The Soviet Far East

The Far East is the largest economic region in the Soviet Union. Enormous, isolated and sparsely populated, it possesses mineral and other resources consistent with its size. Exploiting these resources is one of the key challenges facing the Soviet Union, and Gorbachev has announced a major programme of investment in the region which is to run until the year 2000.

This book looks at the likelihood of this programme succeeding and concludes that it will have to improve significantly on earlier development projects if substantial progress is to be made. Seemingly intractable problems are posed by the combination of geographical remoteness, the inhospitability of the climate and the inefficiency of the administration. Attracting an adequate labour force to the region remains a problem despite the inducements that are offered. Major projects to improve the transport infrastructure, including the construction of the Baykal–Amur mainline railway, have been dogged by delays and poor standards of performance. Even joint ventures involving the Japanese have not been successful.

Yet the region remains central to Soviet regional development. This book explores it in all its geographical and economic complexity. Chapters on the current state of its development are supplemented by examinations of the history of its settlement; analysis of its unique environment and the threats which economic growth might pose for it; and of the region's vital strategic significance. The result will be of interest to all Soviet specialists.

The editor

Allan Rodgers is Professor of Geography at Pennsylvania State University. He has researched and published extensively on questions of economic development with a particular focus on the Soviet Union, Italy and China.

The Soviet Far East
Geographical perspectives on development

Edited by Allan Rodgers

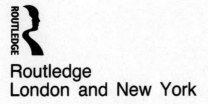

Routledge
London and New York

First published 1990
by Routledge
11 New Fetter Lane, London EC4P 4EE

Simultaneously published in the USA and Canada by Routledge
a division of Routledge, Chapman and Hall, Inc.
29 West 35th Street, New York, NY 10001

© 1990 Allan Rodgers

Phototypeset by Input Typesetting Ltd, London
Printed and bound in Great Britain by Mackays of Chatham PLC, Kent

British Library Cataloguing in Publication Data

The Soviet Far East : geographical perspectives on development
  1. Soviet Union. Regional Economic Development
  I. Rodgers, Allan
  330. 947

  ISBN 0–415–02406–4

Library of Congress Cataloging in Publication Data

The Soviet Far East : geographical perspectives on development / edited by
  Allan Rodgers
      p.   cm.
  Includes bibliographical references.
  ISBN 0–415–02406–4
  1. Soviet Far East (R.S.F.S.R.)—Economic conditions.   I. Rodgers,
  Allan, 1922–
  HC337.R852F3764   1990   89–70214
  330.957'0854—dc20   CIP

Contents

Figures   vii
Tables   viii
Contributors   xi
Preface   xiv

1 Introduction   1
  Allan Rodgers

2 Settling the Far East: Russian conquest and consolidation   5
  Gary Hausladen

3 Environmental constraints and biosphere protection in the Soviet   36
  Far East
  Philip R. Pryde and Victor L. Mote

4 Population and labour force   58
  Ann C. Helgeson

5 Resources   83
  Craig ZumBrunnen

6 Forest and fishing industries   114
  Brenton M. Barr

7 The South Yakutian territorial production complex   163
  Victor L. Mote

8 The Far Eastern transport system   185
  Robert N. North

9 Commodity movements and regional economic development   225
  Allan Rodgers

Contents

10 **Soviet Far Eastern trade**   239
   Michael J. Bradshaw

11 **Economic and strategic position of the Soviet Far East: develop-** 269
   ment and prospect
   Leslie Dienes

12 **Conclusions and recent developments**   302
   Elisa B. Miller and Allan Rodgers

Index   311

Figures

1.1 The Soviet Far East economic region 2
2.1 The Far East 7
2.2 Colonizing the Russian Far East 9
2.3 The Trans-Siberian Railway 15
2.4 The Far East settlement system, 1911 21
2.5 The Far East settlement system, 1939 30
3.1 Average January temperatures in the Soviet Far East 38
3.2 Average date of spring ice break-up in Far East rivers 41
3.3 Depth and continuity of permafrost 42
3.4 Seismicity regions in the Soviet Far East 44
3.5 Areas of endangered fauna in the Far East 47
3.6 Concentration of endangered species, Primorskiy Kray 48
3.7 Location of nature reserves (*zapovedniki*) 49
3.8 Route of the Baykal–Amur mainline 54
5.1 Selected energy resources of the Soviet Far East 86
5.2 Selected mineral resources of the Soviet Far East 100
7.1 South Yakutian territorial production complex 167
7.2 The Amur–Yakutsk mainline 168
7.3 Amur–Yakutsk mainline projected freight flows (c. 2000) 169
7.4 South Yakutian coal and iron deposits 172
8.1 The Soviet Far East transport network 186
10.1 The growth of Soviet foreign trade, 1965–85 244
10.2 The commodity structure of Soviet foreign currency trade, 246
     1985
10.3 Japanese involvement in Soviet Far Eastern development 251

Tables

1.1	The size and population of the major administrative divisions of the Soviet Far East as of 1989	3
2.1	Population of the Far East, 1897–1911	20
2.2	Population of Far Eastern cities, 1897–1911	22
2.3	Urban growth in the Far East, 1911–39	25
2.4	Percentage of urban population in the Soviet Far East compared with Russia/USSR	27
2.5	Urban growth in the Far East by city-size categories, 1911–39	28
2.6	Major cities of the Soviet Far East, 1926–39	29
3.1	Average summer and winter temperatures in cities	37
3.2	Seasonal solar data for Soviet Far East cities at various latitudes	39
3.3	State nature reserves (*zapovedniki*) of the Soviet Far East	50
4.1	Population growth in intercensal periods, 1926–89	60
4.2	Far Eastern urban hierarchy in mid–1980s	62
4.3	Crude estimate of net migrants in intercensal periods, 1926–89	64
4.4	Working-age population in Economic Regions, 1975–2000	76
5.1	The coal reserves of the Soviet Far East	88
5.2	Selected Far East iron-ore deposits	99
6.1	Regional distribution of forests and population	116
6.2	Calculated annual allowable cut in reserve and inaccessible forests	117
6.3	Distribution of forested area: characteristics of industrial stands under central state forest management	122
6.4	1973–4 regional costs and prices for various forest products	124
6.5	Profitability of logging by region, 1967–74	124
6.6	Land use composition of state industrial forests administered by central state forestry agencies	125
6.7	Species composition of mature mountain forests	128
6.8	Mature timber	129

6.9	Species composition of forested areas	130
6.10	Mature growing stock and mean annual increment/ha of mountain forest	131
6.11	Forests of the Soviet Far East	133
6.12a)	Lumber output, 1960, 1975, 1985	135
6.12b)	Plywood output, 1960, 1975, 1985	135
6.12c)	Paper output, 1960, 1975, 1985	136
6.12d)	Paperboard output, 1960, 1975, 1985	136
6.13	Total roundwood equivalents, 1960, 1975, 1985	137
6.14	Change in regional distribution of roundwood equivalents, 1960–85	138
6.15	Regional composition structure of major forest industries, 1960	138
6.16	Regional composition structure of major forest industries, 1975	139
6.17	Regional composition structure of major forest industries, 1985	139
6.18	Mix-and-share: summary of three effects, selected regions, 1960–85	141–2
6.19	Volume and mix of consumed timber, 1973	147
6.20	Volume and mix of harvested timber, 1973	148
6.21	Regional surplus deficit in mix of harvested timber, 1973	148
6.22	Regional shift in industrial removals of timber, 1960–85	149
6.23	Regional origin of exported commodities	151
7.1	Composition of the South Yakutian labour force, 1975 and projected	179
8.1	Provision of transport routes in the Soviet Far East	191
8.2	Spatial distribution of freight traffic in the Soviet Far East, 1966	193
8.3	Modal distribution of freight traffic in the Soviet Far East, 1970	195
9.1	Volume and value of interregional commodity movements for the Soviet Far East, by sector, 1966	226
9.2	Volume of interregional commodity flows for the Soviet Far East, by product, 1965 and 1970	227
9.3	The volume of interregional commodity movements for the Far East, by region, 1950–70	227
9.4	Volume and value of interregional commodity movements for the Soviet Far East, by region, 1966	228
9.5	The volume of interregional movements, by commodity, 1966	229
9.6	Origin and destination of freight shipment to and from the Far East	232

9.7   Siberian linkages with the Far East, by rail, sea and river,   232
        as a share of total flows
9.8   Sectoral linkages of the Soviet Far East, in value terms   233
10.1   Geographical distribution of Soviet foreign trade, 1965–86   244
10.2   Commodity structure of Soviet foreign trade, 1970–86   247
10.3   Soviet–Japanese development projects in the Soviet Far   252
        East
11.1   The share of extensive and intensive factors in the growth   278
        of industrial output in the Soviet Far East, 1960–80
11.2   Dynamics of industrial growth in the Far East by time   279
        period
11.3   Workers and employees in the economy of Sakhalin   291
        Oblast
11.4   Workers and employees in the economy of Kamchatka   291
        Oblast

Contributors

Brenton M. Barr is a Professor of Geography at the University of Calgary in Alberta. He is the Editor of *The Canadian Geographer* and Past President of the Canadian Association of Geographers. His research has focussed primarily on the Soviet forest industry, having published several books and numerous articles on this specialty. His most recent book, written with Kathleen Braden, is on *The Disappearing Soviet Forest: A Dilemma in Soviet Resource Management* (1988).

Michael J. Bradshaw is a 1987 PhD from the University of British Columbia. His dissertation was on *East–West Trade and the Regional Development of Siberia and the Soviet Far East*. He taught for a year at Miami University in Ohio. Currently Dr Bradshaw is a Research Fellow in the School of Geography at the University of Birmingham, United Kingdom. His specialties are Siberian trade, west–east technology transfer, and Soviet regional development. He has published numerous articles and chapters in scholarly journals and books in his field of interest.

Leslie Dienes is Professor of Geography and of the Soviet and Eastern European Studies Program at the University of Kansas. His research interests focus on issues of regional development and energy policy issues, primarily in the USSR. Since 1980, he has twice been on an IREX–USSR Academy of Sciences exchange programme, spent ten months at the Slavic Centre of Hokkaido University of Japan and was Senior Fellow at the Harriman Institute for the Advanced Study of the Soviet Union. He is the author of numerous articles on Soviet regional and energy topics. His latest book, *Soviet Asia: Economic Development and National Policy Choices* was published in 1987.

Gary Hausladen is an Assistant Professor of Geography at the University of Nevada-Reno. He is a specialist on Russian and Soviet settlement. His recent contributions include a guest editorship of a special issue of *Soviet Geography* on 'Settling the Russian Frontier', which contained two of his articles. Additional recent publications in *Soviet Geography*,

Geoforum and *Cities* have focused on socialist and Soviet urbanization. Professor Hausladen received an NCEER research grant (1989–90) for research on *Siberian Urbanization since Stalin*.

Ann C. Helgeson is an Assistant Professor of Geography at the University of Texas in Austin. She received her doctorate from the University of California, Berkeley in 1978 where her dissertation was on *Soviet Internal Migration and its Regulation Since Stalin: The Controlled and the Uncontrollable*.

Elisa B. Miller teaches Soviet Business and Trade at the Graduate School of Business Administration of the University of Washington. Her research publications have focused on US–USSR trans-Pacific commerce and trade. She has served on several trade missions to the Soviet Far East and spent several months in that region in the Spring of 1988 consulting with Soviet authorities in Khabarovsk and Vladivostok.

Victor L. Mote is an Associate Professor of Geography at the University of Houston in Texas. During the academic year 1988–9 he was a Visiting Research Scholar at the Slavic Research Centre of Hokkaido University in Japan. Much of his research has focused on Siberia, particularly his co-authorship with the late Theodore Shabad of *Gateway to Siberian Resources: The BAM*. All told, he has published over sixty other publications in scholarly articles and books, in his field of specialization.

Robert N. North is a Professor of Geography at the University of British Columbia. He is a specialist on Soviet and East European transport and communications policy on which he has written widely. Aside from numerous articles and chapters in scholarly journals and books, his major study on *Transport in Western Siberia: Tsarist and Soviet Development* appeared in 1979.

Professor Philip R. Pryde is a specialist on land-use planning, water resources and environmental impact analysis in the Department of Geography at San Diego State University. His regional specialization is the USSR, and like the other contributors he has travelled widely within that country. He is currently a member of the Advisory Committee for the journal *Soviet Geography*. In addition to numerous articles and chapters in scholarly journals and books, he is the author of *Nonconventional Energy Resources and Conservation in the Soviet Union*. He is currently working on a new book entitled *Environmental Management in the Soviet Union*.

Allan Rodgers is a Professor of Geography at the Pennsylvania State University. His specialties are industrial location and regional economic development. He has worked on the USSR for over three decades with the support of the InterUniversity Committee on Research Grants, the

National Research Council and IREX. That research has resulted in a number of articles and chapters in scholarly journals and books on the Soviet Union.

Craig ZumBrunnen is an Associate Professor of Geography at the University of Washington. He has a PhD in geography from the University of California at Berkeley where he wrote his dissertation on water pollution problems in the Soviet Union. He served for five years in the Geography faculty at the Ohio State University and is currently on the staff at the University of Washington. He is the co-author of *The Soviet Iron and Steel Industry* and editor of *Urban Geography in the Soviet Union and the United States*. His research publications have focussed on Soviet environmental, energy, natural resources and regional-development related topics. His current research interests include the development of GIS data bases for the Soviet Union and the relationships between *glasnost, perestroyka*, and Soviet natural resource and environmental management.

Preface

The major stimulus for this book was the publication of Mikhail Gorbachev's speech delivered at Vladivostok, in the Soviet Far East, in August of 1986. In that address and subsequent pronouncements, he promised the people of this remote area, far from the centre, an investment programme that by the year 2000 would transform this strategic resource frontier into a far more mature economic region. In addition to economic growth, the Far East would have a greatly expanded social infrastructure and a far higher standard of living. Not only would the region be more highly integrated with the national economy, but it would also develop sharply expanded ties with nations on the Pacific Rim. Gorbachev's promises were particularly exciting for me because of my research on that segment of the Soviet Union in the early 1980s.

With the encouragement of Mr Peter Sowden, social science editor of Routledge, I enlisted the participation of a group of Canadian and American geographers, most of whom had researched Siberia in the recent past. It should be stressed that no rigid framework was imposed on each author other than the focus of regional economic development. Three chapters were added; these were designed to provide an historical, environmental and demographic setting for the economic chapters.

In each instance, the contributors agreed to write materials in their own specialties, keying them, of course, to the Soviet Far East. All authors were faced with the problem of data limitations, yet managed remarkably well in overcoming the obstacles to acquiring up-to-date information.

Finally, it is hoped that this book will fill a virtual void in the western literature on economic development in this strategic region.

Allan Rodgers
Pennsylvania State University

Chapter one

Introduction

Allan Rodgers

The Soviet Far East with an area of over 6 million square kilometres is the largest planning region in the USSR (Figure 1.1). It, alone, accounts for roughly 29 per cent of the nation's territory, more than double the size of Kazakhstan. Unlike the latter, which is a union republic, the Far East serves no administrative function. Nor is it a separate and distinct physical region, for its western borders with Eastern Siberia are clearly arbitrary. It is nevertheless, a planning region and it is now a far more unified economic entity. In the past, the Yakut ASSR, its largest sub-region, was linked to Eastern Siberia and in fact, was a part of that areal unit until 1963. However, with the completion of the BAM (Baykal–Amur Mainline) railway, Yakutian ties have strengthened with the Far Eastern 'core'. These links will be reinforced with the completion of the AYAM (Amur–Yakutsk Mainline) railway after the turn of the century.

The region is remarkably diverse for it stretches about 3,000 miles north and south from the Arctic to the North Korean frontier and 2,500 miles in breadth from the Pacific to the borders with Eastern Siberia. That western limit runs from the Argun river, along the flanks of the Stanovoy mountains to the Kolyma lowlands. The Far East's constituent political division, its area and population are demonstrated in Table 1.1 and mapped on Figure 1.1.

Roughly three-quarters of the Far East lies in its three northern administrative units: Magadan and Kamchatka Oblasts and the Yakut ASSR. The remainder lies in the south in Maritime and Khabarovsk Krays; and Amur and Sakhalin Oblasts.

The Far East is sparsely populated containing roughly 8 million inhabitants in 1989 or 2.8 per cent of the national total. However, in spatial perspective, there are vast differences in regional densities. Three-quarters of the population is concentrated on the southern margins of the territory and half live within eighty kilometres of the sea. Vast areas, particular in the 'Far North' are either unpopulated or thinly so (less than one per square kilometre), while the more favoured pockets of the south east have densities well over thirteen per square kilometre. The

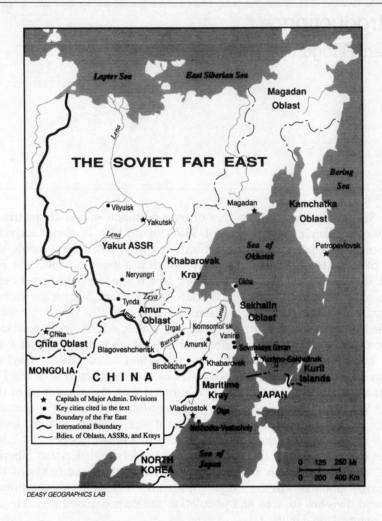

DEASY GEOGRAPHICS LAB

1.1 The Soviet Far East economic region

region's population grew by over 16 per cent from 1979 to 1989 compared to a national increase of 9 per cent. The Yakut ASSR grew by almost 29 per cent. The Far East is more urbanized than the national average with a mean of 75 per cent living in urbanized places and highs of 82–3 per cent compared to a Soviet urban average of 66 per cent.

This huge region plagued by environmental problems and remote from the Soviet heartland has remained largely underdeveloped. These obstacles interdict economic interactions with other regions of the Soviet Union and even constrain commodity movements within its sub-regions. Inaccessibility and sheer distance result in excessive transport costs and

Table 1.1 The size and population of the major administrative divisions of the Soviet Far East as of 1989

Political units*	Area thousands sq. km	Population thousands	Population density per sq. km
Maritime Kray	165.9	2,260	13.6
Khabarovsk Kray	824.6	1,824	2.2
Amur Oblast	363.7	1,058	2.9
Kamchatka Oblast	472.3	466	0.9
Magadan Oblast	1,199.3	543	0.5
Sakhalin Oblast	87.1	709	8.1
Yakut ASSR	3,103.2	1,081	0.3
Soviet Far East	6,216.1	7,941	1.3

Source: *Pravda, 29 April 1989, p. 2.

intervening opportunities from Soviet regions closer to the centre frustrate the Far East in internal markets. Access to Pacific Rim markets presents some of the same problems because of the location of key resources in the interior, in some instances far from conceivable centres of export demand and the rise of new resource frontiers, particularly in mainland China and western Australia.

A discussion of remoteness and the 'friction of distance' must focus on the extraordinary distances involved in rail shipment from the 'West' to Far Eastern ports coupled with trans-shipment to coastal freighters destined for ultimate destinations like Magadan, Petropavlovsk or Yuzhno-Sakhalinsk (see Figure 1.1). However, these are extreme cases which involve only a fraction of the flows to or from the Far East. More realistically, the focus should be on distances to the centre or to the Urals from the more developed parts of the Far East, such as Khabarovsk. The rail distances to the Urals (Chelyabinsk) and the centre (Moscow) are 6,694 km and 8,505 km respectively. Overcoming these distances clearly involves the expenditure of time and money. Despite limited resources, conflicting sectoral and regional demands and the problems of the Far East, cited above, the central authorities have devoted a comparatively large share of capital investment funds to the region in the 1970s, particularly when compared to its share of the nation's population (2:1). That priority has reflected the nation's strategic concerns initially *vis à vis* Japan and particularly since the 1960s with the People's Republic of China. From 1974 on, there were massive investments in the building of the BAM, not comparable with those in the West Siberian oil and gas fields, but clearly significant. Our authors have evaluated the degree to which there were major economic developments in the BAM zone and the resulting volume of freight traffic.

Finally, as we shall see, Gorbachev's speech in August of 1986 on the future development of this region and succeeding pronouncements have kindled renewed interest in prospects for its further development though these must be tempered by the stagnation of the national economic growth from 1987 to 1989.

In July 1987 the Central Committee approved 'the Long Term Programme for the Comprehensive Development of the Far Eastern Region and Transbaykalia [Chita Oblast and Buryat ASSR] till the year 2000'. A month later, Gosplan laid out some details of this plan for the region's development. For the Far East alone an investment of 198 billion roubles was proposed over this twelve-year period. Another 34 billion rouble investment was to be allocated to Transbaykalia. Thirty per cent of this huge overall outlay was to be devoted to the development of its extremely backward social and cultural infrastructure. There was to be self-sufficiency in the supply of key food products. In the energy sectors, major growth was anticipated in coal, natural gas, petroleum and electricity production. Very ambitious growth rates were also proposed for the wood product industries. In an overall sense, the value of industrial output was projected to increase roughly 2.5 fold. More significant perhaps than the mining and industry projections was to be the transformation of the Far East from its previous role as predominantly a supplier of raw materials to a processor of those goods. To accomplish this, the machinery and metallurgy industries were to be overhauled and re-equipped and the construction material industry would show a significant growth. This move towards a more balanced and comprehensive industrial structure would require notable in-migration. Thus the population of the Far East would grow by 20 per cent to about 10 million inhabitants, while that of the BAM zone would increase by 70 per cent.

These extraordinarily ambitious goals are analysed by a number of contributors to this book and their views are summarised in the final chapter. The achievement of those goals would, of course, change the face of the region, but their feasibility, as we shall see, is in serious question.

Chapter two

Settling the Far East: Russian conquest and consolidation

Gary Hausladen

The colonization of the Far East, the 'tattered edge of Siberia', marked the final stage of a saga of Russian imperial expansion eastwards that had begun in the late sixteenth century. By the middle of the seventeenth century this onslaught engulfed the Pacific coastal regions from the Bering Sea and Kamchatka in the north to the Amur in the south. It was not until the nineteenth century, however, that Russian settlement and treaties truly consolidated Russia's hold on its Pacific periphery. The incorporation of the Far East region provided the Russian empire with a vast territory, stretching from above the Arctic Circle to the borders of China, from the Lena river to the Pacific Ocean, accounting for almost one-twelfth of the landmass of the earth.

The Far East, of course, was more than just a vast territory. It was a treasure trove of natural wealth, first with furs, later with minerals. It served as a place of exile for prisoners and revolutionaries, and as a refuge and land of opportunity for dissidents, malcontents and peasants. It also served as an outlet to the Pacific, providing access to China, Japan and America. Throughout its history, the relationship between the Far East and tsarist Russia represented a classic case of a resource-rich periphery and a dominant, European-based core.

For the Far East, the years immediately following the Revolution were years of relative autonomy. The Bolsheviks were preoccupied securing power. Thus, pragmatic considerations dictated that the Far East be allowed to exercise a great deal of local control. This relative autonomy ended in the late 1930s as Stalin consolidated his control and Soviet power was extended over the entire territory that had previously comprised the Russian Empire.

Theoretically, the basic nature of the relationship between the Far East and the European core changed with the establishment of communist control. The new Soviet state was ideologically committed to abolishing imperial relationships and to integrating peripheral regions into the greater Soviet polity – extending the benefits of modernization to all regions of the country. Many have suggested that there has been more

5

continuity than change from tsarist to Soviet times and that the Far East continued to serve as a resource hinterland for the European core even after the Revolution. As it did for the tsars, the Far East also served as a Soviet gateway to East Asia and the Pacific basin. To both the Russian and Soviet empires, it represented a periphery, the integration of which was as difficult as the potential wealth was promising, a periphery which had a pivotal strategic location for the Soviet state.

One characteristic that remained constant throughout Russian and Soviet consolidation was the importance of the settlement system. Throughout the colonial history of the Far East, the settlement system has been the skeleton for conquest, consolidation, development and integration. Whether it represented exploitation or integration, the focus of Far East development has traditionally been found in the settlement system. This continued during the post-Revolutionary period, when the Soviet Far East remained one of the most highly urbanized regions in the USSR. Cities, most of which were originally Cossack settlements and important towns during the tsarist period, still functioned as conduits for implementing central government development policies.

This chapter examines the settling of the Far East, which approximates the boundaries of the present-day Far East economic region. It includes Yakutia and the territories of the Amur and Ussuri basins and along the Pacific coast (Figure 2.1).[1] After briefly describing the history of Russian colonization of the Far East, this chapter details the pattern of economic activities that had developed by the latter stages of tsarist rule. This treatment heavily emphasizes the importance of the settlement system. The chapter then examines the transition from tsarist to Soviet control of the region through the 1930s, when the Far East finally became fully incorporated within the USSR. The role of the Far East within the Russian and Soviet empires is compared, with particular attention to those features of its relationship to the European core that remained the same as well as those that changed. In this manner, the groundwork is laid for better understanding more recent trends, which are described and analysed in the following chapters.

Russian colonization of the Far East

The Russian conquest of the Far East was part of the overall colonization of Siberia, which began with the famous exploits of Yermak in 1581. Early Russian expansion was slowed during the turbulent years between dynasties, known as the 'Time of Troubles' (1605–13). However, the consolidation of the Romanov dynasty, under Tsar Mikhail, precipitated an 'explosion of explorations', beginning in 1518, which extended the Russian empire into the Far East.

By the seventeenth century, when Russian explorers first penetrated

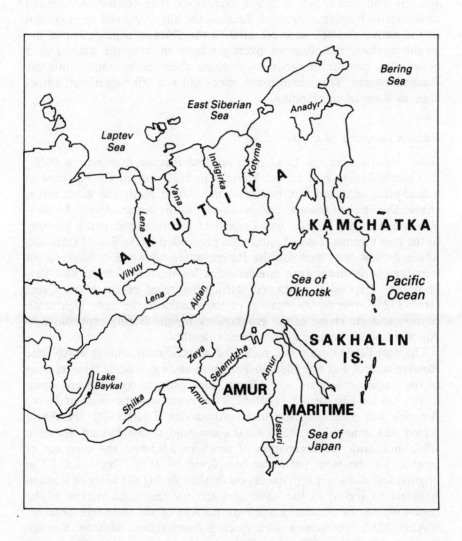

2.1 The Far East

the Far East, there were numerous indigenous tribes throughout the region.[2] For the most part, however, they were not a major barrier to Russian expansion eastwards. The relative ease of Russian conquest of the region resulted from the fact that not only were the indigenous tribes sparsely distributed, but, with few exceptions, they offered only limited resistance to Russian expansion. Because the allegiance and co-operation of the native peoples were essential to the Russian exploitation of the wealth of furs, the Russians overtly sought to treat the natives in a benevolent manner in order to promote their incorporation into the Russian system. Thus, indigenous tribes did not offer significant resistance to Russian colonization.

Russian conquest of the Far East

Initial reports from the Lena river reached Russian explorers in 1620.[3] The establishment of a fort at Yakutsk in 1632 opened the Far East to a four-pronged assault: north-east to the Bering Strait and Kamchatka; east to the Sea of Okhotsk; and south-east to the Amur. Along the way, Russia settled the Yenisey and Lena river basins, established a gateway to the east through Yakutsk, and then proceeded to the Sea of Okhotsk, where its first settlement on the Pacific was established in 1639. Other important settlements were established at Vilyuysk, Olekminsk and Verkhoyansk. Expansion continued during the reign of Peter the Great (1682–1725), with the annexation of Kamchatka in the early eighteenth century and the laying of the groundwork for the Bering expeditions to America, which occurred after Peter's death.

The post-Petrine era was a period of consolidation, during which time Russian control was extended further south-eastwards while those regions of the empire that were too difficult and costly to control were eliminated. Under Elizabeth I (1741–62), the north-western coast of North America was colonized; and under Alexander II (1855–81), the Amur region was annexed in 1858. Alaska was sold to the United States in 1867, and, with the occupation of southern Sakhalin, the conquest of Siberia, for the most part, was completed by 1876 (Figure 2.2). The original fortresses and settlements established during the years of Russian colonization served as the basic skeleton for the consolidation of the region within the Russian empire on the eve of the twentieth century.

After 1637, Siberia was administered directly from Moscow through the Siberian Office (*Sibirskiy Prikaz*), which was responsible for all activities in Siberia. For the four Siberian districts established in 1693, the administrative centre for the Far East was Yakutsk. Because of its location on the Lena, providing access to the northern reaches of the Far East, Yakutsk became the main gathering point for furs to be shipped back to European Russia. It was also an agriculturally productive region.

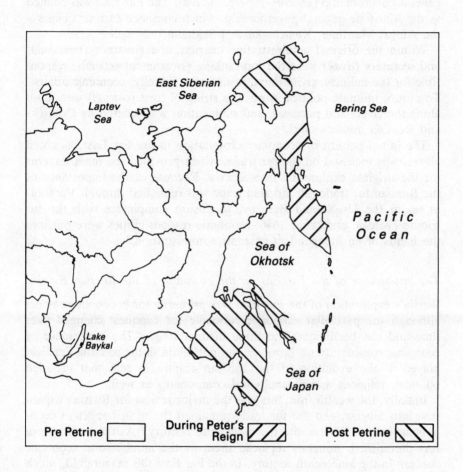

2.2 Colonizing the Russian Far East

As a result, Yakutsk maintained its importance throughout the period of colonization and the evolution of the administrative system.

By the twentieth century, Yakut Oblast, with its capital in Yakutsk, was administratively part of East Siberia, subordinate to the Irkutsk general governorship (*general'-gubernatorstva*). The Far East was defined as the Maritime general governorship, which consisted of four oblasts – the Amur, Maritime, Kamchatka and Sakhalin.[4]

Within the original administrative districts, the governor (*voyevoda*) and secretary (*dyak*) were the top-ranking government officials, responsible for the military, civil affairs, justice and, initially, economic affairs.[5] To ensure ultimate political control, a string of forts (*ostrogi*) were built along the rivers and portages, and subjugation was secured by Cossacks and Russian military units.[6]

The initial pattern of economic exploitation in the Far East, as noted above, was focussed on the fur trade, which provided the main impetus for the original exploration into Siberia. Because of the importance of the Russian fur trade, all Siberian trade was funnelled through Verkhotur'ye, in the Urals, and Berezov, to ensure compliance with the fur tribute (*yasak*), and, after 1646, economic responsibilities were put into the hands of an independent customs administration.[7]

The importance of the Far East in the evolution of the Russian Empire

Russia's exploration of the Far East was primarily for economic reasons, although the particular economic rationale for conquest changed over time and can be divided into three distinct stages. The importance of economic considerations, however, did not diminish the role the Far East played in the evolution of the Russian empire, a role that included political, religious and psychological components as well.

Initially, the wealth from furs was the major reason for Russian expansion into Siberia, and the fur trade remained the most important economic activity in Siberia into the nineteenth century.[8] As the wealth from furs diminished, minerals replaced them as the number one economic concern in the nineteenth century. In the Far East this meant gold, which replaced furs as the primary source of Far Eastern wealth. Mineral resources continued to be the primary source of wealth from Siberia till the Revolution. A third stage in the development of the Far East, however, accompanied the construction of the Trans-Siberian Railway.

In the late-nineteenth and early-twentieth centuries the promise of freedom and land brought massive numbers of Russian peasants into the region. As a result, agriculture became an important economic activity, not because of the revenues it produced or wealth it generated for the empire, but because it provided the stimulus for the true consolidation of the Far East into the Russian empire, consolidation effected by the migration of Russian peasants into the region. And, in the nineteenth century, during the period of consolidation, Russian interest was further stimulated by the potential for increased trade relations with China.[9] The southern reaches of the Far East provided access to China, as well as to other countries of the Pacific, particularly Japan and the United States.

As important as these economic considerations were, political motives also stimulated expansionist tendencies as early as Tsar Mikhail. Even in the early seventeenth century, with the establishment of the Romanov dynasty and through the imperial period of Peter the Great, Siberia and the Far East became symbols of empire to both the Russian state and the Orthodox Church.

The Orthodox Church was a great supporter of Russian expansion into Siberia. The Metropolitan of Moscow still held out hope for Moscow to become the 'Third Rome'. Challenged by the Catholic Church in the west and Islam to the south, Holy Russia presented the Church with an extensive realm of control. From the earliest tsars, it was impossible to separate Russia's political and religious missions. 'To the Tsar belonged not only care for the interests of the State but also care for the salvation of souls.'[10]

This mission affected the manner in which Russia treated the indigenous populations they encountered in the Far East. Because of the small numbers of Russian colonizers, support and allegiance from the indigenous peoples was essential to the success of Russia's colonization. Although not always carried out to the letter, official Russian policy towards the natives was one of conciliation. As Harrison points out:[11]

> By order of the tsar, Russians were not to harm natives or penetrate native hunting reservations. Once a tribe had taken an oath of loyalty to the tsar, remained pacific, and delivered its tribute, they were regarded as friends and wards.

This benevolent approach towards the aborigines notwithstanding, the basic character of Russian colonization fits the classic scenario of a core–periphery relationship within the framework of European-based imperialism. We see in Russia's expansion the 'aggressive encroachment of one people upon the territory of another, resulting in the subjugation of the latter people to alien rule'.[12] Clearly, by the latter stages of tsarism, the Far East, the 'tattered edge of Siberia', was a true colony of European Russia.

The Far East on the eve of Revolution

By the twentieth century, the Far East was perceived as an important part of the Russian empire both economically and strategically. The tsarist government pursued policies designed to consolidate its hold on the region, thus producing an increasingly complex and productive economic geography, built upon the foundation of the urban settlement system.

Tsarist policies towards the Far East

The last sixty years of tsarist rule had a profound effect on the nature of development in the Far East and on the resultant economic geography one found there in the early twentieth century. From 1855 until the Revolution, Russia was ruled by the last three of the Romanov tsars – Alexander II (1855–81), Alexander III (1881–94) and Nicholas II (1894–1917). The policies of each of these monarchs directly and indirectly changed the basic nature of the role of the Far East within the Russian empire. Only several of the more important events are touched upon here, and only as they affected the Far East.[13]

Alexander II, the 'Tsar Liberator', came to the Russian throne as an enlightened and politically liberal tsar. He was keenly aware of the debilitating effects of serfdom, appreciated the value of borrowing from the west, and understood the need to maintain Russia as a great European power, an aspiration somewhat tarnished by the Crimean War (1854–6), in which Russia's defeat at the hands of Britain and France had diminished Russia's stature as a 'great power' in Europe. Alexander II was responsible for the emancipation of the serfs in 1861, which freed the serfs from the land. Tight restrictions, however, prevented massive migrations of peasants out of European Russia. The increase of migrants to Siberia increased only from an annual average of 19,000 in the 1850s to 25,000 in the 1860s.[14] It was not until additional migration laws were passed in the 1880s that the number of migrants would increase substantially.

Alexander II also promoted industrialization, which he viewed as the key to future greatness. The building of railways, primarily in European Russia, greatly benefitted from this outlook. This would eventually carry over to the building of the Trans-Siberian in the 1890s.

After his father's assassination in 1881, Alexander III ascended to the throne as a political reactionary and slavophile. He was anti-western in outlook and pursued policies that promoted the greatness of Russian culture. In fact, under Alexander III 'russification became an official policy'.[15] Yet, he did not go so far as to rescind those of his father's policies that encouraged industrialization. He continued to believe that industrialization was necessary for Russia to be a great power.

Additionally, it was under Alexander III that the government identified the importance of populating the Far East in order to consolidate the empire's hold on this distant territory. As one indication of a 'changed attitude' towards the Far East, the building of the Trans-Siberian Railway was initiated.

It was also during the reign of Alexander III that much of European Russia experienced a great famine. Although the reasons for the crop failures were for the most part beyond the control of the government, its policies greatly exacerbated the situation.[16] Because of the heavy emphasis on industrialization under the two Alexanders, agriculture had not received adequate support from the government and had remained relatively backward; this was not the case in most industrializing countries, where a modernized, efficient agriculture was essential for development. Also, previous heavy taxation of the peasants to pay for industrialization meant that the peasants had little or no surpluses to fall back on during the famine.

The Trans-Siberian was continued and completed under Nicholas II. Whereas various other policies provided the push and pull factors – emancipation, famine, overcrowded conditions in European Russia, along with government incentives for migration – the Trans-Siberian provided the means for migration. Nicholas's interest in the Far East was increased by the prospects of increased trade and the revenues from rail trade between Asia and Europe.[17]

The drive for further expansion in the Far East, especially in Manchuria at the expense of the decaying Manchu dynasty, was finally halted by the Japanese. Japan handed Russia a humiliating defeat in the Russo–Japanese War of 1904–5. Russian losses were estimated at 400,000 killed, wounded or captured and a total cost of approximately 2,400 million roubles, not including 500 million roubles worth of vessels sunk and equipment captured by the Japanese.[18] Although Russia skilfully secured its position in the region at the Treaty of Portsmouth and in the years that followed, the defeat had a devastating effect on national morale and confidence in the government. Many have interpreted the defeat of Russia by Japan as a major contributor to the revolutionary movements that would eventually overthrow the monarchy.

As the role of the Far East changed in the eyes of the European overlords, it became important to consolidate its union with Mother Russia. The Far East was no longer simply a treasure trove of rich furs and gold, a place of exile, or a region of Russian and Orthodox dominion for the conversion and salvation of pagan souls. In the second half of the nineteenth century, Russia's leaders, especially under Alexander III, came to understand its strategic value to Russia. The Far East became increasingly important to the Russian heartland as it came to be perceived as a region with political and economic access to China, Japan and the United States.

Russian consolidation of the Far East: migration and settlement in the late-tsarist period

The true consolidation of the Far East with the European Russian heartland would have been difficult without the building of the Trans-Siberian Railway, which provided the means for the massive migration of peasants that occurred in the late-nineteenth and early-twentieth centuries. Although various political and economic interests had long pressed for the building of a railway to Siberia, it was not until 1891 that Tsar Alexander III finally became committed to the project.[19] Once begun, the building of the various lines that would eventually link together to form the Trans-Siberian proceeded at a rapid pace.

To a great extent the success of the Trans-Siberian resulted from the organizational and financial skills of the man primarily responsible for its construction – Sergei Witte.[20] As Minister of Finance from 1892 to 1903, Witte directed the entire Russian economy; the construction of the Trans-Siberian became one of his top priorities. Despite political opposition, lack of finances and the enormity of the task, Witte was largely responsible for building 'one of the greatest enterprises of the century in the entire world'.[21] The project, excluding the Chinese Eastern line across Manchuria, resulted in the laying of 4,400 miles of track at a total cost of 384.6 million roubles, or approximately 87.4 thousand roubles per mile.[22] For the decade of the 1890s, the Trans-Siberian accounted for approximately 30 per cent of all track laid in Russia and almost 15 per cent of the total government expenditure in the 'nation's railway business'.[23]

The first thrust into Siberia had already linked Yekaterinburg, in the Urals, with Tyumen' in 1885. In the 1890s various parts of the route were begun simultaneously. The line from Chelyabinsk to the Ob' river (Krivoshchekovo) was completed in 1896 and the Ussuri line from Khabarovsk to Vladivostok a year later. From the Ob' the Trans-Siberian reached Irkutsk (Innokentiyevskaya), with a branch line to Tomsk, in 1898, and the Irkutsk to Baykal line was opened in 1899. From Lake Baykal (Mysovaya) the line reached Sretensk in 1900. The initial route between Sretensk and Vladivostok, the Chinese Eastern Railway, crossed Manchuria and was completed in 1904. In just over a decade the Russian heartland had been linked by rail with its Far Eastern territories; Europe had been linked with East Asia (Fig. 2.3).[24]

The construction of the Trans-Siberian Railway greatly facilitated the final stage in the saga of Russia's colonization of the Far East, when large numbers of free peasants migrated to Siberia and the Far East from European Russia. Even prior to this period, as Semenov points out:[25]

. . . the core of the colonization was provided not by deportation or

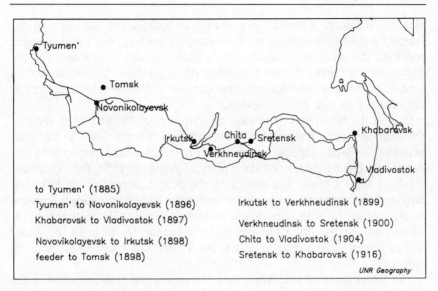

to Tyumen' (1885)

Tyumen' to Novonikolayevsk (1896)   Irkutsk to Verkhneudinsk (1899)

Khabarovsk to Vladivostok (1897)   Verkhneudinsk to Sretensk (1900)

Novovikolayevsk to Irkutsk (1898)   Chita to Vladivostok (1904)

feeder to Tomsk (1898)   Sretensk to Khabarovsk (1916)

UNR Geography

2.3 The Trans-Siberian Railway

'official' resettlement, but by free colonization, which did not worry much about official regulations. It is thanks to this that by the end of the eighteenth century there were over a million Russian settlers in Siberia.

In fact, most scholars agree that free colonization was the primary source of Russian colonists throughout the entire colonial period, although Treadgold suggests that the number of Russians did not reach one million until early in the nineteenth century.[26] In either case, the importance of peasant migration cannot be denied. It provided a base, and the nineteenth century provided the political and economic rationale along with the technological means for greatly stimulating the flow of migrants eastwards. The large numbers of Russian settlers in the nineteenth century truly effected the consolidation of the Far Eastern realm as part of the Russian empire.

Emancipation in 1861, coupled with economic depression in the 1870s and famine in the early 1890s provided the impetus for emigration from the European Russian countryside. Treadgold estimates that the total number of migrants to all of Siberia from 1801 to 1914 was somewhat less than 7 million.[27] There were approximately 375,000 migrants between 1801 and 1850, and the rate of migration increased from 8,000 per year prior to 1850 to 42,000 by 1890.[28] Then came the impact of the Trans-

Siberian Railway. Although sea voyage had shortened the time of the journey considerably earlier in the nineteenth century, the Trans-Siberian provided the stimulus necessary for a truly massive migration. For the decade of the 1890s the total number of migrants to Siberia soared to almost 1.4 million and for the first decade of the twentieth century it skyrocketed to just over 3 million.[29]

During the fifty years of intense migration from 1860 to 1910, most of the migrants were destined for West and East Siberia. But the Far East benefited as well, especially the southern reaches and particularly the Amur and Ussuri basins. At the Treaty of Aigun in 1858, the Amur was regained from China. The return of the Amur, according to Bakunin, provided the government with the rationale for encouraging migration to this region to help consolidate Russian control. 'Through the Amur Siberia has for the first time acquired a meaning, for through the Amur it was linked to the Pacific and is no longer a wilderness without an outlet.'[30]

The importance of the Amur and Ussuri regions was reflected in the special treatment given the regions in 1861. Although emigration from the European core to Siberia was officially discouraged, special state funds were provided to encourage settlement of the Amur and Ussuri regions.[31] It has been estimated for the thirty years between 1863 and 1893, that over 19,000 migrants arrived in the Far East. The Trans-Siberian helped during the latter stages, so that by 1907 20 per cent of all those crossing the Urals were destined for the Far East.[32] Those who migrated were not only from European Russia, many were from other parts of Siberia. All in all, by the end of the nineteenth century the population of the Far East, including Yakutia, was 641,000, of whom 265,000 were Russian.[33] By 1911, these numbers had grown to a total of 1.1 million total, including 651,000 Russians.[34]

The Far East economy

By the twentieth century, the migration of peasants, the increasing value of the Far East to the Russian economy and its strategic location in East Asia had clearly defined the region as an integral and important append-age to the Russian empire. As its role changed and as its consolidation as part of the empire became vital, the complexity of its role increased, producing a more diverse economic geography by the eve of the Revolution.[35] Although there had been continual changes during the two and a half centuries since it had first been opened by Russian explorers, the final years of tsarist rule left a dramatic imprint on the geography of the region as it was about to pass from the rule of tsars to commissars.

The initial stimulus for settling Siberia and the Far East had come from the desire for wealth from the furs. By the turn of the twentieth century much of this wealth had been exhausted. In fact, the sea otters had been so overhunted and the net revenues so negligible, that Russia gave up on its American colony and sold Alaska to the United States in 1867. In Siberia and the Far East, furs continued to play an important, if not leading, economic function, although their relative contribution to the region's wealth was diminishing rapidly. As important as its actual contribution to the wealth of the region was the fact that furs were a major source of revenue from trade with the Chinese, thus further strengthening the government's perception of the Far East's increasingly attractive geographic position *vis-à-vis* China and other Asian countries. By the twentieth century, the most profitable furs were of sable, squirrel and, to a lesser extent, tiger.[36]

During the nineteenth century, gold and coal replaced furs as the primary stimulus to Siberian and Far Eastern colonization. By 1910 the Vitim, Zeya, Bureya, Olekminsk, Amursk, Maritime and Ussuriysk gold regions were producing over 53,800 pounds of gold yearly, worth nearly 290 million roubles.[37] This accounted for one-third of Russia's entire production of gold. Although of lesser value, the production of coal had also increased greatly and was vital to the construction and operation of Siberian railways. Two major fields emerged in the early twentieth century. The mines in the Ussuriysk region were producing up to 358.6 million pounds yearly, while the newer mines on Sakhalin produced almost 72 million pounds.[38]

Fishing had been a traditional activity of the indigenous peoples before the arrival of the Russians. Russian colonization only intensified its importance as fish were an important source of food for the ever-increasing numbers of colonists and the exportation of fish and fish products became an important source of revenue for local economies.

The great varieties of fish in the Far East included herring, cod, flounder, seal and whale. But of greatest importance were the various kinds of salmon, especially Siberian salmon and sig, which were also responsible for the production of caviar. By 1910, the salmon catch was estimated in the tens of millions of fish, which accounted for hundreds of thousands of pounds of caviar.[39]

The exportation of fish and fish products proved to be a major source of revenue for local economies in helping them to build infrastructure and improve the standard of living. Fish exports exceeded 215 million pounds yearly, of which half went to Japan, and the remainder was divided between European Russia and Europe.[40] The government clearly saw that increased foreign interest in the Far Eastern fishing industry would eventually make this activity as important politically, in linking Russia and the Far East to European and Pacific-Rim countries, as it was economically.

Reindeer herding was also a traditional activity that retained its value under tsarist rule. Unlike fishing, however, its importance was primarily local. The domesticated reindeer herds of the Far East, which totalled almost 0.5 million by the twentieth century, provided draft animals, milk and meat, and skins and antlers for the region.[41]

As settlements were built and expanded for the increasing numbers of settlers, the forests of the Far East became important sources of lumber. In fact, most of the lumber cut in the Far East, primarily cedar, pine and larch, was put to local use. Major regions of forestry included the Amur and Ussuri regions along with the coastline along the Sea of Okhotsk.[42] The Russians did export some timber products, mostly to Australia. Here again, the government believed that Far Eastern forest products held the potential for increased economic and political relations with timber-deficit Pacific-Rim regions, particularly Japan and Manchuria.

The last half century of tsarist rule greatly changed the economic geography of the Far East. Just as mineral resources had replaced furs as the primary economic rationale for the settling and consolidation of the region, the great influx of peasant farmers resulting from the 'great migration' produced a third stage in the region's economic history. Although agriculture had been encouraged since the original opening up of Siberia, emancipation, untenable conditions in European Russia, the promise of freedom, and the construction of the Trans-Siberian Railway propelled farmers into the region. Along with the traditional agricultural areas in the Vilyuy and Lena river basins, farmlands were developed along the route of the Trans-Siberian in the south. Major agricultural regions in the south came to include the basins of the Amur and Ussuri rivers and their tributaries, which were estimated to have almost 0.5 million peasants early in the twentieth century.[43]

Even though grain production in the southern regions reached 1,400 million pounds of grain yearly, it was not sufficient to meet the growing demand from the massive influx of settlers.[44] As a result, the Far East imported about 395 million pounds of grain per year from Manchuria, a situation which the government saw as another opportunity eventually to open up Manchurian markets for Russian goods.[45]

Industry had also come to the Far East by the turn of the twentieth century, although most of it was related to the primary economic activities of the region. Wood products, fish products and flour were the main industrial products of the Far East. There was some light metallurgy associated with the Trans-Siberian which was located primarily in Vladivostok, Nikol'sk-Ussuriysk and Khabarovsk.[46]

Yet, as important as economic considerations were to the continued Russian presence in the Far East, one should not diminish the importance of the region's strategic location and the role of the military. As noted above, the growing value of the Pacific realm had been identified during the reign of Alexander III. Even when its economic contribution to the empire waned, its strategic importance dictated that Russia consolidate its south-eastern periphery as a bridge to the Pacific Rim. It might be too narrow to suggest that the Far East was simply Russia's 'military bastion on the Pacific',[47] but clearly the strategic–military rationale for Russian occupation cannot be overlooked.

One sees in Russian occupation of the Far East the eternal struggle between the potential and the actual. The potential wealth of the region has never been questioned, unlike its actual contribution to the economic health and development of the Russian state. The enthusiasm for investing in the Far East in order to hasten the day when potential and actual merge has, in fact, changed over time. It is when this enthusiasm for economic potential has been low that the strategic value has provided central planners with the necessary rationale for continued domination of the Far East. Clearly, when other colonies of the empire were given up for economic reasons, the Far East was retained not solely for its economic potential, but for its strategic location as Russia's pathway to the Pacific.

Patterns of settlement

As the role of the Far East changed in the eyes of Mother Russia, government policies changed as well. As mentioned above, the government, initially concerned about the potential loss of labourers and renters, had steadfastly opposed emigration until the late-nineteenth century, when it began actively to encourage migration. As a result, large numbers of migrants moved to populate the Far East, especially along the southern boundary, which was the most strategic and most vulnerable. In fact, it was the massive colonizing of the Far East by Russian peasants that truly consolidated Russian control of the region.

Population increases between 1897 and 1911 are indicative of the impact of the Trans-Siberian and the magnitude of the explosion of settlement that the Far East experienced during the last years of tsarist rule (Table 2.1).[48] During the fourteen-year period, 1897–1911, the population of the Far East almost doubled from 641,000 to over 1.1 million. During the same period, the ethnic Russian population more than doubled from 265,000 to over 650,000. This increased the proportion of Russians from 41 per cent to 60 per cent. In other words, Russians became the largest ethnic group in the Far East during this period. There was, however, great regional variation.

The more remote and inhospitable regions of the Yakutsk Oblast showed only a slight increase in total population from 269,000 to 277,000, while experiencing a decrease in the number of Russians from 30,000 to 18,000. The Kamchatka region experienced slight increases in both total

Table 2.1 Population of the Far East, 1897–1911

	1897	1911
	(000s)	(000s)
Yakut oblast	269	277
Amur oblast	120	286
Maritime oblast	189	524
Kamchatka oblast	35	36
Sakhalin Island	28	9
Total	641	1,132

Source: *Aziatskaya Rossiya*, vol. 1 (reprint: (1974) Oriental Research Partners, Cambridge, Ma.), pp. 241–2.

and Russian populations. The total population rose by 1,300 to just over 36,000, with an increase of 320 ethnic Russians to 4,200. Sakhalin Island, primarily as a result of Russo–Japanese conflicts during the period, experienced the greatest losses of people. Population under Russian control dropped by over 19,000 to just under 9,000, of which 5,600 were Russian, a total that represented a loss of almost 13,000 from 1897.

The areas of greatest appeal were along the Trans-Siberian in the Amur and Maritime oblasts. These regions showed exceptional growth (138 per cent and 177 per cent, respectively) from a combined population of 309,000 to 810,000. Russians provided the greatest share of this increase. In the Amur, their numbers increased from 103,500 to 242,300; in Maritime oblast, from 109,800 to 380,400. In these two oblasts, the total increase of ethnic Russians amounted to 409,400 people, thus increasing their proportion of the total population from 69 per cent to 77 per cent.

At the same time that the Far East was experiencing a massive in-migration of settlers, especially along the route of the Trans-Siberian, it was also experiencing migration into its towns and cities. From 1897 to 1911 the number of people living in urban areas in the Far East grew from 93,000 to 262,000, an increase in the proportion of people living in cities from 15 per cent to 23 per cent, a very high proportion for Russia, which had a total urban population of only 13 per cent in 1913.[49]

By 1911 an economically diverse Far East, linked to the European Russian heartland, had taken on the basic profile that would serve as the foundation for Far Eastern development into and during the Soviet period. Russians had become the largest and dominant ethnic group, and the relatively high level of urbanization served as the skeletal structure for future growth. Dominant regional urban centres had emerged, many of which had grown on the sites of early Cossack forts and some of which had flourished as a result of the changing economic geography of the Far East in the twentieth century.

By the twentieth century, the basic urban character of the Far East reflected the colonial relationship between the region and the European heartland. Several large cities dominated the settlement system of the Far East, although by 1911 smaller urban centres were beginning to fill in the network (Table 2.2, Figure 2.4).

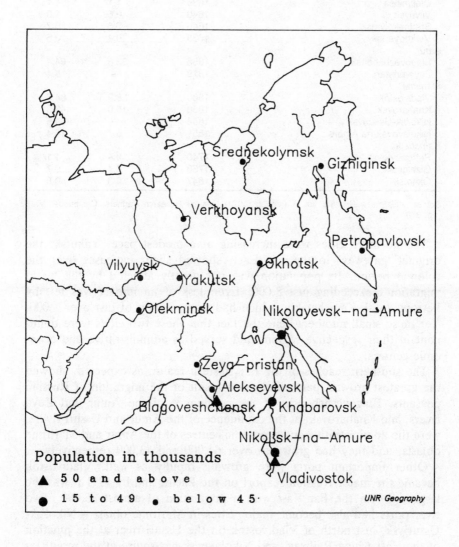

2.4 The Far East settlement system, 1911

21

Table 2.2 Population of Far Eastern cities, 1897–1911

	Date of founding	Population	
		1897 ('000s)	1911 ('000s)
Yakut			
Yakutsk	1632	6.5	8.2
Olekminsk	1635	1.1	1.1
Vilyuysk	1640	0.6	1.0
Srednekolymsk	1646	0.5	0.7
Verkhoyansk	1638	0.4	0.5
Amur			
Blagoveshchensk	1856	32.8	64.4
Zeya-pristan'	1879	–	5.4
Maritime			
Vladivostok	1860	28.9	84.6
Khabarovsk	1858	15.0	43.3
Nikol'sk-Ussuriyskiy	1858	–	34.6
Nikolayevsk-na-Amure	1851	5.7	16.4
Kamchatka			
Petropavlovsk	1740	0.4	1.1
Gizhiginsk	1753	0.4	0.7
Okhotsk	1647	0.3	0.6

Source: Aziatskaya Rossiya, vol. 1 (reprint: (1974) Oriental Research Partners, Cambridge, Ma.), p. 350.

The northern cities were increasing at a modest pace. Yakutsk, the original 'gateway' to the Far East, showed slight increases over the colonial period. Its population was not greatly affected by the 'great migration', exceeding just 8,000. Even less of an impact was felt by Petropavlovsk-Kamchatskiy, which had a population of just over 1,000. Yet these small numbers belie the fact that these two cities were dominant in their respective regions and served as administrative and economic centres.

The southern reaches of the Far Eastern realm, as expected, showed the greatest growth, particularly as a result of the migration of Russian peasants. Blagoveshchensk, at the confluence of the Amur and Zeya rivers, and Khabarovsk, at the confluence of the Amur and Ussuri rivers, were the administrative and economic centres of the Amur and Maritime oblasts, and they had grown to over 64,000 and 43,000 respectively.

Other important ports were growing rapidly as well. Vladivostok became the major Russian seaport on the Pacific and by 1911 it was the largest city in the Far East with a population of almost 85,000. Two river ports had also become major cities in Maritime oblast – Nikol'sk-Ussuriysk, just north of Vladivostok on the Ussuri river at the junction of the East China Railway, and Nikolayevsk-na-Amure at the mouth of the Amur river.

The fourteen cities of the Far East accounted for a total urban population of 262,000, of which 179,000, or 68 per cent, was attributed to the ports of Maritime oblast, reflecting the increasing connectivity of the region not only to Mother Russia but to the entire Pacific region, especially China and Japan.

The Far East after the Revolution: continuity and change

The eventual fate of the Far East was anything but certain during the period 1918–22. The Civil War brought widespread fighting and bloodshed. Red forces battled White forces, which were in turn supported and challenged by Allied interventionist forces: Britain, France, Japan and the United States. For a time, 1918–20, the White Admiral Kolchak declared himself to be 'supreme ruler' of the Far East.[50] After his execution in 1920, an independent and democratic Far Eastern Republic was established, 1920–22, greatly supported by the Soviet government, which saw the republic as a means to thwart Japanese intentions in the region.[51] At one point, economic and political relations were established between the Soviet government and the Far Eastern Republic and economic and military aid were provided by the Soviets.

The capture of Vladivostok by the Red Army in 1922 effectively put an end to hostilities in the region and, with strong encouragement from the United States, the Japanese withdrew their forces in the same year. As the Civil War came to a close and the Soviet government gained control of the territory once under the control of the Russian Empire, the Far Eastern Republic was subsumed within the Russian republic (RSFSR or Russian Soviet Federative Socialist Republic) and in 1923 came under the control of the newly created Union of Soviet Socialist Republics.

Even then, the Far East enjoyed a great degree of local autonomy. From 1926 to 1937–8, the Far Eastern Territory (FET), with its own semi-independent army organization, was established. In many instances, particularly with regard to the implementation of the first two five-year plans, it actually opposed and blocked policies from the centre. From 1937 to 1938 the central government carried out military and civilian purges in the FET that effectively brought the region under Stalin's control and the political entity was dissolved in 1938.[52]

The Bolshevik state that emerged after the Revolution exhibited characteristics that represented both a continuation of the basic features of tsarism and a radical change from the past.[53] The blending of these seemingly disparate factors provided for a complex and ambiguous role for the Far East within the new Soviet state. Political and economic pragmatism balanced ideological considerations in determining the nature of the relationship between the Far Eastern periphery and the Russian core.

The Far East continued to be an important region in the Soviet Union for many of the same reasons that had made it desirable to the Russian empire. Although the specific kinds of resources changed in importance over time, the Far East still represented a repository of real and potential wealth of untold value. Additionally, Siberia and the Far East still provided the Soviet Union with a land link between Europe and Asia and, thus, played an increasingly important role economically and strategically.

It is, in fact, virtually impossible to separate the economic from the strategic considerations when discussing the importance of the Far East to the Soviet Union. Strategically, the Far East provided the Soviet state a presence in Asia and access to the Pacific realm. The Pacific presence afforded the Soviet Union by controlling the Far East served to confront traditional and potential enemies, particularly Japan, China and the United States at the same time that it encouraged greater integration with these countries. This presence might not have been so important if it were not for the economic value of the region.

Yet, even given the dominance of economic and strategic considerations for the continued Soviet interest in the Far East, one cannot totally ignore the political and psychological factors. As this chapter argued earlier, the accumulation of a vast empire was not solely an economic exercise for imperial Russia. Also of importance were the political and psychological motivations – the prestige afforded European powers as a result of having great empires. One might expect such a rationale to diminish, if not disappear, under communist ideology. One might, but one should not. Apparently, the national pride associated with controlling a large empire or state, ideologically anathema to a true communist state, continued into the Soviet period. In the 1930s, as Stalin was purging the oppositionist tendencies of the FET as a prelude to its extinction, a 'Russian national "mystique" ' was being created to emphasize the role of Russian heroes in the development of the Far East.[54]

To a great extent these factors show the great continuity between tsarist and Soviet perceptions of and approaches to the role of the Far East. These economic, political and political–psychological considerations, which were part of the Far East's allure in tsarist times, were balanced by an ideological commitment to spreading the benefits of socialism and development to all regions of the country, to providing remote regions with some measure of equality in economic development, and to creating a true 'Soviet nation-state' that would eventually supersede and replace antagonisms and inequalities based on class and ethnicity.

According to communist ideology, the Far East should no longer serve as a second-rate periphery to the Russian heartland, but should attain equal status. The socialist state was ideologically committed to eliminate the dichotomy between standards of living and the qualities of life in the core and periphery. It is in this regard that Soviet and tsarist perceptions differed most. And although it may be argued that this was simply communist rhetoric and the Far East has, in fact, retained its traditional role as a resource colony, this difference should not be so easily dismissed. One can detect a sensitivity to ideological considerations in the treatment of the Far East region from one five-year plan to the next.

Settlement system to the 1930s: the skeleton for Soviet development and integration

For the first decades following the Revolution, the Far East experienced sustained population growth, thus continuing growth trends begun in the latter years of tsarist rule (Table 2.3). From 1911 until the first Soviet census of 1926, the total population of the region grew from 1.1 million to 1.5 million, an increase of 36 per cent. From 1926 to 1939 the population almost doubled to just under 3 million. These trends were all the more dramatic when one considers that this growth occurred during periods of war, civil war, collectivization, and military and political purges.

Table 2.3 Urban growth in the Far East, 1911–39

	Population			Population increase			Relative increase		
	1911	1926	1939	1911–26	1926–39	1911–39	1911–26	1926–39	1911–39
	(000s)	(000s)	(000s)	(000s)	(000s)	(000s)	%	%	%
Far East total	1,132	1,580	2,976	448	1,396	1,844	40	88	163
urban	263	377	1,385	114	1,008	1,122	43	267	427
Russia/ USSR total	139,313[a]	147,028	170,557	7,715	23,529	31,244	6	16	22
urban	24,820	26,314	56,125	1,494	29,811	31,305	6	113	126

Sources: Aziatskaya Rossiya, vol. 1 (reprint: (1974) Oriental Research Partners, Cambridge, Ma.), pp. 241–2 and 350; E.N. Pertsik (1980) Gorod v Sibiri, Moscow, p. 84; Itogi vsesoyuznoy perepisi naseleniya 1959 goda, (1963) Moscow, pp. 24–9; F. Lorimer (1946) Populations of the Soviet Union, Geneva, pp. 69–70.
Notes: [a]1913 data for 1938 boundaries.

At the same time, the total population of tsarist and Soviet Russia, as defined by the borders that existed in 1938, grew from 139 million to 147 million during the period 1911 to 1926, an increase of 6 per cent. From 1926 to 1939 the total population of Soviet Russia increased 16 per cent to 171 million. Thus, the growth rates for the Far East were substantially greater than for the country as a whole.

To a great extent, the differences in growth rates between the Far East and the country as a whole can be explained by examining the tsarist and Soviet components of this period separately. During the last years of tsarist rule the primary factors affecting growth, as discussed above, were related to the stimulus of the Trans-Siberian Railway, which provided for large numbers of immigrants, especially during the early years of the twentieth century prior to the Revolution. On the other hand, differences for the two decades after the Revolution resulted from the fact that the European regions of the country experienced greater losses as a result of the First World War, the Civil War, collectivization and purges. There was also some transferring of population as a result of forced migration. For the most part, then, greater rates of increase during the tsarist period, and the Soviet period, accounted for the dramatic population increases in the Far East during the first four decades of the twentieth century.

At the same time that the total population of the Far East was increasing, so were the numbers of ethnic minorities, although their proportion of the total population decreased because of the large gains made by Russians and other Slavs.[55] In 1911 there were 671,000 Russians, or 61 per cent of the population of the Far East. By 1926 the number of Russians and other Slavs had increased to over 80 per cent of the population. Although indigenous tribes were able to maintain, and in most cases, increase their numbers during the transition to Soviet rule, by the Second World War, the Soviet Far East, especially in its cities, was clearly a Russian land.

Urbanization

Development strategies in the Far East during the Soviet period continued to be concentrated in cities, which served as centres for resource exploitation and as the framework for an integrated settlement network. In addition to providing the means to develop resources and to link European Russia with Asia and the Pacific, urbanization was also designed to improve standards of living and lessen regional inequalities in a socialist state.

The importance of improving standards of living and lessening regional differences was greatly de-emphasized during the Stalin years. He effectively put an end to comprehensive urban planning at the 1931 Plenum of the Central Committee.[56] Yet, this did not adversely affect growth trends in the urban Far East. Migration policies were still supported and the numbers of migrants were supplemented by large numbers of convicts, ex-military and Komsomols, all of whom made a major contribution to growth during the Stalin years.[57] (Komsomols are members of the All Union Leninist Union of Youth, the final stage before selection for Communist Party membership.)

From 1911 to 1926 the urban population of the Far East increased by 43 per cent to 377,000, and from 1926 to 1939 it more than tripled to 1.4 million (see Table 2.3). Although the absolute increase of 1.1 million new urban dwellers between 1911 and 1939 represented only about 4 per cent of the total urban increase for the country as a whole, the relative increase of 427 per cent showed a substantially greater rate of growth than the national average, which produced a relative increase of 126 per cent. From 1911 to 1939, the Far East averaged an increase of 6 per cent in the urban population annually, which was double the national average annual urban increase of 3 per cent.

When one examines the periods 1911–26 and 1926–39 separately, very different trends can be identified. The 1911–26 urban increase of 114,000 comprised 25 per cent of the total population increase for the Far East. This apparently resulted from the fact that most of the growth in the Far East during the last years of tsarist rule was attributable to the migration of peasant farmers. Greater levels of increase that might have occurred in the cities were most likely mollified by war and purges.

The effects of industrialization and urbanization during the first five-year plans became apparent during the period 1926–39. The 1 million-plus increase in the number of people living in cities accounted for 74 per cent of the total population increase for the region. Whereas the proportion of urban dwellers actually decreased from 23 to 20 per cent from 1911 to 1926, it showed a healthy increase to 47 per cent by 1939. These trends paralleled national trends during the earlier period and greatly exceeded them for the period to 1939, when the national urban proportion of the population increased to 33 per cent (Table 2.4). These data show that as the newly-emerged Soviet state was becoming an urban society, the Far East was becoming even more highly urbanized. This was particularly true of the late 1920s and 1930s as Stalin's economic blueprint was being put into operation.

Table 2.4 Percentage of urban population in the Soviet Far East compared with Russia/USSR

	1911	1926	1939
Soviet Far East	23.0	24.0	47.0
Russia/USSR	18.0	18.0	33.0

Sources: Aziatskaya Rossiya, vol. 1 (reprint: (1974) Oriental Research Partners, Cambridge, MA.), pp. 241–2 and 350; E.N. Pertsik (1980) Gorod v Sibiri, Moscow, p. 84; and Itogi vsesoyuznoy perepisi naseleniya 1959 goda (1963) Moscow, pp. 24–9; F. Lorimer (1946) Population of the Soviet Union, Geneva, pp. 69–70.

One characteristic of urbanization in the Far East was clearly established during these early years – the dominance of the settlement system by relatively few large cities. This, of course, reinforced the pattern established during the tsarist colonization (Table 2.5). In 1911 there were a total of fourteen cities in the Far East. Of these, three had populations between 15,000 and 50,000, and another two had populations of more than 50,000. These five cities accounted for 89 per cent of all people living in urban settlements.

Table 2.5 Urban growth in the Far East by city-size categories, 1911–39

City-size categories ('000s)	No. of cities	1911 Pop. ('000s)	% of urban	No. of cities	1939 Pop. ('000s)	% of urban
100+	0	0	0	2	413	30
50–99	2	149	57	4	255	18
15–49	3	84	32	14	386	28
Total		233	89		1,054	76

Sources: Aziatskaya Rossiya, vol. 1 (reprint: (1974) Oriental Research Partners, Cambridge, Ma.), pp. 241–2; and Itogi vsesoyuznoy perepisi naseleniya 1959 goda (1963) Moscow, p. 38.

By 1939, after two decades of Soviet rule and one decade of five-year plans, the number of cities had greatly increased and to these had been added a substantial number of urban-type settlements.[58] Between 1917 and 1940, fifteen cities and forty-two urban-type settlements were officially designated in the Far East. Yet, the largest cities continued to dominate. In 1939, the twenty cities with populations of 15,000 and above accounted for 76 per cent of the entire urban population of the region, and of these, cities of 50,000 and above accounted for almost half of all urban dwellers (Table 2.6; Figure 2.5).

Thus three characteristics, highlighted urban growth during the early years of the Soviet state. A large proportion of the dominant cities in the region were pre-Revolutionary in origin, highlighting the degree of continuity between pre- and post-Revolutionary urban policies and the importance of historical advantage. All but one of the cities with populations of 50,000 and above in 1939 were pre-Soviet in origin. Yet, new Soviet cities were built and they also became important parts of the urban settlement system in the Far East. Of the twenty cities with 1939 populations of at least 15,000, twelve had been founded since the Revolution. Third, at the same time that dominant cities were increasing their status within the hierarchy, there was also a dispersion of population to a greatly increasing number of smaller cities and urban-type settlements.

Growth of major cities

The growth of individual cities had been modest up to 1926.[59] Only Vladivostok, which had increased its population by 23,000, had shown substantial growth. From 1926 to 1939, however, all major cities, with the exception of Blagoveshchensk, which had been bypassed by the Trans-Siberian, showed substantial gains. Khabarovsk and Vladivostok reigned as dominant regional centres with populations in excess of

Table 2.6 Major cities of the Soviet Far East, 1926–39 (with populations of 15,000 and above)

	Population 1926	Population 1939	Population increase 1926–39	Relative increase 1926–39
	(000s)	(000s)	(000s)	%
Yakut				
Yakutsk	11	53	42	382
Maritime				
Vladivostok	108	206	98	91
Ussuriysk	35	72	37	106
Suchan	–	37	–	–
Artem	–	35	–	–
Lesozavodsk	–	24	–	–
Spassk-Dal'niy	–	23	–	–
Khabarovsk				
Khabarovsk	50	207	157	314
Komsomol'sk-na-Amure	–	71	–	–
Birobidzhan	–	30	–	–
Nikolayevsk-na-Amure	7	17	10	140
Bikin	–	15	–	–
Amur				
Blagoveshchensk	61	59	–2	–3
Svobodnyy	10	44	34	340
Belogorsk	–	34	–	–
Skovorodino	–	20	–	–
Kamchatka				
Petropavlovsk-Kamchatskiy	2	35	33	1650
Magadan				
Magadan	–	27	–	–
Sakhalin				
Aleksandrovsk-Sakhalinskiy	–	25	–	–
Okha	–	20	–	–

Sources: *Itogi vsesoyuznoy perepisi naseleniya 1959 goda* (1963) Moscow, p. 38; Harris, Chauncy (1970) *Cities of the Soviet Union*, Chicago: American Association of Geographers, pp. 256–67.

200,000. Four other cities – Ussuriysk, Komsomol'sk-na-Amure, Blagoveshchensk and Yakutsk – had populations above 50,000, and fourteen others had populations between 15,000 and 50,000.

If any one city in the Far East benefited from Soviet industrialization and urbanization policies in the 1930s it was Khabarovsk. The four-fold increase in population from 1926 to 1939 established it as the region's largest city, a status it held for three decades. At the confluence of the Amur and Ussuri rivers, Khabarovsk had long served as the crossroads to the Russian and Soviet Pacific realm. It was the administrative and economic centre of the Russian Maritime region. It later developed into an important trans-shipment point for the Trans-Siberian Railway. During Soviet industrialization, it also became a multi-functional manufacturing centre, producing agricultural, timber and fish products, along with machinery and ships.

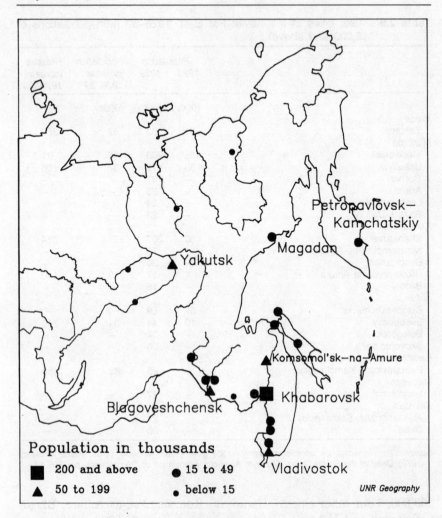

2.5 The Far East settlement system, 1939

Vladivostok also had a population of over 200,000 in 1939, although its growth, which almost doubled the population between 1926 and 1939, had not been nearly so spectacular as Khabarovsk's. In fact, it had barely lost its status as the largest city in the Far East during this period. Within a few years of its founding in 1860, Vladivostok had established itself as Russia's queen city of the east. Not only did Vladivostok become Russia's major Pacific port and window to the Pacific, it also provided access to Russia for a great variety of foreign influences, especially American, Japanese and Chinese. It was the cosmopolitan nature of Vladivostok that worried Stalin and led to the purges of the 1930s, which may have been partly responsible for it not having even greater rates of growth, comparable to Khabarovsk, during the period. It was often referred to politically and ideologically as the eastern equivalent of Leningrad.

The geographical advantage of Vladivostok lay in its location on a well protected bay. Its port can be kept open all year with ice breakers, and it became the main Pacific port for the Soviet navy. It also developed a range of industries closely associated with the processing and exporting of the region's natural resources, e.g. fish canning and processing, and timber processing. But its importance and growth were primarily dependent on its strategic role as a port and greater economic integration in the Pacific basin.

The increasing importance of the zone between Khabarovsk and Vladivostok along the Ussuri river and rail line also stimulated the growth of smaller towns and cities. The most noteworthy of these was Ussuriysk, which developed into a regional centre for the processing of natural resources, particularly agricultural, timber and fish products.

Komsomol'sk-na-Amure, initially founded in 1932, was a truly Soviet new town, built with the enthusiasm and idealism of the communist youth league, Komsomol. It soon developed into one of the most diversified cities in the Far East. It was a major iron and steel centre, with oil refining facilities, machinery works and a variety of other industrial activities. In just over seven years it had already become the region's fourth-largest city.

Blagoveshchensk had long been a major Far Eastern entrepôt and was the region's second-largest city into the twentieth century. It remained the administrative centre of Amur oblast and the economic centre for the Zeya-Bureya agricultural region into the Soviet period. As such its major economic functions were related to food processing and transportation. Located at the confluence of the Amur and Zeya rivers, Blagoveshchensk was bypassed by the Trans-Siberian and connected only by a branch line. This greatly contributed to its relative stagnation during the early Soviet period, showing small declines in population from 1911 to 1939.

Yakutsk had been the 'gateway to the Far East' since the seventeenth century. It remained the capital of the largest administrative unit in the Soviet Union, outside of the RSFSR. Its status as a regional economic centre continued to grow during industrialization because of the discovery of new natural resources, especially in southern Yakutia.

The administrative centres of three remotest regions of the Far East – Kamchatka, Magadan and Sakhalin oblasts – all had populations under 40,000. Petropavlovsk-Kamchatskiy had been a very small regional centre from tsarist times, whereas Magadan and Yuzhno-Sakhalinsk were new

Soviet cities, both officially founded in 1939, although Yuzhno-Sakhalinsk was not included in the 1939 census.

In addition to being the administrative and economic centres of two of the most remote regions of the Soviet state – Kamchatka and Sakhalin oblasts, respectively – Petropavlovsk-Kamchatskiy and Yuzhno-Sakhalinsk also developed into important ports for the Soviet navy, although more limited than Vladivostok because of long periods of icing. The industries that developed were closely tied to the processing of sea foods, timber products and the building and repair of ships. Magadan, in addition to serving as the administrative centre for Kamchatka oblast, developed as the economic centre for the nearby Kolyma gold fields.

Conclusion: the Soviet Far East as periphery to the Russian core

For three centuries, Russian colonization of the Far East was directed and controlled through a select number of geographically well-situated settlements. Building upon the original Cossack forts, these settlements grew into cities, which served as administrative and economic centres for vast regions of the Russian empire's eastern periphery. By the time of the Revolution, the Far East possessed a settlement system that was linked by rail, river and sea. This settlement system not only served as control points for the exploitation of natural resources by the European core, but also as a source of access to the countries of the Pacific rim. As Russian presence in the Far East became increasingly important, the cities of the Far East became increasingly necessary for the integration of the periphery to the European core.

During the first decades of Soviet rule, the new Bolshevik state also appreciated the value of the Far East. Rather than destroying the vestiges of pre-Revolutionary capitalist development, as embodied in the settlement system, the new rulers in the European core built upon the tsarist system, which it used as the basis for their own policies towards the periphery. Pragmatic considerations, namely the need for the Far East's natural resources and its strategically important location *vis-à-vis* the Pacific, modified ideological considerations as the settlement system built by the tsars was maintained to help retain Far East dependence on the European core. Particularly during the Stalin years, when actual investment and the growth of the settlement system were highly concentrated in specific regions for those projects of value to the Soviet economy as a whole, the relationship that developed was similar to the pre-Revolutionary relationship – one between a dominant European core and a Far Eastern periphery.

The administrative and economic centres inherited by the Soviet regime continued to grow because they had a more developed infrastructure and an existing concentration of goods, people and services. To these were added 'new cities' to enhance the settlement system to integrate the region more fully into the Soviet economy. By 1939, the settlement system for the Far East had increased its role in the region and, although the Second World War temporarily set back the course of Soviet development and urbanization, the skeletal structure for future exploitation and development of the region was firmly in place.

Notes

1. Most historical treatments of the Far East exclude Yakutia because it was always considered part of East Siberia, and it did not officially become part of the Far East until 1963. This study includes Yakutia throughout for historical continuity.
2. For more detailed treatments of indigenous peoples of Siberia see: *Aziatskaya Rossiya*, vol. 1 (2 vols, originally published in St. Petersburg, 1914; reprint published in 1974 by Oriental Research Partners, Cambridge, MA.); Walter Kolarz (1954) *The Peoples of the Soviet Far East*, New York City: Praegar; M.G. Levin and L.P. Potapov, eds (1964) *The Peoples of Siberia*, trans. by Stephen Dunn, Chicago: University of Chicago Press; and Ronald Wixman (1984) *The Peoples of the USSR: an ethographic handbook*, New York: Sharpe.
3. This brief summary is based on detailed treatments of the Russian conquest of Siberia and the Far East and can be found in *Aziatskaya Rossiya* (2 vols); John Harrison (1971) *The Founding of the Russian Empire in Asia and America*, Coral Gables: University of Miami Press; A.P. Okladinov, ed. (1968) *Istoriya Sibiri*, 5 vols, Leningrad; Y.N. Semyonov [Semenov] (1963) *Siberia: Its Conquest and Development*, trans. from the German by J.R. Foster, London: Hollis & Carter; and Donald Treadgold (1957) *The Great Siberian Migration*, Princeton: Princeton University Press.
4. *Aziatskaya Rossiya*, op.cit., vol. 1, p. 44.
5. Harrison, *The Founding of the Russian Empire*, op.cit., pp. 65–8.
6. George Lantzeff (1972) *Siberia in the Seventeenth Century*, New York: Octagon.
7. ibid.
8. Mark Bassin (1988) 'Expansion and colonialism on the eastern frontier: views of Siberia and the Far East in pre-Petrine Russia', *Journal of Historical Geography* 14, p. 8.
9. Semyonov, *Siberia*, op.cit., p. 104.
10. Nicolas Berdyaev (1962) *The Russian Idea*, Boston: Beacon, p. 9.
11. Harrison, *The Founding of the Russian Empire*, op.cit., p. 69.
12. Donald Meinig (1982) 'Geographical analysis of imperial expansion' in A. Baker and M. Billinge (eds), *Period and Place*, Cambridge: Cambridge University Press, p. 71.
13. A more detailed analysis of tsarist policies can be found in Hugh Seton-Watson (1967) *The Russian Empire 1801–1917*, Oxford: University of Oxford Press.
14. Treadgold, *Great Siberian Migration*, op.cit., p. 33.
15. Seton-Watson, *The Russian Empire*, op.cit., p. 485.
16. ibid., pp. 512–13.
17. ibid., p. 548.

18. *Great Soviet Encyclopedia* (1950–58) *Bol'shaya Sovetskaya entsiklopedia*, 2nd edn, Moscow, vol. 22, p. 502.
19. *Aziatskaya Rossiya*, op.cit., vol. 2, pp. 515–16.
20. For an excellent account of Witte's role as Minister of Finance see Theodore von Laue (1963) *Sergei Witte and the Industrialization of Russia*, New York: Columbia University Press.
21. ibid., p. 87.
22. Lyashchenko, P.I. (1949) *History of the National Economy of Russia*, trans. by L.M. Herman, New York: Macmillan, p. 585.
23. ibid., pp. 534 and 585.
24. *Aziatskaya Rossiya*, op.cit., vol. 2, pp. 513–26.
25. Semyonov, *Siberia*, op.cit., p. 317.
26. Treadgold, *Great Siberian Migration*, op.cit., p. 32.
27. ibid., p. 35.
28. ibid., p. 33.
29. ibid., p. 34.
30. Semyonov, *Siberia*, op.cit., p. 281.
31. Treadgold, *Great Siberian Migration*, op.cit., pp. 69–71.
32. ibid., p. 70.
33. *Aziatskaya Rossiya*, op.cit., vol. 1, pp. 82–6.
34. ibid.
35. Detailed discussions of the economic geographies of late-tsarist Far East can be found in the various sections of *Aziatskaya Rossiya*, vol. 1, 'Lyudi i poryadki za Uralom', and Okladnikov, *Istoriya Sibiri*, vol. 3, 'Sibir' v epokhu kapitalizma'.
36. *Aziatskaya Rossiya*, op.cit., vol. 1, p. 509.
37. ibid., vol. 2, pp. 186–7.
38. ibid., vol. 1, pp. 501–2.
39. ibid., vol. 1, p. 507.
40. ibid.
41. ibid., vol. 2, p. 327.
42. ibid., vol. 1, p. 505.
43. ibid., vol. 1, p. 511.
44. ibid.
45. ibid.
46. Okladnikov, *Istoriya Sibiri*, op.cit., vol. 3, p. 340.
47. Bassin, Mark (1986) 'The Russian Far East: an enduring national periphery', paper presented to the American Association of Geographers' annual meeting, Minneapolis.
48. *Aziatskaya Rossiya*, op.cit., vol. 1, pp. 241–2.
49. Harris, Chauncy (1970) *Cities of the Soviet Union*, Washington, DC: American Association of Geographers, p. 232.
50. Ulam, Adam (1976) *A History of Soviet Russia*, New York: Praeger, pp. 36–9.
51. Kolarz, *Peoples of the Soviet Far East*, op.cit., pp. 2–3.
52. ibid., pp. 4–10.
53. For an overview of the various theories of the Soviet state, see Osborn, Robert (1974) *The Evolution of Soviet Politics*, Homewood, IL.: The Dorsey Press.
54. Kolarz, *Peoples of the Soviet Far East*, op.cit., p. 22.
55. *Aziatskaya Rossiya*, op.cit., vol. 1, and Levin and Potapov (eds), *The Peoples of Siberia*, op.cit.

56. For an overview of urban planning policies during the Soviet period see Jensen, Robert (1984) 'The anti-metropolitan syndrome in Soviet urban policy', in George Demko and Roland Fuchs (eds), *Geographical Studies on the Soviet Union* (The University of Chicago, Department of Geography, Research Paper No. 211), pp. 71–91.

57. For a brief identification of the various kinds of migrants see Kolarz, *Peoples of the Soviet Far East*, op.cit., pp. 12–22.

58. Soviet classification procedures distinguished between cities (*goroda*) and urban-type settlements (*poselki gorodskogo tipa*); although definitions vary, cities are generally defined as places of more than 10,000 to 12,000, while urban-type settlements have a minimum of 2,000 to 3,000 inhabitants and between 60 and 85 per cent of the labour force is non-agricultural; cities and urban-type settlements are combined to make up the total urban population.

59. The descriptions of individual cities were compiled from various sources, especially useful were *Aziatskaya Rossiya*, op.cit., vol. 1; *Istoriya Sibiri*, op.cit., vol. 3; and Harris, *Cities of the Soviet Union*, op.cit.

Chapter three

Environmental constraints and biosphere protection in the Soviet Far East

Philip R. Pryde and Victor L. Mote

An expanded rate of economic growth has been predicted for the Soviet Far East for some time. This has been assumed in part on the expectation of accelerated Siberian development generally, and in part on the expansion of Soviet trade with other countries in the Far East. The most promising recent catalyst has been the construction of the new Baykal–Amur Mainline (BAM) railway.

The potential of the Far East was well summarized in the title of a recent book chapter which referred to Siberia in its entirety as 'a country in reserve' (Komarov 1980, p. 112). However, much of the environment of Siberia and the Soviet Far East is simultaneously very harsh and very fragile, and among Soviet authorities there is an increasing appreciation that these factors need to be better assimilated into the planning process. The region's severe climatic limitations are probably the best known and best reported constraints to development (Borisov 1965; Lydolph 1977a), but the numerous potential adverse environmental impacts that await the careless developer of Siberian resources have been pointed out also by numerous Soviet writers (Komarov 1980; Rasputin 1985; Vorob'yev 1984).

The ensuing two sections of this chapter will review the primary natural constraints, and environmental protection factors, that might bear on the development of the vast expanses of the Soviet Far East.

Natural constraints to development

The natural constraints to the development of the Soviet Far East include a variety of climatic factors, as well as considerations involving soils, permafrost, seismicity, volcanism and other physical factors. The basic nature of these constraints will be summarized in this section. For a more detailed treatment, the interested reader is referred to Chapter 3 of *Soviet Natural Resources in the World Economy* (Mote 1983).

Topography

Topography was not an important initial constraint to the development of the Far East, because transportation through this region (initially horse caravans and later the Trans-Siberian Railway) could easily follow the water-level route of the Amur river. Any effort to develop the interior, however, would have to confront a formidable array of mountain ranges and, especially in the north, vast swampy or boggy areas in the low-lying river valleys and Arctic coastal plains.

The problems posed by the interior topography quickly impressed themselves on the planners of the Baykal–Amur railway, resulting in the preparation of a number of environmental studies on conditions along the BAM corridor. Even so, the rugged topography has seriously delayed the completion of tunnels through mountain ranges, and has greatly increased the project's cost. These high construction costs, which can run two to four times the cost of construction in the western USSR, are an economic fact of life in the majority of the Soviet Far East.

Climate

The general impression of the inhospitable nature of the winter climate over most of this region is accurate, and probably represents the dominant physical constraint (Figure 3.1). However, it also should be noted that summers here are relatively mild, and that the southernmost portion of Primorskiy Kray (Vladivostok area) has a climate no worse than Moscow or Minnesota (Table 3.1).

Table 3.1 Average summer and winter temperatures in cities

City	Ave. July T (°C)	Ave. Jan. T (°C)	Record Minimum
Vladivostok	17	−18	−31
Komsomolsk	19	−25	?
Khabarovsk	22	−23	?
Blagoveshchensk	21	−26	?
Zeya	16	−29	?
Petropavlovsk	12	−16	−34
Magadan	11	−21	?
Yakutsk	17	−42	−64
Anadyr	11	−22	−51
Oimyakhon	13	−48	−71

Source: Tochenov, U.V. *et al.*, eds. (1983) *Atlas SSSR*, Moskva: Glav. Uprav. Geodezil i Kartografii, p. 99.

The harsh winter cold creates not only physical discomforts and health hazards for humans, but also continuous mechanical problems for machinery (Kolyago 1970). The winter cold frequently results in ice fogs,

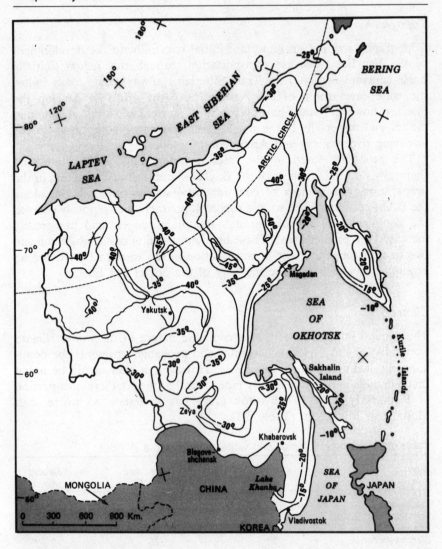

3.1 Average January temperatures in the Soviet Far East

especially near large rivers, that can paralyse air transportation. All types of transportation equipment must be kept running or otherwise warmed, lest critical parts freeze up, crack, or in any way suffer from the cold. Work productivity is necessarily reduced markedly.

In winter, lack of sunlight also becomes a constraint on normal activities. A predictable consequence is the need for greatly increased rates of energy consumption. One unusual effect of inadequate sunlight is a need for vitamin D supplements during winter months. The decrease in daylight hours at progressively more northerly habitations is illustrated by Table 3.2.

Table 3.2 Seasonal solar data for Soviet Far East cities at various latitudes

Latitude	Representative city close to given latitude	Hours sun is above horizon[a]		Noon sun elevation	
		June 21	Dec. 22	June 21	Dec. 21
64°N	Anadyr'	21:04	4:13	49.5°	2.6°
62°N	Yakutsk	19:45	5:09	51.5	4.6
60°N	Magadan	18:52	5:53	53.5	6.6
58°N	Aldan	18:10	6:27	55.5	8.6
56°N	Ust'-Kamchatsk	17:37	6:57	57.5	10.6
54°N	Zeya	17:08	7:23	59.5	12.6
52°N	Lazarev	16:43	7:45	61.5	14.6
50°N	Blagoveshchensk	16:22	8:05	63.5	16.6
48°N	Khabarovsk	16:02	8:23	65.5	18.6
46°N	Yuzhno-Sakhalinsk	15:45	8:39	67.5	20.6
44°N	Vladivostok	15:29	8:53	69.5	22.6

Note: [a]From US Naval Observatory ephemeris.

Storms and high winds aggravate the problems caused by low temperatures. Major storms are most common near the coast, where regional precipitation in winter (and summer) is at a maximum. These harsh, raw conditions impede both human and economic activities along the Sea of Okhotsk, and remind us why the only large city there, Magadan, has been best known historically as a penal colony (major place-names are located on Figure 3.8).

In the interior, snowfall totals are relatively low in many areas, due to the dominance of high pressure over eastern Siberia in winter. Most interior locations south of 58°N receive 60–75 per cent of their annual precipitation from April to October. However, at the low temperatures found in the interior, extreme wintertime wind-chill conditions are possible, further imperilling health and reducing work productivity.

Because interior Siberia is dominated by high pressure in winter, atmospheric temperature inversions are common, especially in valleys and lowland areas where cold air drainage adds to the problem. These inversions not only contribute to the severity of the ice-fog problem alluded to earlier, but also create ideal conditions for potentially severe air pollution. 'Smokestack industries' should avoid such valleys in northern latitudes.

Snowfall is extensive only in areas closer to the coast, and although there is snow cover on the ground (except in the Vladivostok area) for 150 to 280 days depending on latitude and elevation, it is not usually a major problem *per se*. It becomes a problem in severe winters, when winds produce 'whiteout' conditions or cause the snow to drift across transportation corridors; it is particularly dangerous in mountainous areas where avalanches become an occasional hazard.

Rivers

The Soviet Far East is drained by two major rivers, the Lena and the Amur, together with their major tributaries (Aldan, Bureya, Olekma, Ussuri, Zeya), and a few other major streams that drain into the Arctic (Indigirka, Kolyma, Yana). These rivers tend to be very wide, since downcutting is inhibited by permafrost. The sheer size of these rivers, especially the Amur and Lena, necessitates very costly bridges at well-chosen locations if motorized surface transport is to be accommodated.

The usefulness of these streams as transportation arteries is limited in two other ways. First, they are frozen much of the year. The Amur freezes on average around mid-November; the Lena freezes in late October at Yakutsk and around 1 October near its mouth. Spring break-up can be quite late: early May on the Amur and early June on the lower reaches of the Lena (Figure 3.2). While frozen, however, the ice-covered rivers themselves become convenient roads.

Equally a problem is the seasonal flooding, usually occurring in late spring. The normal spring flooding produced by snow-melt is greatly exacerbated on north-flowing rivers such as the Lena by the fact that when heavy run-off is occurring in the southern headwaters, the river's mouth (and surrounding land) is still tightly frozen. Consequently, huge lakes can form behind ice dams, making riverside settlement and transportation routes highly vulnerable. The widespread standing water in spring and summer also should not be overlooked as an ideal breeding ground for mosquitoes and other insects.

Soils and permafrost

Permafrost (permanently frozen ground), together with its effects on soils and groundwater, is the most critical consequence for building that results from the very cold climate which dominates most of the Soviet Far East. In the north, permafrost is deep and continuous, towards the south it becomes discontinuous or sporadic (Figure 3.3). Only the lowlands near the Ussuri and lower Amur rivers and the coast along the Sea of Japan are free of it.

The construction implications of permafrost are the most serious. Buildings are safe from destruction only if they are built on pilings sunk deep into permafrost well below the seasonal thawing zone, thus providing them with a solid anchor. This, of course, increases construction costs. If this is not done, structures quickly sink into a summer quagmire of irregularly thawed earth, and can be totally destroyed in a few years.

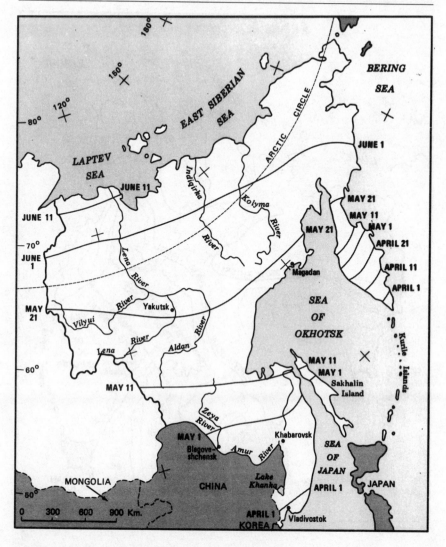

3.2 Average date of spring ice break-up in Far East rivers

The cost of roads and railroads is greatly increased by the far more elaborate road-bed preparation that is needed. Even telephone poles can be thrust out of the ground by frost-heaving.

The seasonal thin layer of thawed ground is especially fragile. Vehicle tracks pressed into it quickly become permanent linear rivers, creating a number of secondary problems. The soils over permafrost tend to be thin, nutrient-poor, waterlogged, and in general unstable and hence

41

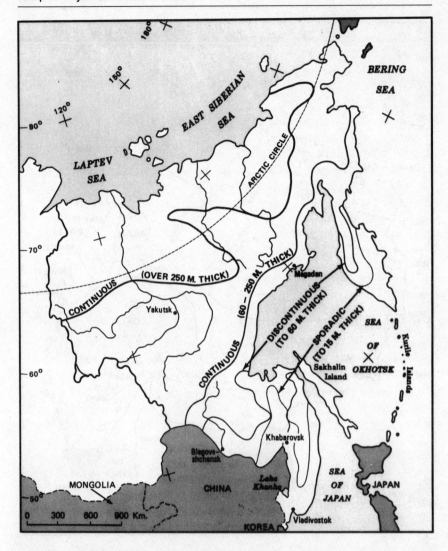

3.3 Depth and continuity of permafrost

unsuitable for either agriculture or construction (Nefedova 1978). The vast amounts of standing water near the surface tend to be stagnant, acidic and easily polluted.

Ironically, with all this surface water, most of the permafrost zone has little or no natural potable water supply. The surface water is generally undrinkable without treatment, and the permafrost precludes the use of wells. Small streams may freeze solid in winter. The larger rivers represent the best source of domestic water, despite being frozen at the surface for many months. The problems of transporting water in such cold climates can be easily imagined.

Seismicity

Earthquakes are to be expected throughout much of the Soviet Far East. By far the most intensive area is south-eastern Kamchatka and the Kuril Islands, where colliding continental plates along a zone of subduction create conditions similar to those along the west coast of North America. The city of Petropavlovsk lies within one of the most intense seismic regions (Figure 3.4).

Other areas where earthquakes have a high potential for damage are near the mouth of the Lena river, in the Cherskiy range, and especially the area along the west boundary of the study region near the Aldan and Olekma rivers. This latter area is the eastward extension of an intensive seismic zone that is centred on Lake Baykal and the Stanovoy range. This zone has the most potential of any in the Soviet Far East for disrupting economic activities, for it includes a portion of the route of the new BAM railway, as well as several proposed mining complexes (Shabad and Mote 1977). Earthquakes measuring from 6.5 to 7.9 on the Richter scale have been recorded in this area in recent decades.

Elsewhere in the Far East, earthquakes measuring between 4 and 5 (Richter scale) have been recorded on Sakhalin Island and along the coast of Primorskiy Kray. More ominously, a quake measuring 5.8 occurred not too far from the new Zeya dam (Mote 1983). However, most of the Ussuri river valley region has a low earthquake probability.

Landslides and avalanches are common consequences of earthquakes, even small ones. Their frequency of occurrence is enhanced both by the mountainous nature of much of the Soviet Far East, and by permafrost which serves to destabilize surface geology, especially during maximum freeze-thaw periods. The disruptive implications for economic activities and transportation routes throughout much of the Far East are clear.

A final consequence of earthquakes is the tsunami, or seismic sea wave, commonly called a tidal wave. Tsunamis are infrequent and unpredictable after-effects of earthquakes, but on occasion they can be devastating. Severe tsunamis smashed against south-eastern Kamchatka in 1904, 1923 and 1952, and against the southern Kuril Islands in 1958. The eastern coast of Sakhalin Island is also a vulnerable area.

Volcanism

The Soviet Far East contains one of the most active zones of volcanism to be found anywhere on earth, together with some of the most impressive

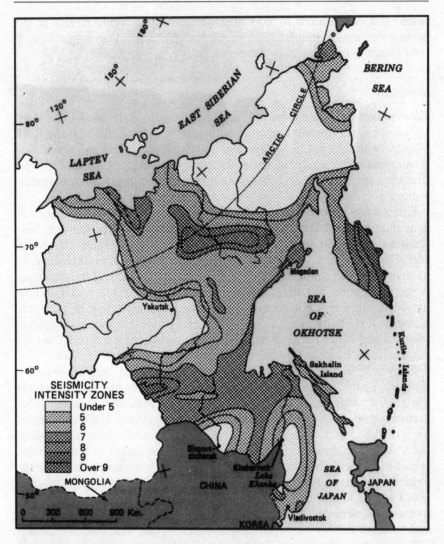

3.4 Seismicity regions in the Soviet Far East

volcanic cones and one of the world's four great geyser basins. This volcanic zone, however, is largely conterminous with the lightly populated Kamchatka Peninsula and Kuril Islands.

On both Kamchatka and the Kurils, volcanic eruptions are a common occurrence. There are several dozen active volcanoes, and many of their cones are very symmetrical, proof of their recent and continuing activity. The largest of them, Klyuchevskaya, extends 4,750 m into the Kamchatka sky, one of the tallest volcanic cones on the planet. It has erupted dozens of times in the nineteenth and twentieth centuries.

Some of these eruptions have been of enormous size. In 1956, a close neighbour of Klyuchevskaya named Bezymyannyy (3,085 m) erupted violently in a manner similar to Mount St Helens. The top was blown off the mountain and the landscape was denuded for vast distances around its base. Fortunately, because of the low population in the area, the many eruptions on Kamchatka and the Kurils have taken few lives. However, the only major city in the region, Petropavlovsk, is only 25 km away from the active Avachinsk volcano.

Of the world's four great geyser basins, one is located on Kamchatka, in the Kronotskaya valley (location no. 7 on Figure 3.7). Because of its remoteness, few non-Soviets have ever seen it, but it is described as second only to Yellowstone in its variety and extensiveness. It has been designated as a state nature reserve since 1967.

The Kurils consist almost entirely of active or dormant volcanic islands. Only the larger southernmost islands, which are still claimed by Japan, have any significant habitations. Elsewhere in the Soviet Far East, older volcanic calderas and plugs occur on the plateau just west of the lower Lena river, and a few inactive cones are found in the Kolyma mountains west of Anadyr'.

Protective constraints to development

The natural zones of the Soviet Far East vary from the temperate southern portions of Primorskiy Kray to the most climatically inhospitable areas of the entire country. These include mixed broadleaf-coniferous forests in Primorskiy Kray and along the Amur lowland, north of which is a broad belt of conifer forests (*taiga*), and still further north lies Arctic tundra (Knystautas 1987). This pattern is extensively broken by mountain ranges, on whose higher elevations montane tundra is found.

These geobotanical regions all possess inherent degrees of tolerance (or intolerance) for human activity, as well as varying degrees of biological resilience. Constraints of nature on human activities have been discussed already; in this second part of the chapter constraints placed by humans on economic development of the natural landscape will be examined.

The previous sections indicated that the further north one goes, the less tolerant the natural environment becomes to developmental activities (Sokolov and Chernov 1983). Both the United States (in Alaska) and the Soviet Union (in West Siberia) have discovered this in the process of extracting rich hydrocarbon deposits. Other natural zones in the Soviet Far East having inherently little tolerance for disturbance include high alpine ecosystems, virtually all wetlands and riparian systems, and most coastal areas.

In areas where such disturbance has already taken place, some species of plants and animals may have been greatly depleted or even pushed to the edge of extinction. The extent of this phenomenon in the Far East, and what the USSR is doing about it, are discussed in the ensuing sections.

Biosphere protection: endangered species

The protection of endangered species of flora and fauna has become an issue of world-wide importance, and is one the Soviet Union has also embraced. The main focus of this effort is the compilation of information on rare, threatened and endangered species into volumes called 'Red Books' (in the USSR, *Krasnaya kniga SSSR*). The most recent Soviet compilation as of this writing appeared in 1985; it included two volumes that dealt separately with fauna and flora (Borodin 1985). Volume One lists 463 species or sub-species of animals, of which 70 are designated as endangered. Out of 603 species of plants that appear in Volume Two, 135 are considered endangered.

A study on the distribution of the habitats of the 70 species of endangered fauna throughout the country revealed a very uneven pattern (Pryde 1987a). There is a marked concentration in the southern regions of the country, from the Ukraine to the southern portion of the Soviet Far East. These areas and the coastal seas account for the vast majority of all 70 species. A generalized map of this distribution within the Far East is presented as Figure 3.5.

The southernmost portion of the Soviet Far East, principally Primorskiy Kray and the immediate offshore waters, is one of the most critical areas in the Soviet Union for endangered species. Figure 3.5 is at too small a scale to show this clearly, so the Primorskiy Kray region has been re-drafted at a much larger scale for this purpose (Figure 3.6). The most significant areas of endangered species are concentrated around Lake Khanka, in the region of the south-eastern coast, and within the thin arm of land extending southwards towards Korea.

Does the presence of these endangered species pose any major hindrances to the development of this region? Probably not, even though the next section will note a corresponding concentration of nature reserves here. Most of these reserves are not large, and cumulatively represent only a small fraction of even the southern part of Primorskiy Kray. Further, the Soviet Union presently has no equivalent of the United State's Endangered Species Act, which can delay or (in theory at least) even halt development projects.

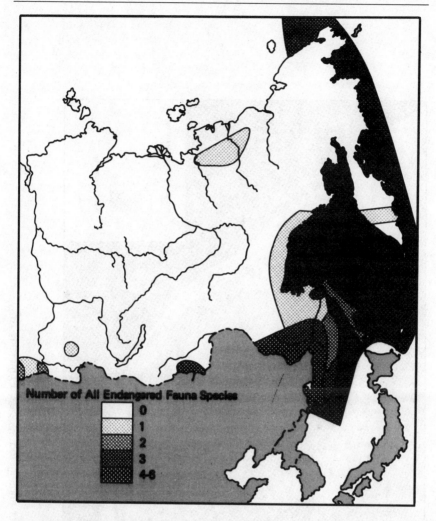

3.5 Areas of endangered fauna in the Far East

Biosphere protection: preserved areas

In the effort to conserve significant landscapes and ecosystems, and the flora and fauna within them, the Soviet Union has created various types of preserved territory. These include state nature reserves (*zapovedniki*), partially preserved areas (*zakazniki*), hunting preserves, national parks and natural monuments. Of the foregoing, the *zapovedniki* are the most important. They function as biological preserves and scientific research stations; few have any tourism function (Pryde 1972).

3.6 Concentration of endangered species, Primorskiy Kray

There are presently about 150 *zapovedniki* in the USSR (Fischer 1981; Borodin and Syroyechkovskiy 1983), of which sixteen are in the Far East (Figure 3.7 and Table 3.3). All of the largest of the Far East reserves (greater than 500,000 ha) are located in relatively hostile climatic zones. Table 3.3 reveals that over half of the protected area in this region has been set aside since 1982.

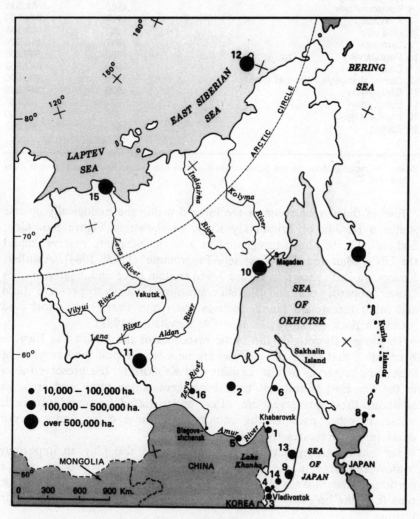

3.7 Location of nature reserves (*zapovedniki*)

Table 3.3 State nature reserves (*zapovedniki*) of the Soviet Far East

Name	Created	Hectares
1 Bol'shekhekhtsirskiy	1963	44,928
2 Bureiskiy	1985	350,000
3 Dal'nevostochnyy	1978	64,360
4 Kedrovaya Pad'	1916	17,897
5 Khinganskskiy	1963	82,186
6 Komsomol'skiy	1963	61,208
7 Kronotskiy	1967	964,000
8 Kuril'skiy	1984	63,365
9 Lazovskiy	1957	116,524
10 Magadanskiy	1982	869,200
11 Olekminsk	1984	847,102
12 Ostrov Vrangelya (Wrangel Island)	1975	795,650
13 Sikhote-Alin'	1935	347,532
14 Ussuriyskiy	1932	40,432
15 Ust'-Lenskiy	1986	1,433,000
16 Zeyskiy	1963	82,567
	Total:	6,179,951

Source: International Union for the Conservation of Nature and Natural Resources Protected Areas Data Unit.

Five of the sixteen reserves are located within the biologically diverse southern portion of Primorskiy Kray, as shown in Figure 3.6. One, Sikhote-Alin, has been designated as a world biosphere reserve, under the UN's 'Man and The Biosphere Programme' (Pryde 1984). Another, Kedrovaya Pad', shown in Figure 3.6 to contain three endangered species (Amur leopard, goral and Blakiston's fish-owl), works to preserve these and other threatened fauna, such as the Amur tiger, black stork and mandarin duck (Vasil'yev *et al.* 1984; Knystautas 1987).

To protect thoroughly the biotic resources of the Soviet Far East, it is probable that more *zapovedniki* are needed. Despite the existence of such huge preserves as Ust'-Lenskiy and Kronotskiy, the preserved area in the Far East is still less than 1 per cent of the total land area. In addition, these preserves are often under-staffed and under-funded; worse, in other parts of the country some have come under severe development pressures from other ministries.

The *zakazniki*, or partially preserved areas, also play an important conservation role in the Soviet Far East. These may protect only a portion of the biota within them, or for only a proscribed period of time (e.g. the breeding season), while various economic activities occur nearby.

Of 2,912 *zakazniki* in the USSR in 1985 having a total area of 48,075,500 ha, 109 embracing 14,340,800 ha (30 per cent of the total area in *zakazniki*) were in the Soviet Far East (Shalybkov and Storchevoy 1985). This indicates that these preserves in the Far East are characterized by very large average size. One very critical *zakaznik* of 16,500 ha exists around Lake Khanka on the Chinese border. The importance of Lake Khanka for endangered species can be seen clearly from Figure 3.6.

Although other types of protected area exist in the USSR, they are of little importance in the Far East. Since 1971 the Soviet Union has created about a dozen national parks, but as of 1987 none had yet been established east of Lake Baykal. The USSR has also identified several areas of 'wetlands of international importance' under the Ramsar Convention, but only one (Lake Khanka) is in the Soviet Far East.

Due to the relatively small percentage of the Far East encompassed by all these preserved areas (about 3 per cent), and especially given that all the largest ones are in northerly areas, they should have little detrimental effect on the economic development of the Soviet Far East. However, a general concern for biotic conservation exists (and is statutorily mandated) even outside the preserves.

Other conservation considerations

In addition to endangered species and preserved areas, a number of other conservation considerations weigh on economic development in the Soviet Far East. These include various aspects of fishing, timber harvesting, mining and air quality.

The waters off the Far East coasts have always been among the major fisheries of the Soviet Union. Yet Soviet fishing endeavours have long come under foreign scrutiny, especially in the United States. Criticism from abroad was clearly one factor, though certainly not the only one, in recent reductions in the scope of Soviet whaling. Both international agreements and international opinion can be expected to play a continuing role in defining the breadth and intensity of Soviet fishing and crustacean-harvesting activities on ships operating out of Far East ports.

The nature of the Far East forests, especially in the more northerly regions, also sets inherent limits on the scope of timber harvesting. The sparse forests that border on the Arctic tundra are designated as Class I (protected) forests, which virtually prohibits logging in them. Cutting on steep slopes has been banned in the Lake Baykal basin, and similar limitations could be imposed in sensitive areas of the Far East as well. Indeed, overcutting of easily accessible forests (i.e. those along rail lines, roads and rivers) has been criticized for years (Pryde 1972), but to do otherwise clearly implies higher production costs. Forest fires, such as ravaged Khabarovsk Kray in 1976, are also common in these remote *taiga* areas (Kostyrina 1986). Boreal forests are also slow-growing and fragile. One writer refers to Soviet studies that suggest that in order to preserve Siberian/Far East forest ecosystems, not more than 20 per cent of the trees should be removed (Komarov 1980, p. 118).

One official of the timber industry, A. S. Isayev, has stated: 'I am against an overenthusiastic view of the BAM zone's forest riches and think they must be used very judiciously . . . in a number of districts of Amur Province the forests are of rather low yield, and a significant proportion of them are situated in the permafrost zone. The commercial exploitation of such forests could lead to irreversible ecological processes' (Druzenko 1984a). Thus, the development of the Far East timber industry should, if knowledgeable counsel is accepted, proceed slowly and with care.

Great care must also be taken in the extraction and processing of mineral resources, if waste, landscape despoilation and severe air pollution are not to result. This is true in all parts of the country, but especially so in the Far East where biological limitations and potential for frequent atmospheric inversions compound the problems. Two areas where adequate care appears not to have been taken are around the cities of Noril'sk and Chul'man (Bond 1984; Mote 1985). In both locations, a serious air quality problem has resulted from natural resource processing. Both areas lie just west of the Soviet Far East region, but in each case their emissions undoubtedly flow periodically into it.

The effects of all the above are compounded by the low tolerance of arctic vegetation to pollution. Thus, economic development of all types must proceed with care in the fragile environments of the Soviet Far East.

Environmental impact analysis

In order to ensure that environmental considerations are taken into account in the course of economic construction, the Soviet Union, like the United States, utilizes a process of environmental impact analysis. In this process, adverse environmental changes brought about by development are predicted and evaluated, and wherever possible, mitigation measures are suggested that would reduce or eliminate the severity of these effects.

Although environmental impact analyses are now frequently prepared in the Soviet Union, the USSR has no specific law that states when such analyses must be prepared and what type of issue they should discuss. Instead, the USSR relies on broad enactments which govern the use of land, water, wildlife, air and other natural resources. These acts mandate conservation and the prevention of harmful actions, and Article 21 of the 1980 wildlife law calls for 'the organization of scientific studies aimed at substantiating measures for the protection of the animal world' (Anon., *Law of the USSR*). This article, however, does not say that such studies are mandatory, that they must be done prior to the start of the project, or that mitigation measures are required.

Within the Soviet Far East, the project currently receiving the most environmental study is the new Baikal–Amur mainline railway (BAM). This major new railway will function as a second transportation corridor across the Soviet Far East, running to the north of the existing Trans-Siberian Railway (Shabad and Mote 1977). The BAM enters the Far East near the town of Khani (on the border of the Yakut ASSR and Amur Oblast), and then continues eastwards to Komsomol'sk-na-amure, where it connects with an existing rail line to Sovetskaya Gavan' on the Sea of Japan (Figure 3.8).

The environmental studies being conducted within the BAM corridor are intended to assemble baseline data on such features as geology and vegetation that can be used to assess changes brought about by future economic development (Belov 1983; Belov and Krotova 1981; Sochava 1977). However, such studies seem to lack appropriate mitigation measures to help overcome the expected environmental problems. *Izvestiya* described the situation in these words:

We became acquainted with a project . . . known as the Territorial Comprehensive Plan for Environmental Protection Along the BAM. Its many volumes provide a comprehensive analysis of the territories and resource potential of the BAM zone and forecast potential pollution and environmental damage. But we failed to find in the plan specific proposals as to how to protect primary topographical features and basic ecosystems . . . and what kind of organizational measures are needed to protect the environment. Yet such proposals are essential.

(Druzenko 1984b)

Because of these omissions, and because these environmental studies were begun after the start of work on the railway, they do not meet the requirements of American environmental impact statements. Nevertheless, the baseline data they provide is vital. The status of these studies was the subject of a national conference on the BAM project in 1984 (Naprasnikova 1984).

The BAM project, as one of the Soviet Union's largest current construction projects, has received considerable environmental review by competent scientists. However, it is less clear whether thorough environmental studies are routinely carried out on smaller projects in the Far East, on military construction projects, or on non-construction projects (such as clearing land for agriculture). There is no available evidence that they are. The Soviet Union appears to need a law explicitly mandating advance environmental impact reports and mitigation measures on *all* significant projects, and calls for this have already appeared (Kolbasov 1983, 123ff).

Although environmental impact analysis is becoming a standard and accepted procedure in the Soviet Union, the bureaucracy appears at

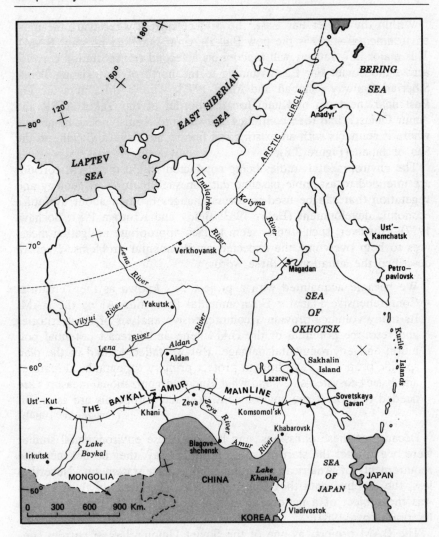

3.8 Route of the Baykal–Amur mainline

present to lack an effective administrative enforcement mechanism capable of halting environmentally questionable construction projects.

Summary

At first glance, it would seem that there exist a number of natural constraints to the economic development of the Soviet Far East. In addition to severe cold, the region contains significant earthquake and landslide hazards, belts of active volcanoes, and large areas containing permafrost, poor drainage and infertile soils.

In practice, however, none of the foregoing constraints are absolutely limiting in their economic impact. The most serious earthquake and volcanic zones are in regions whose relative location gives them limited economic potential in any event. Although a vast majority of the entire Soviet Far East is subject to extreme winter cold and is underlain with permafrost, the portion that is not so encumbered is still quite large. Further, since this latter portion is located in the southern part of the region, it also lies close to neighbouring countries, to other centres of economic activity and to major trade routes.

This more favoured portion of the Soviet Far East – the general lowland areas along the Amur and Ussuri rivers in Primorskiy and Khabarovsk Krays – has few natural constraints to being developed to a level considerably above what exists today. Indeed, the major natural constraint to its development lies mainly outside its boundaries, and that is simply the vast continental distances that separate this region from the heartland of the Soviet Union. These distances make this an expensive part of the country to develop and provision, even in the aftermath of half a century of efforts to promote regional self-sufficiency.

In summary, then, although it is probable that natural constraints will prevent the vast majority of the Soviet Far East from ever being very highly developed (except for local natural resource extraction), and although prudence is essential in transforming the natural environment in all of the Far East biomes, the portion that is amenable to construction has considerable potential. The southernmost part of the Soviet Far East can be developed to play a considerably larger role in the future regional economy of the Pacific Rim, and there are no significant natural constraints, other than sheer distances, to preclude this. Whether the USSR wishes to accept the higher developmental costs of expanding the economic base of this region is another question.

Bibliography

In the citations that follow, *CDSP* refers to *Current Digest of the Soviet Press*; *SG* refers to *Soviet Geography*; and *SGRT* refers to *Soviet Geography: Review and Translation*.

Anon. (1980) *Law of the USSR on the Protection and Utilization of the Animal World. CDSP*, vol. 32, no. 29, pp. 10–14, 24.

Aseyev, A. and Korzhuyev, S., eds. (1982) *Dal'niy Vostok i berega morey, omyvayushchikh territoriyu SSSR*, Moskva: Izdat. 'Nauka'.

Barr, B. M. (1983) 'Regional Dilemmas and International Prospects in the Soviet Timber Industry', Ch. 17 in R. Jensen, T. Shabad and A. Wright, (eds.) *Soviet Natural Resources in the World Economy*, Chicago: University of Chicago Press.

Belov, A. V. (1983) 'Ecological Problems in Economic Development of the BAM Zone' (paper presented at Soviet–American Meeting on the Social-Geographic Aspects of Environmental Change), Irkutsk: Akademia nauk.

Belov, A. V. and Krotova, V. M. (1981) 'Geobotanicheskoe rayonirovanie Amurskoi Oblasti', *Geografiia i prirodnie resursy*, no. 4, pp. 34–43.

Bond, A. (1984) 'Air Pollution in Norilsk: A Soviet Worst Case?', *SG*, no. 25, pp. 665–80.

Borisov, A. A. (1965) *Climates of the USSR*, Chicago: Aldine.

Borodin, A. M. (ed.) (1985) *Krasnaya Kniga SSSR*, Moskva: Izdat. Lesnayaprom.

Chenov, Yu. (1985) *The Living Tundra*, Cambridge: Cambridge University Press.

Dienes, L. (1987) *Soviet Asia: Economic Development and National Policy Choices*, Boulder: Westview Press.

Druzenko, A. *et al.* (1984a) 'Man in the Taiga', *Izvestiya*, 24 August, as translated in *CDSP*, vol. 36, no. 34, pp. 20–21.

Druzenko, A. *et al.* (1984b) 'The Law of Restitution', *Izvestiya*, 7 October, as translated in *CDSP*, vol. 36, no. 40, p. 23.

Fischer, D. (1981) 'Nature Reserves of the USSR: An Inventory', *SG*, vol. 22, no. 8, pp. 500–522.

Knystautas, A. (1987) *The Natural History of the USSR*, New York: McGraw-Hill.

Kolbasov, O. S. (1983) *Ecology: Political Institutions and Legislation*, Moscow: Progress Publishers, pp. 111ff.

Kolyago, V. A. (1970) 'A Classification and Regionalization of the Harsh, Cold Climate of Siberia and the Far East in Relation to Problems of Cold Resistance of Machines', *Soviet Geography: Review and Translation*, no. 11, pp. 38–47.

Komarov, B. (1980) *The Destruction of Nature in the Soviet Union*, White Plains, NY: M. E. Sharpe.

Kostyrina, T. B. (1986) 'Meteorologicheskiye i sinopticheskiye usloviya formirovaniya periodov pozharnoy opasnosti v lesakh yuga Dal'nego Vostoka', *Geografiya i prirodnyye resursy*, no. 2, pp. 78–85.

Luchitski, I. V., Vorob'ev, V. V. and Yermikov, V. D. (1983) 'The "Siberia" Comprehensive Programme and Environmental Protection', in V. P. Maksakovski, (ed.) *The Rational Utilization of Natural Resources and the Protection of the Environment*, Moscow: Progress Publishers.

Lydolph, P. E. (1977a) *Climates of the Soviet Union*, vol. 7 of H. Landsberg, (ed.) *World Survey of Climatology*, New York: Elsevier Scientific Publishing Co.

Lydolph, P. E. (1977b) 'Some Characteristics of the Climate of the USSR With a Direct Bearing on Human Activity', *SGRT*, vol. 18, pp. 145–62.

Mote, V. L. (1983) 'Environmental Constraints to the Economic Development of Siberia'. Ch. 3 in R. Jensen, T. Shabad and A. Wright, (eds) *Soviet Natural Resources in the World Economy*, Chicago: University of Chicago Press.

Mote, V. L. (1985) 'A Visit to the Baikal–Amur Mainline and the New Amur–Yakutsk Rail Project', *SG*, vol. 26, no. 9, pp. 691–716.

Naprasnikova, Ye. V. (1984) 'Vsesoyuznaia nauchno-prakticheskaia konferentsiia "chelovek i priroda na BAMe" ', *Izv. vsesoyuznogo geograficheskogo obshchestva*, no. 3, pp. 279–80.

Nefedova, V. B. (1978) 'Research on the State of the Environment of North Western Siberia in Connection with Economic Development and Conservation', *SGRT*, vol. 19, no. 1, pp. 25–31.

Nikol'skaya, V. and Khomentovskiy, A., (eds) (1969) *Yuzhnaya chast' Dal'nego Vostoka*, Moskva: Izdat. 'Nauka'.

Pryde, P. R. (1972) *Conservation in the Soviet Union*, Cambridge: Cambridge University Press.

Pryde, P. R. (1984) 'Biosphere Reserves in the Soviet Union', *SG*, vol. 25, no. 6, pp. 398–408.

Pryde, P. R. (1987a) 'The Distribution of Endangered Fauna in the USSR', *Biological Conservation*, vol. 42, no. 1, pp. 19–37.

Pryde, P. R. (1987b) 'The Soviet Approach to Environmental Impact Analysis', in F. Singleton, ed. *Environmental Problems in the Soviet Union and Eastern Europe*, Boulder: Lynne Rienner.

Rasputin, V. (1985) 'Let Siberia Serve the Fatherland', *Izvestiya*, 3 November, p. 3, as translated in *CDSP*, vol. 37, no. 44, p. 19.

Shabad, T. and Mote, V. (1977) *Gateway to Siberian Resources (The BAM)*, New York: John Wiley & Sons.

Shalybkov, A. M. and Storchevny, K. (1988) 'Nature Reserves: A Reference Guide', *Soviet Geography*, vol. 39, no. 6, pp. 589–98.

Shilo, N. A., ed. (1970) *Sever Dal'nego Vostoka*, Moskva: Izdat. 'Nauka'.

Sochava, V. B. (1977) 'The BAM: Problems in Applied Geography', Ch. 8 in Shabad and Mote, *Gateway*, op.cit., pp. 163–75.

Sokolov, V. E. and Chernov, V. I. (1983), 'Arctic Ecosystems: Conservation and Development in an Extreme Environment', *Nature and Resources*, vol. 19, no. 3, pp. 2–9.

Sokolov, V. E. and Syroyechkovskiy, Ye. (eds) (1985) *Zapovedniki Dal'nogo Vostoke SSSR*, Moskva: 'Mysl' '.

Suslov, S. P. (1961) *Physical Geography of Asiatic Russia* (Part III), San Francisco: W. H. Freeman.

Tochenov, V. V., *et al.*, (eds) (1983) *Atlas SSSR*, Moskva: Glav. Uprav. Geodezii i Kartografii.

Vashjev, P. V. (1971) 'Forest Resources and Forest Economy' in I. Gerasimov, D. Armand and K. Efron, (eds), *Natural Resources of the Soviet Union*, San Francisco: Freeman.

Vorob'yev, V. V. (1984) 'Problems of Protecting the Environment in Siberia', *Geoforum*, vol. 15, no. 1, pp. 105–11.

Wood, A. (1987) *Siberia: Problems and Prospects for Regional Development*, New York: Barnes & Noble.

Chapter four

Population and labour force

Ann C. Helgeson

Introduction

The Far East is the most sparsely populated region of the Soviet Union. Its location on the eastern periphery of the USSR, its severe environment and the cumulative effects of inadequate investment in social and economic infrastructure have led to its being stereotyped as a warehouse of natural resources since long before 1917. The Far Eastern economy seems always to have been controlled by outsiders – foreign investors in the nineteenth century and All-Union ministries in the twentieth – which have been primarily interested in the contents of the warehouse. Of the factors of production the region is immensely rich in natural resources; capital can be directed here by central authorities or foreign investors. What is lacking is labour. In an earlier era involuntary migrants did much of the work in the region, especially in its less inviting northern parts. Even today a need for convict labour is felt, although the sources for this form of manpower are no longer unlimited.[1]

The sudden realization that the Pacific Ocean is the centre of a huge and developing part of the global economy and not the end of civilization has found the Soviets ill prepared to exploit their 6,000 mile Pacific coastline.[2] It is arguable that the most serious obstacles to Far Eastern development are the lack of a sizeable labour force and the proven ability of the Far East's rugged living standards and lack of amenities to inhibit the permanent settlement of migrants from other parts of the country.

Demographic description of the region

In 1987 the Far Eastern Economic Region accounted for about 2.8 per cent of the Soviet population on 28 per cent of its territory.[3] Even this description understates the sparsity of population in most of the region. Large tracts are empty. A recently compiled map of unsettled territory in the USSR, defined as those tracts of at least 1,000 sq. km which are more than 60 km from an urban place and 15 km from a rural settlement, shows that 79 per cent of the Far East in unsettled.[4]

The fifth of Far Eastern territory that is populated is largely south of about 55° N latitude or in scattered settlements at coastal locations, along river routes, in the middle Lena valley or at mining and forestry sites. It is probable that three-quarters of the region's population of 7.8 million inhabit less than 10 per cent of the total land area, concentrated in the Amur and Ussuri valleys, the plain stretching between Lake Khanka and Vladivostok and the southern parts of Sakhalin Island. This population of 5 million stands in sharp contrast to more than 60 million in the two Koreas, over 120 million in Japan and another 90 million in the three northern Chinese provinces of Heilongjiang, Jilin and Liaoning.[5]

Although this is a small population it has shown continuous growth over the whole of Soviet history. It is five times what it was in the 1926 census, while that of the USSR as a whole has not quite doubled. Over the entire period since 1926 there has been consistent growth – never population loss – in the Far East, although the growth rates have varied among the constituent sub-regions and periods. The single exception is Sakhalin which seems to have slightly declined during the 1960s. The 1960s was also the only decade when several Far Eastern regions grew at rates less than the national average. The region has grown steadily in the last two decades. Only three oblasts in the RSFSR grew at a greater rate[6] than any of the sub-regions of the Far East[7] over the period 1970–87. The Far East grew 34 per cent, in contrast to West and East Siberia which grew at rates more comparable to the USSR average of 16 per cent. Table 4.1 shows estimated annual population growth in the sub-regions of the Far East over the intercensal periods of the twentieth century. The growth rates are consistently higher than those for the USSR as a whole, again with the exception of Sakhalin and the 1960s.

In 1987 77 per cent of the Far Eastern population lived in cities and urban-type settlements, which is high when compared with the Soviet average (66 per cent), but something we would expect in a region where agriculture is not widespread. However, despite the high proportion of urbanites one is struck by the large proportion of urbanites living in very small towns.[8] Over half of the urban population lives in towns with less than 100,000 inhabitants. Of the 1987 urban population as a whole, 43 per cent (2,575,000 people) lived in 263 places with fewer than 10,000 people in them and 67 places with between 10,000 and 50,000 people in them. Another 10 per cent lived in 9 cities of 50–100,000.[9] Table 4.2 shows what can be reconstructed of the urban hierarchy in the region, based on statistical and cartographic sources.

Table 4.1 Population growth in intercensal periods, 1926–89

	Population in 1926	Increase 1926–39	Population in 1939	Increase 1939–59	Population in 1959	Increase 1959–70	Population in 1970	Increase 1970–79	Population in 1979	Increase 1979–89	Population in 1989	Increase 1926–89	Pop growth 1926–89
	'000	%	'000	%	'000	%	'000	%	'000	%	'000	%	'000
USSR	147,028	30	190,678	10	208,827	16	241,720	9	262,436	9	286,717	95	139,689
Far East Economic Region	1,572	89	2,976	62	4,834	20	5,781	18	6,819	16	7,941	405	6,369
rural	1,204	32	1,591	−1	1,569	5	1,649	5	1,738	10	1,920	59	716
urban	368	276	1,385	136	3,265	27	4,132	23	5,081	19	6,021	1,536	5,653
Khabarovsk	147	273	549	78	979	20	1,174	17	1,375	17	1,608	994	1,461
rural	87	136	205	21	248	−1	245	8	263	22	321	269	234
urban	60	473	344	113	731	27	929	20	1,112	16	1,287	2,045	1,227
Jewish A.O.	36	203	109	50	163	6	172	10	190	14	216	500	180
rural	27	37	37	24	46	17	54	13	61	20	73	170	46
urban	9	700	72	63	117	1	118	9	129	11	143	1,489	134
Maritime	637	39	888	56	1,381	25	1,721	15	1,978	14	2,260	255	1,623
rural	464	−6	436	4	453	3	467	3	479	6	509	10	45
urban	173	161	452	105	928	35	1,254	20	1,499	17	1,751	912	1,578
Amur Ob.	414	53	634	13	718	11	793	18	937	13	1,058	156	644
rural	308	12	345	−16	289	5	303	8	328	4	342	11	34
urban	106	173	289	48	429	14	490	24	610	4	716	575	610
Kamchatka	9	856	86	124	193	33	257	34	344	24	427	4,644	418
rural	7	629	51	14	58	−18	48	−6	45	40	63	800	56
urban	2	1,650	35	286	135	55	209	43	299	22	364	18,100	362

Koryak A.O.	10	130	23	22	28	11	31	10	34	15	39	290	29
rural	10	130	23	−4	22	−5	21	0	21	14	24	140	14
urban	0	0	0	0	6	67	10	30	13	15	15	150	15
Magadan Ob.	7	2,071	152	24	189	33	251	32	333	16	385	5,400	378
rural	7	1,671	124	−80	25	132	58	5	61	−2	60	757	53
urban	0	0	28	486	164	18	194	41	272	19	325	1,061	325
Chukchi A.O.	13	62	21	124	47	115	101	32	133	19	158	1,115	145
rural	13	38	18	11	20	20	31	29	40	8	43	231	30
urban	0	0	3	800	27	55	70	33	93	24	115	3,733	115
Sakhalin Ob.	12	733	100	549	649	159	616	6	655	8	709	5,808	697
rural	9	456	50	222	161	−5	133	−13	115	9	125	1,289	116
urban	3	1,567	50	878	489	−18	483	12	540	8	584	19,367	581
Yakut ASSR	287	44	414	18	487	−1	664	26	839	29	1,081	277	794
rural	272	11	302	−18	248	36	290	12	325	11	360	32	88
urban	15	647	112	113	239	57	375	37	514	40	721	4,707	706

Sources: Naselenie SSSR (1973); *Izvestiya*, 28 April 1989; TsSU SSSR (1984) *Chislennost' i sostav naseleniya SSSR: Po dannym Vsesoyuznoy perepisi naseleniya 1979g*, Moscow: Statisticheskiy sbornik.

Table 4.2 Far Eastern urban hierarchy in mid–1980s (numbers of settlements)

	Cities >100,000	Cities 50–100,000	Cities <50,000	Urban-type >10,000	Urban-type <10,000
Far East ER	10	9	36	56	230
Khaberovsk Kr.	2	1	4	23	8
Jewish A.O.	0	1	1	4	8
Maritime Kr.	3	3	3	10	38
Amur Ob.	1	3	5	4	26
Kamchatka Ob.	1	0	2	2	6
Koryak A.O.	0	0	0	0	5
Magadan Ob.	1	0	1	2	32
Chukchi A.O.	0	0	2	2	16
Sakhalin Ob.	1	0	10	1	33
Yakut ASSR	1	1	8	8	56

Sources: Naselenie SSSR (1987) pp. 34–42. Places with fewer than 50,000 are estimated from the dot size in Atles SSSR (1985) Moscow. Population data in this atlas refer to 1984.

These are the troubled small towns written about by Boris Khorev and others.[10] Many of them are of the 'company-town' type – places with only one major employer, often a mining, timber or fishing enterprise. The special problems in providing housing and social infrastructure in these places is treated later in the chapter.

The Far Eastern labour force

The size of the labour force in the region, or even that of the labour-age population, is not currently published in the USSR. We have to rely therefore on intelligent estimates. Using several methods to estimate the working-age population we conclude that it is somewhere between 4.5 and 5 million. If we assume that the age structure in the Far East is identical to that recently published for urban and rural populations in the USSR as a whole,[11] then there are about 4.5 million[12] persons of labour-age in the region. This is quite probably an underestimate, because all indications are that a larger proportion of the population of the Far East are in the labour ages than for the USSR as a whole. An alternate estimate of about 5 million[13] distributes the 1987 Far Eastern population according to the last published age structure of the region in the 1970 census. Baldwin's estimate of the working-age population for both 1985 and 1990 is about 4.4 million.[14] This latter figure is a projection from the 1970 census and does not include any migrants.

We also know little about participation rates among the working-age population, although some qualitative statements are made often enough in Soviet research to lend them widespread acceptance. For instance, it is said that the female participation rate is lower than average, especially in one-industry towns. This is accounted for by, among other things, poor child-care facilities. Data from the 1979 census on household structure indicate that the stereotype of whole settlements filled with young single males may be misleading.

Neither do we have detailed data on the sectoral structure of the Far Eastern work-force. Again, qualitative statements are repeated frequently. The importance of the extractive industries – fishing, mining and forestry – clearly results in a larger proportion than elsewhere in these industries. There are a limited number of machine-building plants, all of which are in the southern areas of Khabarovsk Kray.[15] Construction is probably more heavily represented than in other parts of the USSR. The construction industry is weak in the region. Over 40 per cent of labour is unmechanized. Annual labour turnover in construction is up to 60 per cent.[16] Transport is more numerically important in the region than in other parts of the USSR. Dienes reports that 19 per cent of workers and employees are in the transport sector.[17] We cannot even single out the agricultural portion of the labour force or use the size of the rural population as a surrogate because many who participate in agriculture live in urban-type places.

Although the military presence in the region would not be included in labour force figures, even if they were published, it is useful to estimate the extent of the military manpower in the region. Dienes estimates that one-quarter of the men between the ages of 18 and 60 in the border provinces east of Lake Baykal are in uniform.[18] This would be well over 500,000. He estimates total uniformed naval personnel in the Far East at 180,000, citing a rare source which reveals that sailors make up 10 per cent of the permanent population in Vladivostok and more than 40 per cent in Nakhodka.[19] These uniformed personnel could be utilized in the civilian labour force with a lessening of political tensions in the region.

Complaints about labour deficits in the region are common. To make things worse there is an extremely bad record in labour productivity. Duyring the 10th Five Year Plan only 66 per cent of production increases were due to labour productivity (the figure is 81 per cent for the RSFSR as a whole). In the 11th Five Year Plan this figure went down to 59 per cent.[20] The labour redistribution which is going on all over the Soviet Union today will be particularly problematic in the Far East. With such low levels of labour productivity the region is not at the stage of development where even a single sector can stand redundancies in order to transfer labour to developing sectors.[21]

Migration to the region

The explanation for the more or less continuous population growth in the Far East is connected with migration and not a particularly high rate of natural increase. Table 4.3 shows average annual net migration during intercensal periods. The estimates are calculated using a crude residual method which assumes that the USSR total population growth rate can be considered as natural growth (the balance of births and deaths) only and that Far Eastern natural growth was at the same as the national rate. Thus the residual left over when the Far Eastern populations were subjected to the national growth rates represents net migration.

Table 4.3 Crude estimate of net migrants in intercensal periods, 1926–89

	1926–39	1939–59	1959–70	1970–79	1979–89
Far East ER	937	1,575	183	541	491
rural	29	−173	−168	−52	21
urban	908	1,748	351	594	470
Khabarovsk	358	378	40	100	106
rural	92	24	−43	−3	34
urban	266	354	83	103	72
Jewish A.O.	62	44	−17	3	8
rural	2	5	1	2	6
urban	60	38	−17	1	2
Maritime	62	409	122	108	99
rural	−166	−24	−58	−28	−14
urban	228	433	180	137	113
Amur Ob.	97	24	−38	76	34
rural	−54	−89	−31	−1	−16
urban	152	113	−7	78	50
Kamchatka	74	99	33	65	51
rural	42	2	−20	−7	14
urban	32	97	53	72	37
Koryak A.O.	10	3	−1	0	2
rural	10	−3	−4	−2	1
urban	0	6	3	2	1
Magadan Ob.	143	23	33	60	21
rural	115	−111	29	−2	−7
urban	28	133	4	62	28
Chukchi A.O.	4	24	47	23	13
rural	1	0	8	6	−1
urban	3	24	39	17	13
Sakhalin Ob.	84	540	−136	−14	−7
rural	38	106	−54	−29	−1
urban	46	434	−83	15	−6
Yakut ASSR	42	34	100	118	165
rural	−51	−83	2	11	5
urban	93	116	98	107	159

Source: Calculated from Table 4.1

According to these rough calculations the 1926–39 period saw a net increase of nearly a million people with more than a half of this in Khabarovsk Kray and Magadan Oblast. One thinks of the construction of Komsomol'sk-na-Amure beginning in 1932, the first in a long series of Komsomol-sponsored construction projects. The name Magadan sends a chill up the spine of many even today and one wonders how many of those migrants were involuntary.

The dramatic migration growth of the 1939–59 period was concentrated in the earlier part of this period and connected with wartime evacuations of population and industry, military activities themselves and the postwar reconstruction period. The Maritime province grew rapidly at this time, as did Sakhalin in connection with the Soviet settlement of the southern half of the island after the war. Magadan's migration growth in the earlier period did not continue.

The decade of the 1960s stands out in this table as having been a period of considerably reduced migration growth when compared to the decades before and after. The net migration residuals are small, even negative for several regions, and probably not even large enough to be statistically significant in many cases. My earlier research[22] shows that over the 1960s the Far East achieved a near balance in the numbers of out-migrants and in-migrants, which represented a more successful performance than West or East Siberia over that period. Within the net balance a large number of young adults (aged 20–29 in 1970) counterbalanced considerable net out-migration among older age groups. This was particularly striking in the southern Khabarovsk and Maritime Krays.

The out-migration of the over–30 age groups signals the persistent migration problem of the Far East. Young persons have always been attracted here by such things as high wages and sometimes romantic notions of the frontier, but they have not settled permanently. The older out-migrants recorded for the 1960s may have moved to the region in their 20s.

Detailed data for independent analysis of migration trends in the Soviet Union have not been available for periods later than the 1960s. Even published Soviet research has limited itself to broad conclusions for the country as a whole, probably based on the same data available to non-Soviet researchers, or to small case study regions. Table 4.3 shows net migration to have achieved nearly the pre-war levels in the 1970s and 1980s, exceeding them in the BAM regions. Yakutia for the first time becomes noteworthy in the migration tables.

Kulakov reports a positive migration balance of 700,000 to Siberia and the Far East in 1981–6.[23] But this is the very inefficient result of a turnover in urban places of around 14 million. Among the out-migrants at least 60 per cent had been there less than three years. In the 1970s, he reports, this kind of irrational migration cost about 6 billion roubles in the Far Eastern region alone.[24] Recently it was pointed out that the cost of resettling a worker from the central regions to the Far East was 17–20,000 roubles. This cost includes the creation of a job and the provision of necessary housing and social infrastructure.[25]

Migration channels

Most migration policies, either the official migration channels treated here or other incentives for independent migrants discussed later, have focussed on the initial move to the region rather than inducing migrants to stay once they have arrived. State-organized migrations have always played an important role in recruiting labour for the Far East and they are expected to do so in the future.[26] Three government channels of directed migration, all of them with roots in the 1930s, have sent large numbers of young people to the Far East in recent years. These three channels, the distribution of university graduates, Orgnabor and the Komsomol 'appeals' as well as the agricultural resettlement programme treated later are probably numerically more significant in the Far East than in other regions of the USSR. All are aimed at young adults and none seem to have been particularly successful at getting these young people to settle permanently in the region.

The most numerically significant official channel of migration is the distribution of graduates of higher educational institutions to compulsory three-year work assignments. Every year over 2 million young people are placed in jobs all over the Soviet Union this way. The Far East, Siberia and the North have been singled out as preferred destinations for graduates in all the recent official statements on the programme.

Many graduates leave as soon as their duties are fulfilled. Over the years evidence has collected to demonstrate that a large proportion of the distributed graduates does not even serve its full three years. Various strategies to address this problem, such as the withholding of degrees until the end of the three-year period, have not been successful. Many of those who do not fulfil their obligations leave because of low standards of housing and living conditions generally, despite the enterprises' legal obligations to provide them. This is a particularly acute problem in the fishing industry. Consider the case of Magadanrybprom which did not build any housing or social infrastructure for its employees during the entire 11th Five Year Plan and only one kindergarten is in the plan for 1986–90. They, and other fishing associations, have avoided the necessity of building housing for their specialists with the use of 'fleet residence permits' (*propiska po flotu*), which allow the residence registration of its employees without the usual necessity of demonstrating that a person has the requisite housing space at their disposal. A trawler captain with Magadanrybprom said that the specialists distributed to them invariably left after three years due to the housing problem.[27]

The graduate distribution programme is the single example in the Soviet labour market of command-administrative methods, so we should not be surprised to see criticism surfacing in the press of this compulsory assignment of jobs.[28] Changes in the system, not its abolition, are going to exacerbate further the problems of the Far East in getting qualified specialists. Enterprises are now required to pay 3,000 roubles from enterprise development funds to the education authorities for each graduate assigned to them. When self-financing becomes a reality throughout the Soviet economy the hiring of graduates may decrease as enterprises calculate the value to them of what used to be a gift from the state. It is planned that the 3,000 rouble payment will be replaced in the future by variable sums negotiated by the education authorities and the enterprises concerned on a contract basis. Already the payments are financing 10 per cent of the cost of higher education, which is planned also to become self-financing. Over the years more regulations regarding the housing and social facilities which enterprises must provide for newly employed specialists have also made the hiring of large numbers of specialists an expensive proposition.

Orgnabor, born in 1931 as a mechanism for moving unskilled rural workers to the construction sites of the 1st Five Year Plan, exists still. It has been a major source of seasonal and year-round workers for the extractive industries of the Far East. Although the latest model contracts for Orgnabor recruiting stress the provision of housing and other social facilities for the recruits there is much anecdotal evidence that these are seldom provided. A 1986 description of the awful working and living conditions for a fishing organization in the Kurils quotes its director: 'These people don't understand. This is not Moscow. In a word – it's Orgnabor. . . . They even want warm toilets!'[29] The author of this newspaper exposé compares the costs of continuous hiring of seasonal Orgnabor recruits unfavourably with that of providing facilities for permanent workers.

The numbers of recruits moving in this channel are vastly reduced when compared to its heyday in the 1930s and the 1940s.[30] As late as 1950 36 per cent of all hirings in construction were of Orgnabor recruits.[31] Even 13 per cent of industrial hirings were of Orgnabor recruits at the mid-century,[32] although by 1967 industrial hirings through Orgnabor were down to 2.2 per cent.[33] After this there seems to have been a slow but steady growth to 5.6 per cent of industrial hirings in 1981.[34] Many have said the time has come to abolish Orgnabor altogether,[35] or at least radically reorganize it.[36] The mentioning of the programme in the 1988 decree on Labour suggests that it will limp on.[37] As has been customary in official statements on Orgnabor since the mid–1960s the eastern and northern regions of the country are singled out as primary destinations for recruits.

Another official migration channel embraces the practice by the Komsomol and occasionally other 'social' organizations of sponsoring large prestigious construction projects and recruiting young workers for them. The first 'Komsomol appeal' was the construction of Komsomol'sk-na-

Amure beginning in 1932 and since then many young people have come to the region in this way. About 34 per cent of the BAM construction workers arrived via this channel of migration in the second half of the 1970s.[38] About 100,000 young workers were placed in jobs throughout the USSR this way in the 1970s. Although numbers are not available for the 1980s the following Far Eastern projects were included on the list of major Komsomol construction projects for the 11th Five Year Plan: Kolyma GES, the BAM, Vostochnyy port, the South Yakutian coal complex at Neryungri, civilian construction in Amursk and Komsomol'sk-na-Amure, a metallurgical plant in Dal'negorsk and the development of irrigated rice cultivation in the Maritime Kray.[39]

The appeals are organized by local 'social' organizations in the sending regions, predominately the Komsomol, including Komsomol organizations in the military, but also the Party and trade union local organizations. Those moving in this way are usually young[40] single working adults, recent school-leavers and those demobilized from compulsory military service.[41] When applicants receive permission to join a Komsomol brigade they receive a *putevka*, which entitles them to a job at the designated construction project and various privileges and material benefits. Most volunteers get jobs in construction in industry, the railroads, road-building and the construction of housing, although since the 1970s there have also been Komsomol volunteers going into the service industries.

There are few strings attached to this sort of placement. The volunteer does not have to agree to work for a specific term in most cases and the employer is only 'encouraged', but not obliged, to provide employment in a particular job. Although these loose arrangements encourage volunteers, they have led to high levels of job-changing. Zhelezko who has studied the BAM construction workers in depth concludes that the Komsomol volunteers are considerably less likely to settle down than workers who have come independently. Among those Komsomol volunteers surveyed by Zhelezko and his colleagues in the 1970s no more than 30 per cent intended to stay for more than five years.[42]

There was, however, some success in assembling the work-force for the construction of the BAM, despite the evidence above on the intentions of the Komsomol volunteers. About a quarter of the BAM workers surveyed by Zhelezko came independently and 30 per cent were intraministerial transfers. Only 3 per cent were Orgnabor recruits. The relative success in recruiting and retaining workers for the BAM project is now being analysed in the search for efficient ways of transferring labour to the Far East. The practice of transferring whole labour collectives with experience of working together in the home regions, similar to the *shefskye otryady* of construction workers for the BAM is singled out as having been notably successful.[43] It is recognized that careful selection of potential members of these brigades will result in enormous savings on transfer costs for those who would otherwise leave after a few years.

But still there is large-scale independent migration of those seeking high wages in seasonal or temporary jobs. Most migrants to the Far East come there independently, outside of the state's official migration channels.

Attracting independent migrants

Since 1932 there have been a series of regulations allowing wage supplements and other monetary advantages to those who choose to work in northern and eastern regions. The emphasis on these wage supplements and their levels have waxed and waned. Khrushchev, in particular, de-emphasized them, stating that wages alone would not populate Siberia and the Far East. The lowering of wage coefficients in 1960 led to labour resource problems in northern and eastern regions.[44] Khrushchev's statement in Vladivostok in 1959 of the necessity to provide superior living conditions (housing, consumer goods, social facilities) in the eastern regions to those in the centre has been echoed down the years. Both Brezhnev and Gorbachev, again both in Vladivostok, repeated a variant of the statement.[45] History seems to have demonstrated the efficacy of wage supplements and the like in eliciting waves of migrants, but not in retaining them.

Currently the wage coefficients vary by region within the Far East from a low of 1.2 to 1.3 in the southern parts[46] to a doubling and more of basic wage rates. Wage supplements, which increase with time spent in the region, are now offered to all the work-force. For many years these were operative only in the northern parts of the region and those with equivalent extreme conditions, but since 1986 they have been available to everyone. In addition there are other entitlements like paid passage to one's place of holiday every three years which supplements the well-being of workers in the Far East, but also, one notes, their attachment to other parts of the country.

Taking all the wage benefits into account the incomes of workers in the Far East have always been considerably higher than those in the central regions of the country. In 1982 the average Far Eastern wage was 230 roubles per month compared to 187 on average for the RSFSR, or 23 per cent higher.[47] In 1984 Far Eastern wages were 48.4 per cent higher than the average for the RSFSR. This figure is apparently highly distorted by wage levels in the northern parts of the region as the comparable statistic for Khbarovsk Kray was 26.4 per cent; for the Maritime Kray it was 18.8 per cent and for Amur Oblast it was 17.4 per cent.[48] Real wages are another matter as is the reduced ability in much if not all the region to spend those wages. The Far Eastern Academy of

Sciences in taking account of living conditions and costs estimates the level of real wages in the Far East as only 110.8 per cent of the RSFSR average.[49]

The tides of migrants passing each other in what Aganbegyan has called the 'railroad terminals'[50] that are Far Eastern settlements, it has been argued, are partly caused by the lure of high wages in the eastern and northern regions. Zaslavskaya, among others, has noted that a raising of wage incentives in the north and east can actually have the perverse effect of shortening the stays of target earners who need less time to acquire the sums they need to establish themselves in more westerly and southerly parts of the country.

For decades, among at least the propagandists, there has been a wistful hope that enthusiastic young migrants attracted to construction work in the more severe environments of the east and north would settle down after construction was over and provide the work-force for the developing economy.[51] One is reminded of all the songs – from Komsomol'sk in the 1930s to the BAM – which reiterate the old story of starting life in a tent, building the city, railroad etc. going to work in it and watching a new city grow before one's eyes. While this kind of notion may have been attractive to the unskilled rural Orgnabor recruit of the pre-war period it is harder to support in the case of the skilled construction workers who now regularly move from one highly paying project to another. It makes more sense to talk about the optimal length of tour for temporary workers rather than permanent settlement, as has been advocated by many.[52]

Paradoxically at least some of the BAM migrants have not only decided to remain in the region as the old song goes, but are vigorously complaining about the slowness of the promised development of the BAM zone. Recently an 'Initiative Group for the Defence and Development of the BAM Zone' published a statement in the local press[53] vociferously criticizing the lukewarm treatment now being given to the railroad and the development of the BAM zone. They have even proposed leasing the entire BAM zone and taking charge of the region themselves.

The problem of social infrastructure

Since the study of interregional migration in the USSR resumed in the mid-1960s, the return migration from the eastern regions has been consistently attributed, at least in large part, to the poor success in acclimatization of the migrants due to the lack of social amenities.[54] From time to time the leadership has affirmed the importance of this problem and piously stated the intention to make the standard of living in the eastern regions even higher than that in the more settled central parts of the country.[55] When Gorbachev visited the region in the summer of 1986 his speeches and conversations with residents inevitably came back to problems of housing and social infrastructure.

But the promises have not been fulfilled. Judging from published data on the level of housing provision and other social facilities, the Far East falls far short of parity with European regions of the USSR and nowhere near the desired superior levels. In 1985 residents of the region had only 84 per cent of the RSFSR average per capita housing space. Eighty-seven per cent of housing had running water (the figure for the RSFSR is 91 per cent). Fifty-six per cent had hot water (the RSFSR figure being 67 per cent); 86 per cent had central heating (the RSFSR figure being 89 per cent).[56] Dienes notes that the March 1985 session of the Yakut Supreme Soviet made it clear that the average level of supply of consumer goods and services in that ASSR is in last place among the provinces of RSFSR.[57] None of the Far Eastern oblasts fulfilled their housing plans during the 11th Five Year Plan. The plans for the provision of child-care institutions in Khabarovsk Kray were fulfilled by 79 per cent; in Sakhalin by 88 per cent. Hospital plans for the region as a whole were fulfilled by 76 per cent.[58]

One suspects from anecdotal evidence in the press that housing conditions are worse in the scattered workers' settlements of the Far North than in the more southerly regions of the Far East. An article about a murder in Susuman in Magadan Oblast in the newspaper *Izvestiya* draws attention to the neighbourhood where the murderer lived – a swamp covered with shanties built on chicken legs which was called 'Shanghai' by local residents. The article points to 600 such houses in Susuman in which 'several thousand' people dwell and says that about 8,000 people in the town of Magadan alone live in these 'unplanned structures'.[59]

Part of the problem is that the organizations that build and provide housing and service facilities have little incentive to do so. Housing and social infrastructure are expensive in all regions of the Soviet Union and grudgingly provided, it seems, by ministries whose priorities are elsewhere. But there are special problems in the Far East which make the building of housing and the provision of other aspects of well-being both more expensive and more problematic for the ministries. There are several reasons for the great expense of providing *sotskul'tbyt* in the Far East. Construction costs are obviously high due to environmental features like permafrost and the cold climate. To add to this the construction industry itself in the Far East is weakly developed and is characterized by higher levels of manual labour, higher labour turnover, lower wages etc. than the construction industry elsewhere. Union Ministries are evaluated on the total volume of such construction regardless of where the construction takes place, so they understandably will build in the 'cheapest' locations, which makes the Far East unattractive.[60]

Particularly in the smaller 'company towns' of the northern parts of the Far East the resource base of the local soviets, the other traditional provider of social facilities, is weak. Milovanov and Singur[61] repeat the traditional complaint about the low priority of territorial planning when compared with sectoral. *Sotskul'tbyt* planned by local soviets is insignificant when compared to that planned by ministries. This is a result of not only the small size of typical Far Eastern urban places but also of the predominance of all-Union sectors in the enterprises of the Far East. Minvostokstroi does only 28 per cent of the construction work in the region. The rest is done by non-construction ministries.[62]

At the same time the demographic structure of migrants to the region makes extra demands on social infrastructure. While the ability to provide social facilities is more limited in the Far East the demographic and socio-economic characteristics of the population would lead one to expect higher than usual demand for social facilities and services. The age structure of the Far Eastern population and even more the age structure of the migrant population is, they say, heavily weighted towards young adults. It will probably be possible to confirm this soon with data more recent than that from the 1970 census.[63] In addition to being young (and maybe more male than in other regions) the population, and again especially the migrants, are relatively highly paid when compared to similar groups in other parts of the USSR. They make demands on social infrastructure that are greater than a population with larger numbers of older and less highly paid people.

The imbalances between money incomes of the population and the availability of consumer goods in the region is legendary. The anti-alcohol campaign has exacerbated the problem. Again, priority distribution of consumer goods to the region has been a pious and unrealized promise for decades.[64] Personal consumption in the region is much more concentrated in the state retail trade network (well over 90 per cent as against a figure of about 80 per cent in the USSR as a whole) which means that little is bought or sold in urban *kolkhoz* markets. Yet retail trade turnover growth in the Far East is lower than that for the USSR as a whole. This began to be noticeable in the 10th Five Year Plan period and was particularly acute in the early 1980s.[65]

At present about 30 per cent of consumer goods purchased in the Far East are of local production. Only 50 per cent of food goods are produced locally[66] and clearly a smaller proportion of non-food consumer goods. There are limited possibilities for the development of local consumer goods production due to the expense of imported raw materials and added labour costs, which, of course, would subvert the labour supply problem further. There is some evidence, however, of low levels of female employment in the region, particularly in small one-industry settlements. Production of consumer goods in the region is to increase by 40 per cent in the 12th Five Year Plan, which is a higher rate of increase than any other Soviet region, but even the achievement of this increase would mean only a decrease of imported consumer goods to a level of 67 to 68 per cent. The necessity of importing the bulk of consumer goods introduces problems of assortment. Rudenko states that the tendency is for distant supply organizations to load whatever is available into freight cars when the deadline comes for shipment to the Far East, resulting in the dumping of unwanted goods in little or no variety and the filling up of Far Eastern warehouses with these goods.[67]

The poor development of the transport infrastructure in the region exacerbates the problem of access to consumer goods and services. The distances to larger towns, even the rayon centres, can be formidable. A transport accessibility norm for rayons which requires that there be a two-hour access for passenger transport from any point in the rayon to any other is satisfied in many western regions of the USSR. In Maritime Kray this norm is 60 per cent satisfied. In the Terneysk rayon the average journey is nineteen hours. There are even more sub-normal rayons in Amur and Kamchatka Oblasts. In the Zeysk rayon of Amur Oblast the average journey is eleven hours.[68] A recent source states that doubling the length of hard surface roads increases the living standards of a rayon by 1.22–1.35 times and results in the decrease of labour turnover of 1.65 times and labour demand by 1.04–1.09 times.[69]

Rural labour force and rural migrants

The size of the rural population has been remarkably stable when compared with the growth of urban population. In the whole period since 1926 the rural population has grown by only 49 per cent. Even this growth rate would be much more modest if we considered only the southern parts of the region where the agricultural population is located. Here there is practically no growth at all over the Soviet period. It is said that the outflow from rural areas of the region to the cities began as early as the 1870s, when it was necessary for the authorities to forbid the hiring of peasants by manufacturers and other urban employers.[70]

The resettlement of agriculturalists from densely populated western parts of the country has a long history. It was particularly intense from the 1880s when transport by sea from ports like Odessa began. Even then, and partly in response to the peasants' fear of the sea passage, incentives had to be offered, including free passage, food for half a year and 100 roubles for seed and inventory and livestock.[71]

In recent decades the state agricultural resettlement programme, administered by Goskomtrud has favoured the Far East as a destination, with larger relocation subsidies available to those willing to move to rural areas in the Far East. In the 1950s and 1960s an average of 20 per cent of agricultural resettlements in the RSFSR were to the Far East.[72] Churakov maintains that a third of RSFSR resettlers went to the Far East in the 1960s. Yet from 1960 to 1968 the number of agricultural workers there actually decreased by 8.4 per cent so that by the early 1970s there were two jobs to every available agricultural worker in the region.[73]

There has been limited success in more recent years in transferring rural residents to the Far East from the relatively overpopulated southern republics. In keeping with the new emphasis on group migrations there has been some success in transferring groups of families from the Turkmen Republic to Amur Oblast. There is evidence of movement from Andizhan Oblast in Uzbekistan, a particularly overpopulated region, to the Far East through the agricultural resettlement programmes, although also evidence of considerable reluctance on the part of the Uzbek authorities to let native Uzbeks go.[74] There have also been movements from Azerbaydzhan to the Amur region.[75] It seems likely that many of these migrants from the southern republics are Slavs. Kulakov reports that among migrants to Siberia and the Far East in 1981–5 only 2 per cent were from any of the Central Asian nationality groups.[76]

The financial benefits and incentives for agricultural resettlement are considerable and include such things as transportation and baggage costs (measured in tons rather than the Orgnabor kilograms), costs for the transfer of livestock, free accommodation after arrival and house-construction loans. The benefits vary depending on the region of settlement. In the 1960s these benefits were highest for Kamchatka and Sakhalin followed by the rest of the Far East. Since 1973 the highest benefits have also been given to resettlers on the border with China. Specialists note that the absolute size of the payments to resettlers has not changed substantially in forty years while those volunteering for the programmes are much younger and more qualified than those typically involved in agricultural resettlement in earlier years. Goskomtrud is interested in fulfilling its plans for the resettlers. Farm leaders in the Far East are given little opportunity to select settlers. A survey by Goskomtrud RSFSR of farms showed that 95 per cent of them did not get the kinds of labour they were in need of. The same source says that the agricultural resettlement programme has been exploited by Far Eastern farm leaders as a kind of government subsidy for the funding of housing and social infrastructure. Farms cannot afford to build these facilities for their own young persons in order to encourage them to remain on the farm. But migrants are entitled to new housing immediately upon resettlement. When they leave, the housing remains behind. In partial response to this problem it has been announced that recently demobilized soldiers from the Soviet Army are allowed to be classified as agricultural resettlers, even if they are returning to their home villages.[77]

The relation between labour force growth and migration

It can be argued that differential levels in living conditions among regions in the Soviet Union take on a greater importance in times of overall labour shortage. When there are jobs available everywhere the potential migrant is more alert to the possibilities of improving his standard of living by moving and has plenty of opportunity to do so. The Far East will not gain in these periods. On the other hand, at times when the labour force is growing rapidly (as, for example, in the 1970s) one could equally argue that the geography of job-creation has a greater effect on the regions to which migrants move.

The late 1980s have been a period of slow labour force growth, a demographic fact that is now well known. The changing contours of the labour age population in the country as a whole in the last quarter of the century are shown in Table 4.4. As can be seen, the national labour supply will only begin to come out of its slump at the end of the 1990s having been through the worst of the labour shortages in the later 1980s and early 1990s. In the period around 1990, according to these projections, the industrial areas in the Donets-Dnepr region of the Ukraine will begin to suffer from supply problems, and those in the central and north-western areas of the RSFSR will begin to ease only slightly. There are going to be plenty of job openings in both of these critical industrial regions with their high living standards, making opportunities in the Far East less competitive. In the later 1990s when the situation will have eased considerably in the Ukraine and the Russian industrial areas the labour age population in the Far East will experience a considerable spurt in growth, if Baldwin's medium series of projections turn out to be correct.

Foreign workers for the Far East

The southern Soviet border in the Far East is one of the more dramatic population discontinuities to be observed in the world today. The one factor of production seriously lacking in the region, labour, is potentially available in regions considerably closer and more environmentally similar to the Soviet Far East than the western regions of the USSR. Now that these nations are on better terms politically the possibilities of joint activities involving labour from Korea, China and Vietnam are being explored at a dizzying pace.

There have been foreign workers in the Far East since at least the mid-1960s. The USSR has had joint timber activities with North Korea since 1967 in Khabarovsk Kray and since 1975 in Amur Oblast. The Korean side supplies labour for many of these operations while the USSR supplies the machinery and transport.[78] Korean labour also figures heavily

Table 4.4 Working-age population in Economic Regions, 1975–2000 ('000s)

	Labour-age pop. 1975	Increment 1975–80	% of total USSR growth 1975–80	Labour-age pop. 1980	Increment 1980–85	% of total USSR growth 1980–85	Labour-age pop. 1985	Increment 1985–90	% of total USSR growth 1985–90	Labour-age pop. 1990	Increment 1990–95	% of total USSR growth 1990–95	Labour-age pop. 1995	Increment 1995–2000	% of total USSR growth 1995–2000	Labour-age pop. 2000
Northwest	7,777	353	3	8,130	-185	-5	7,945	-207	-9	7,738	-188	-6	7,550	-21	0	7,529
Central	16,931	520	4	17,451	-590	-16	16,861	-642	-27	16,219	-527	-18	15,692	-55	-1	15,637
Volga-Vyatka	4,780	314	3	5,094	-2	0	5,092	-74	-3	5,018	-60	-2	4,958	116	2	5,074
Cent. Black Earth	4,401	233	3	4,634	-35	-1	4,599	-113	-5	4,486	-97	-3	4,389	73	1	4,462
Volga	10,897	926	8	11,823	188	5	12,011	-41	-1	11,970	-17	-1	11,953	345	5	12,298
N. Caucasus	8,290	691	6	8,981	186	5	9,167	61	3	9,228	97	3	9,325	409	6	9,734
Urals	9,086	614	5	9,700	23	1	9,723	-82	-4	9,641	8	0	9,649	266	4	9,915
W. Siberia	7,421	553	5	7,974	30	1	8,004	-36	-2	7,968	61	2	8,029	274	4	8,303
E. Siberia	4,714	405	3	5,119	98	3	5,217	38	2	5,255	76	3	5,331	184	3	5,515
Far East	4,030	324	3	4,354	50	1	4,404	25	1	4,429	12	0	4,441	440	6	4,881
Ukraine	27,896	1,393	12	29,289	-39	-1	29,250	-13	-1	29,237	-262	-9	28,975	184	3	29,159

Region																
Donets-Dniepr	12,050	504	4	12,554	-122	-3	12,432	-137	-6	12,295	-260	-9	12,035	-65	-1	11,970
Southwest	11,858	660	6	12,518	88	2	12,606	118	5	12,724	22	1	12,746	232	3	12,978
South	3,989	228	2	4,217	-5	0	4,212	6	0	4,218	-24	-1	4,194	16	0	4,210
Baltic	4,563	218	2	4,781	25	1	4,806	-1	0	4,805	-39	-1	4,766	11	0	4,777
Transcaucasus	6,917	1,162	10	8,079	800	22	8,879	519	22	9,398	502	17	9,900	875	12	10,775
Central Asian	10,392	2,291	19	12,683	2,106	58	14,789	2,130	91	16,919	2,569	88	19,488	3,278	45	22,766
Kazakh	7,516	1,148	10	8,664	772	21	9,436	670	29	10,106	674	23	10,780	891	12	11,671
Belorussia	5,276	451	4	5,727	134	4	5,861	25	1	5,886	22	1	5,908	224	3	6,132
Moldavia	2,130	193	2	2,323	87	2	2,410	83	4	2,493	100	3	2,593	147	2	2,740
USSR	143,018	11,788	100	154,806	3,649	100	158,455	2,341	100	160,796	2,932	100	163,728	7,240	100	170,968

Source: Baldwin, Godfrey S. (1979) Population Projections by Age and Sex: For the Republics and Major Economic Regions of the USSR, 1970 to 2000. FDAD, Series P-91, no. 26. September, p. 128.

in the recent proposals by Ota Seizo, president of the Toho Life Insurance Company of Japan to set up companies, hotels, airports, seaports and teleports in the region where the Soviet, Chinese and North Korean borders meet. The USSR is to provide the land, the west the capital and North Korea the manpower.[79] Chinese and Korean farmers are now growing crops at three places in Maritime Kray. The numbers are small so far. There are 66 Chinese at Baranovskiy *sovkhoz*, and about 100 Koreans near the Ussuri and Lake Khanka.[80]

The housing and social facility shortages in the region can be significantly improved by the use of foreign labour. At the meeting of the Inter-Governmental Soviet–Korean Committee for Economic Co-operation in March 1988 decisions were made involving the North Koreans producing clothes for the Far East, North Korean workers growing vegetables and building hotels and other buildings in Nakhodka, Khabarovsk and Vladivostok.[81] A delegation from the Chinese province of Heilongjiang signed agreements in the summer of 1988 for construction of houses, youth facilities, hotels, drainage facilities, airports and surface facilities for a coal mine. Agreements were also signed for joint ventures: a timber processing plant, a concrete parts plant, a wooden parts plant, a base for maintaining and repairing urban drainage facilities and a building materials enterprise.[82] The first group of 50 workers left China in mid-November for Maritime Kray.

Vietnamese workers may increase in the Far East. It was recently announced that Haiphong will send 2,000 young workers to Vladivostok, apparently for work in the consumer goods sector where shoe production is expected to go up from 2.5 million to 8 million pairs annually.[83] It will also open a Vietnamese restaurant there.

In commenting on the new decree on joint ventures[84] V.I. Ivanov of the Pacific Institute of the World Economy and International Relations noted possibilities in construction and agriculture using labour from China, Korea and Vietnam.[85] He and many others have referred to the potentials of the combination of foreign labour, foreign tourist firms and the Far Eastern landscape. Lutsenko, the head of the Maritime Krayispolkom, also waxes eloquent on the hard currency that could be brought in by foreign tourists.[86]

Conclusion

The increased possibilities for the use of foreign labour in the Far East are not the only labour changes liable to affect the region. The possibilities of individual labour and the establishment of co-operatives that have been recently instituted could help significantly to address the social infrastructure problems of the region. At least one commentator has advocated the rapid development of these kinds of economic activities in the region. So far there is little evidence to judge whether these forms have been enthusiastically embraced in the region.

There are those now saying that in the next two or three decades the USSR has to turn its face to the Far East and send millions of energetic and enterprising people there.[87] The announcement of the Far Eastern Development Programme was accompanied by commentary very familiar to those in the eastern regions about the coming developments being as dramatic as the industrialization of the country or the reconstruction after the Second World War.

This time around, though, the concentration seems more focused on the southern parts of the region. There is a good deal of criticism of the emphasis on the natural resources of the region alone, which it is maintained are too expensive to exploit at the present time. The concentration on the southern regions is to focus not only on foreign participation in the development of the region, but also on unleashing the entrepreneurship of the local and migrant population. Free rein should be given to independence, co-operatives, local industry, crafts and small manufacturers.

It all comes back to labour and how the necessary volume of migration is to be elicited. We read again that living conditions in the region have to be made not only equal to the rest of the country, but considerably better. Not only should wages be better, but there should be opportunities to buy things that are not even available in the European parts of the country.

Notes

1. In 1987 a regulation from 1971 allowing first-time offenders to serve their time in their home regions was modified to allow the transport of even these convicts to areas where their labour is needed. See *Izvestiya*, 4 August 1988, p. 3.
2. Klyuchnikov, B.F., 'Sovetskiy Dal'niy Vostok v "tikhookeanskom stoletii". K razrabotke novoy kontseptsii razvitiya regiona.' *Problemy Dal'nego Vostoka*, 3, pp. 3–14. See his repeated assertion that the Soviet Far East is 'geographically' part of the Asia–Pacific region, p. 9.
3. Calculated from *Narkhoz SSSR za 70 let*, pp. 389–91.
4. See Dmitriev, A.V.A.W., Lola, A.M. and Mezhevich, M.N. (1988) *Gdye zhivet sovetskiy chelovek. Sotsial'nye problemy upravleniya rasseleniya*, Moscow, pp. 56–7.
5. *The Population Atlas of China* (1987) Hong Kong, pp. 162–8.
6. Tyumen', Murmansk and the Komi ASSR.
7. With the exception of Sakhalin which grew only 12.5 per cent over the period.
8. Zhelezko, S.N., Morozova, G.F. and Serditykh, B.G. (1976) 'Opyt sotsiologicheskogo issledovaniya migratsii naseleniya Dal'nego Vostoka', *Sotsiologicheskie Issledovaniya*, 2, pp. 75–81. Note that according to data for 1974,

60–80 per cent of the Far Eastern population live in small towns and workers' settlements.

9. The number of places with more than 10,000 population is ascertained by recording the size of the dot for these places in *Atlas SSSR* (1985) Moscow. Population data in this atlas refer to 1984. If we assume, for the sake of argument that the places between 10,000 and 50,000 had on average 20,000 persons then the average for the remaining 361 places is only just over 3,000.
10. Khorev, B.S., (ed.) (1972) *Maliy gorod*. Moscow.
11. *Planovoe khozyaystvo* (1988) no. 3, pp. 121–2.
12. 4,459,181.
13. 4,986,371.
14. Baldwin, G.S. (1979) *Population Projections by Age and Sex: For the Republics and Major Economic Regions of the USSR 1970 to 2000*, FDAD, Series P–91, no. 26, September.
15. Dal'diesel (diesel engines), Dal'selkhozmash in Birobidzhan (agricultural machinery and rice-harvesting combines – largely exported), Energomash (electric power machinery works) and Amurkabel (cable for industry and shipbuilding), among others.
16. Singur, N. (1988) 'Dal'niy Vostok: Kompleksnoe Razvitie proizvoditel'nykh sil', *Planovoe khozyaystvo*, no. 3, pp. 94–5.
17. Dienes, Leslie (1987) *Soviet Asia. Economic Development and National Policy Choices*, Boulder: Westview Press, pp. 217–18.
18. ibid., p. 98.
19. ibid., p. 218.
20. Singur, op.cit., p. 94.
21. Minakir, P.A., Renzin, O.M. and Chichkanov, V.P. (1986) *Ekonomika Dal'nego Vostoka: perspektivy uskoreniya*, Khabarovsk, p. 64.
22. Helgeson, A. (1978) 'Soviet Internal Migration and Its Regulation Since Stalin. The Controlled and the Uncontrollable'. PhD dissertation, University of California at Berkeley.
23. Curiously this is a number often cited as the net out-migration from these same regions in 1959–69. See, for example, Baldwin (1979), p. 7.
24. Kulakov, V. (1988) 'Planirovanie Ispol'zovaniya trudovykh resursov', *Planovoe khozyaystvo*, no. 11, p. 113.
25. Milovanov, Ye. and Singur, N. (1986) 'Planirovanie sotsial'nogo razvitiya dal'nego vostoka', *Planovoe khozyaystvo*, no. 2, pp. 102–8.
26. Khorev, B.S., ed. (1986) *Razmeshchenie naseleniya v SSSR: Regional'nyi aspekt dinamiki i politiki narodonaseleniya*, Moscow, p. 132.
27. *Sovetskaya Rossiya*, 15 May 1987.
28. See *Izvestiya*, 23 November 1988, p. 3.
29. *Sovetskaya Rossiya*, 17 August 1986, p. 2.
30. Matveev, Yu. A. (1973) 'Organizovanniy nabor kak odna iz osnovnykh form planovogo pereraspredeleniya rabochey sily', in A.Z. Maykov, (ed.), *Migratsiya naseleniya RSFSR*, Moscow, p. 67.
31. *Stroitel'stvo v SSSR 1917–67* (1967) Moscow, p. 615.
32. Kolosova, R.P. (1980) *Ekonomika trudovykh resursov*, Moscow, p. 77.
33. Maslova, I.S. (1985) *Mekhanizm pereraspredeleniya rabochey sily pri sotsializme*, Moscow, p. 120.
34. ibid.
35. See Panteleev, N. (1982) 'Vazhnaya forma planovogo obespecheniya kadrami', *Sotsialisticheskiy trud*, no. 4, pp. 15–24 for a particularly outspoken opinion of this type.

36. *Sovetskaya Rossiya*, 2 September 1987, p. 1.
37. *Ekonomicheskaya Gazeta*, no. 4, January 1988, pp. 2 and 9. Decree of Central Committee, Council of Ministers and All Union Council of Trade Unions, 'On guaranteeing effective employment and perfecting the job placement system and strengthening social guarantees for the workers'.
38. Zhelezko, S.N. (1980) *Sotsial'no-demograficheskie problemy v zone BAMa*, Moscow, p. 65.
39. *Adresa odinnadtsatoy pyatiletki* (1982), Moscow, pp. 18–31.
40. One source says 75 per cent of them are under the age of 25. Alkin, I.D. and Zeltyn, M.S. (1978) *Organizatsionnye formy raspredeleniya rabochey sily v narodnom khozyaystve SSSR*, Moscow, p. 31.
41. ibid., p. 30.
42. Zhelezko, op.cit., p. 73.
43. Minakir et al., op.cit., p. 68.
44. Rybakovskiy, L.L. (1961) *O sozdanii postoyannykh kadrov na Sakhaline. In Voprosy trudovykh resursov v rayonakh sibiri*, Novosibirsk, pp. 130–31.
45. The *ukaz* of the USSR Council of Ministers of 10 February 1960 can be found in *Sbornik zakonodatel'nykh aktov o trude*, Moscow, 1970.
46. Minakir, et al., op.cit., p. 78.
47. ibid., p. 79.
48. Singur, op.cit., p. 96.
49. ibid., p. 96.
50. Minakir *et al.*, op.cit., p. 76.
51. ibid., p. 75.
52. ibid., p. 75.
53. *Pravda buryatii*, 2 January 1989, cited in *Izvestiya*, 11 January 1989, p. 3.
54. See, for example, Zayonchkovskaya, Zh.A. (1972) *Novosely v gorodakh*. Moscow.
55. Klyuchnikov, 1988, p. 12.
56. Milovanov and Singur, op.cit., p. 104. As of the beginning of 1986 Moscow and Leningrad had an average of 14.69 sq. m of housing per person while the average for the RSFSR urban areas was 13.76 sq. m. Far Eastern housing provision was at the level of 12.7 sq. m per person. Skorokhodov, Yu.G. (1988) 'Sovetskiy Dal'niy Vostok – problemy i perspektivy', *Problemy Dal'nego Vostoka*, no. 2, p. 9.
57. Dienes, op.cit., p. 215.
58. Skorokhodov, op.cit., p. 9.
59. *Izvestiya*, 1 August 1988, p. 4.
60. See Milovanov and Singur, op.cit., pp. 105–6.
61. ibid., p. 105.
62. Skorokhodov, op.cit., p. 9.
63. Only in 1988 was the age/sex structure of the entire Soviet population published for the first time in nearly twenty years. See *Planovoye khozyaystvo* (1988), no. 3, pp. 121–2.
64. See Minakir, et al., op.cit., p. 74.
65. Rudenko, P. (1988) 'Razvivat' torgovlyu dal'nevostochnogo regiona', *Sovetskaya torgovliya*, 10, p. 16.
66. In Amur Oblast and Maritime Kray it is 70 to 80 per cent, in Yakut and Magadan 12 to 18 per cent, ibid., p. 17.
67. ibid., p. 17.
68. Bugromenko, V.N. (1987) 'Kakim byt' administrativnomu raionu', *Izvestiya*

sibirskogo otdeleniya akademii nauk SSSR. Seriya ekonomiki i prikladnoy sotsiologii, no. 1, Vyp. 1, p. 50.
69. ibid., p. 52.
70. Minakir et al., op.cit., p. 10.
71. Klyuchnikov, op.cit., p. 5.
72. Maykov, A.Z. (1973) 'Effektivnost' sel'skokhozyaystvennogo pereseleniya', in A.Z. Maykov ed., *Migratsiya naseleniya RSFSR*, Moscow, p. 81.
73. Churakov, V.Va. (1976) 'Migratsionnye protsessy i trudoobespechennost' v sel'skoy mestnosti', in L.L. Rybakovskiy, ed., *Territorial'nye osobennosti narodonaseleniya RSFSR*, Moscow, p. 173.
74. *Izvestiya*, 7 October 1987, p. 3.
75. Muradov, Sh. (1988) 'Ratsional'naya Zanyatost' Naseleniya v regione', *Voprosy ekonomiki*, no. 6, p. 107.
76. Kulakov, op.cit., p. 113.
77. *Sovetskaya Rossiya*, 7 October 1987, p. 3.
78. Trigubenko, M.Ye. and Shlyk, N.L. (1987) 'K voprosu ob usilenii neposredstvennogo sotrudnichestva dal'nevostochnykh raionov SSSR s sotsialisticheskimi stranami Azii', *Problemy Dal'nego Vostoka*, no. 4, p. 25.
79. TASS, 28 November 1988, cited in *SUPAR Report* (1989), 6, January, p. 10.
80. *Hokkaido shimbun*, 14 August 1988 cited in *SUPAR Report*, ibid., p. 41.
81. Radio Moscow, 29 June 1988, p. 19. Cited in *SUPAR Report*, ibid., p. 19.
82. *Jingji cankao*, 17 October 1988, p. 3. Cited in *SUPAR Report*, ibid., p. 4.
83. ibid., p. 41.
84. *Izvestiya*, 10 December 1988, p. 2.
85. *Izvestiya*, 16 December 1988, p. 5.
86. *Izvestiya*, 24 January 1989, p. 7.
87. Klyuchnikov, op.cit., p. 11.

Chapter five

Resources
Craig ZumBrunnen

The Far East is traditionally called the country's outpost in the Pacific. This is undoubtedly true. But today this view can no longer be regarded as sufficient. The Maritime Kray and the Far East must be turned into a highly developed national economic complex.

(Gorbachev 1988, p. 205)

As geologists have established, the region abounds with numerous rich deposits of non-ferrous metals, gold, silver and many other valuable elements and minerals.

(Gorbachev 1988, p. 208)

The resources of fuel, hydrocarbon raw material, are colossal here.

(Gorbachev 1988, p. 208)

It is necessary to start from the belief that in the future the Far East must not only supply fuel and power to the nearby areas of the country, but also become a sound exporter of them.

(Gorbachev 1988, p. 209)

The Far East must not be looked upon just as a raw material base. The tremendous raw material potential must be used here for building enterprises with complete cycles, to produce here as a minimum semi-finished products, but better still, end-products.

(Gorbachev 1988, p. 208)

Introduction

The first three of the above excerpts from Gorbachev's now famous July 1986 Vladivostok speech contain the salient facts about the Soviet Far East's vast resource endowment while the final two excerpts express what are probably at best long-term hopes, possibly, and indeed perhaps even probably, grandiose and wistful as Leslie Dienes agrues in Chapter 11 and in his recent comment in *Soviet Geography* on the Politburo's 24 July 1987 approval of the long-term programme for the Far East (Dienes 1988, pp. 420–22). On the one hand, despite Gorbachev's above comments about the role of Far East resources in export markets, it appears that his administration has embarked on economic development policies which both explicitly and implicitly place greater emphasis on the western more developed and populated portions of the Soviet Union (*Ekonomicheskaya Gazeta*, no. 46, Nov. 1985, pp. 3–15). On the other hand, this chapter concludes with recent information about proposed 'free-trade' and/or 'joint-venture' zones which, if implemented as expected, could possibly significantly enhance the prospects for co-operative foreign-Soviet Far East resource development and resource export trade over the next decade. The continuing thaw in Sino–Soviet relations and the recent announcement by China's Premier Li Peng about a possible China–Soviet summit clearly could buoy Soviet Far East trade (Pomfret 1988, p. B–6; 'Soviets who return to China', 1988, p. B–6; 'China–Soviet summit possible, Li says', 1988, p. A–18).

As noted in earlier chapters of this volume, the official Soviet Far East region is enormous in size. Accordingly, it is not surprising that this vast area contains a treasure trove of natural resources. The upsurge in both Soviet and foreign interest in the natural resources of the Soviet Far East during the 1970s and 1980s has been predicated in part on the construction of the Baykal–Amur Mainline (BAM) railway to transport resources either west to the industrial core or east to potential Pacific-Rim markets, especially to natural-resource-poor Japan. In the following chapter Brenton Barr discusses two very important biotic-based resource industries, namely, the Far East's forestry and commercial fishing industries. Victor Mote's chapter on the South Yakutian Territorial Production Complex covers some of the same material as this chapter but in greater regional detail. The focus of this chapter is rather to discuss in turn some of the locational, qualitative and, where possible, quantitative aspects of selected Far Eastern natural resources. For the sake of discussion these selected resources will be grouped into two general categories: energy resources and selected metallic and non-metallic mineral resources.

Energy resources

In general, the energy resources reserves of the Soviet Far East are enormous. Coal, oil, natural gas deposits and hydro-power sites are abundant and widely dispersed. Energy-resource development and electrical-power generation and transmission are still the crucial linchpins of any type of modern economic development. Despite large-scale reserves of energy resources, chronic lags in the development of these resources and power complexes are 'holding up the development of other branches' of the Soviet Far East's economy (Gorbachev 1988, p. 208). To Gorbachev's dismay millions of tons of oil from the regions west of Lake Baykal are still being shipped in annually by railroad tankers to satisfy the fuel needs of the Far East region and a powerful, unified electrical transmission grid is still not completed (Gorbachev 1988, pp. 208–9). It is not surprising then that Aleksandr Semyonov, the USSR's Deputy Minister of Power Development and Electrification, announced on 8 June 1988, that the Soviet Union planned to triple energy generation in the Far East by the year 2000. As part of this programme the Far Eastern Nuclear Plant, utilizing four 440 megawatt water-water type reactors, is to be generating electricity for the United Power Grid of the Far East by 1995. Also reported were plans to build the Primorskaya Nuclear Plant in the Vladivostok area ('Power generation plans for Far East outlined', 1988, pp. 67–8). The Far East currently has one small, 48 megawatt, nuclear plant in the far north-east at Bilibino (Central Intelligence Agency 1985b, p. 57). The post-Chernobyl Soviet anti-nuclear lobby could well terminate all of these plans.

Let us now turn to a systematic discussion of the Far East's energy resources and their potential for economic development.

Coal

In the early 1980s total estimated Soviet geological coal reserves amounted to 6.8 trillion metric tons, about double those of the United States and half of the world's total reserves. Although only 281 billion tons or slightly over 4 per cent of these reserves have been explored, they still represent four times the combined energy equivalent of the Soviet Union's vast natural gas and oil reserves. Unfortunately from the perspective of the geographically developed parts of the country, nearly 75 per cent of the explored reserves lie east of the Ural mountains, thus posing costly and problematic mine-to-market transport bottlenecks. Furthermore, many of these remote coal reserves are of poor quality because of high levels of ash, water and sulphur (*Zapasy Ugley Stran Mira*, 1983, pp. 93–102). While coal-resource maps of the USSR appear to be dominated by the areal extent of the Central Siberia Tunguska and Far East Lena basins, these basins still remain largely unexplored, inaccessible and unexploited. For example, as of 1983 only 2,000 million tons of the Tunguska's estimated geological reserves of 2,299 billion tons, 4,000 million tons of the Lena's estimated geological reserves of 1,647 billion tons, and 4,000 million tons of the South Yakutia basin's estimated geological reserves of 44,000 million tons had been explored (*Zapasy Ugley Stran Mira*, 1983, pp. 93–102). Only a small portion of the Tunguska's hard coal deposits are located within the formal Far East Economic Region, but most of the basin's lower quality brown coal or lignite deposits lie huddled along the western border of the Far East's formal boundary (see Figure 5.1). The massive Lena and Tunguska reserves being largely unexplored, inaccessible, plagued by permafrost conditions and far from market areas, as noted above, will very likely be of little

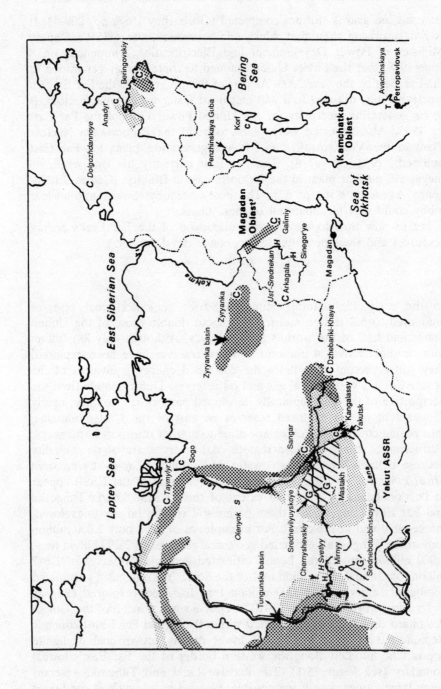

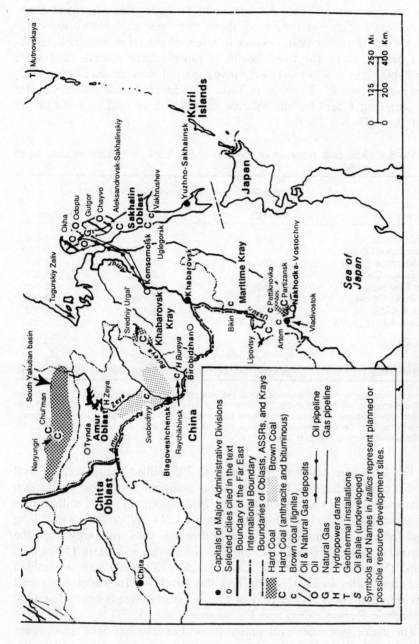

5.1 Selected energy resources of the Soviet Far East

commercial significance during the remainder of this century. Although the reserve data in Table 5.1 appear to be quite old, this is not really significant because as reiterated above little new exploratory work has been done in recent years, except in a few of the more accessible southern basins. Even for these basins if more recent reserve data were available, they would most likely reveal partial conversions from type C_2 reserves to type C_1. The data in Table 5.1, therefore, are still very useful in obtaining a feel for the relative distribution of coal reserves in the Far East (Dienes 1989).

Table 5.1 The coal reserves of the Soviet Far East (in millions of tons as of 1966)

Region	Productive categories $A+B+C_1$	Prospective reserves C_2	Total
Amur Oblast	419	3	422
Kamchatka Oblast	67	60	127
Khabarovsk Oblast	807	1,179	1,986
Magadan Oblast	1,068	3,259	4,327
Maritime Kray	2,450	1,359	3,809
Sakhalin Oblast	2,073	1,496	3,569
Yakut ASSR	4,580	4,469	9,049
(Yakut ASSR (1983)	8,000–10,000	na	na)
Far East	11,464	11,825	23,289
USSR	237,200	235,000	472,200
(USSR (1983)	281,000	na	na)

Sources: Rodgers, A. (1983) 'Commodity flows, resource potential and regional economic development: The example of the Soviet Far East', in Robert G. Jensen, Theodore Shabad, and Arthur W. Wright (eds) Soviet Natural Resources in the World Economy, Chicago and London: University of Chicago Press, p. 201; Central Intelligence Agency (1985b) USSR Energy Atlas, Washington, D.C.: US Government Printing Office, p. 35.

The 1990 Plan calls for an increase to 795 million metric tons in total Soviet coal production. The most recent available Soviet coal production data indicate that total production for 1988 amounted to 772 million tons of gross mine output of which an estimated 211 million tons was coking coal. Siberia accounted for 316 million metric tons, primarily from the Kuznetsk basin (160 million) and the Kansk-Achinsk basin (50 million tons) (Sagers 1989a, p. 329). The residual 110 million tons was mined at various sites in East Siberia and the Far East. In recent years coal has supplied about 50 per cent of the fuel-energy needs of the Far East Economic Region (Shilo and Chichkanov 1981, p. 77).

The major new coal development in the Far East has been the opening up of the South Yakutian coal basin centred at the new town of Neryungri. In contrast to many other local Siberian deposits and mines, this basin contains a high-grade bituminous coal acceptable for both steam-raising and coking uses. Initial commercial development began in the 1970s under an agreement with Japan. Initial steam-coal shipments destined for local and other Soviet Far East power generation began in late 1978 when the north-south traversing 'Little BAM' reached the large open-pit Neryungri mine and linked up with the Trans-Siberian Railway through Tynda. In exchange for loans and technical assistance Japan began receiving deliveries of South Yakutian coking coal in 1985 after the deeper coking coal was uncovered in 1984 (Shabad 1986a, pp. 8, 12–13; Shabad 1986b, p. 271). Prior to shipment to Japan the raw Neryungri coking coal undergoes minehead beneficiation at the USSR's largest coal washery (Sagers 1988a, p. 447). This washery can convert 13 million tons of raw coal into 9 million tons of export-grade coking coal concentrate and 4 million tons of steam coal. Its design capacity was to be expanded by 250,000 metric tons in 1988 (Sagers 1989a, p. 337). Despite their high quality and open-pit mining, these South Yakutian coals are the most expensive coking coals to exploit in the Soviet Union. The cost of the Neryungri project was originally estimated to be 936 million roubles, but by early 1979 the actual investment figure had already grown to over 3,000 million roubles. This three-fold cost escalation can largely be charged to additional mining costs associated with the removal of overburden, the costs of coal washing needed to reduce the high ash content (18 to 20 per cent), the costs associated with coping with a layer of oxidized steam coals, and the costs attributable to labour and equipment problems during Siberian 'cold snaps' when the temperatures drop to less than −62°C (Rodgers 1983, pp. 201–2). Probably for this reason a proposed underground mine in the basin was cancelled shortly after Gorbachev came to power (Shabad 1986b, p. 271).

In 1987 the South Yakutian basin yielded over 14 million tons, 0.7 million tons above its designed capacity. An additional combined 1.5 million tons of hard and brown coal were extracted elsewhere in Yakutia in 1987 at widely scattered mines such as Sogo, Zyryanka, Sangar, Dzhebariki-Khaya, and Kangalassy near Yakutsk (see Figure 5.1) (Sagers 1989a, p. 337).

To the south east in Amur Oblast brown-coal mining activity occurs in the Raychikhinsk basin at Raychikhinsk and is planned in the Nizhnezeysk basin at Svobodnyy. The major current mining in Khabarovsk Kray is in the Bureya basin at Sredniy Urgal where recently new planned capacity did not come on line as scheduled. None the less, hard-coal output in 1987 reached 1.8 million tons or 220,000 tons over plan (Sagers 1988a, p. 447). A second section of the Luzonovskiy mine here should add 2 million tons to the basin's annual capacity (Shabad and Sagers 1987, p. 274). As indicated in Figure 5.1, several local-scale mining activities currently exist in Magadan and Kamchatka Oblasts.

In the south-eastern, most populated regions of the Far East, scattered mines operate in Primorsk (hereafter, Maritime) Kray. At present the kray has nine shaft-mining and four open-pit operations. The Maritime Kray's combined output in 1986 was about 19 million tons and is projected to attain 21–2 million tons by 1990 (*Ugol'*, no. 1, 1988, p. 22). The Artem mine near Vladivostok underwent *perestroyka* in 1986 adding 2.4 million tons to its capacity which should now (1990) be about 5 million tons annually (Shabad and Sagers 1987, p. 274). An additional one million tons of annual capacity was announced at the Pavlovskiy mine in early 1987 (*Stroitel'naya Gazeta*, no. 5, February 1987). Finally, coal-mining operations exist at four locations on Sakhalin Island.

In summary, the Soviet Far East's vast reserves of both hard and brown coal have thus far not been exploited at high enough levels to make the region self-sufficient in coal-based energy consumption. Accordingly, coal is still being shipped into the region from basins west of Lake Baykal. As noted above, the gigantic, but geographically remote, reserves of the Tunguska and Lena coal basins seem destined to remain poorly explored and exploited well into the next century. On the other hand, coking-coal production at the more accessible South Yakutian Basin could undergo significant expansion if a long proposed new integrated iron and steel facility were to be built in the Far East region. Mathematical modelling indicates that Chul'man just 20 km to 30 km north-east of Neryungri and within 100 km to 150 km of the four Aldan iron-ore deposits of Tayezhnoye, Desovskoye, Sivagli and Pionerskoye would be an optimal location for such a ferrous-metal facility (Zum-Brunnen and Osleeb 1986, pp. 185–95).

Oil

The Soviet oil industry over the past two decades has become dominated by the fields of the West Siberian plain which in 1988 produced about two-thirds (419 million metric tons) of total Soviet oil (624.3 million metric tons). This share is planned to increase to 440 million metric tons out of a total of 635 million metric tons in the 1990 Plan (Sagers 1989a, p. 308). By comparison the Far East's oil production of 3 million metric tons in 1988 is negligible. Currently the only producing region is Sakhalin Island off the Pacific coast. Its meagre production, stable since the late 1970s, is in fact planned to decline slightly to 2.7 million metric tons in the 1990 Plan (Sagers 1988a, p. 425). Meyerhoff's study of the history, technology, geology and reserves of Soviet petroleum may be used to supplement this cursory overview (Meyerhoff, 1983, pp. 306–62).

In 1975 the USSR and Japan reached a US $152 million agreement for joint exploration and development of Sakhalin's offshore petroleum resources (Rodgers 1983, p. 203; Central Intelligence Agency 1985b, p. 11). The consortium of Japanese petroleum and trading companies and one US firm, Gulf Oil, is called SODECO. This joint venture involved the use of credits guaranteed by western governments to finance the purchase of western petroleum equipment in exchange for Soviet repayment in terms of oil and gas resources. Plagued by short ice-free drilling seasons, equipment shortages, economic-commercial sanctions and conservative decisions to drill convenient but unproductive geological structures, this venture has been somewhat disappointing. Thus, although two oil and gas fields – Odoptu and Chayvo – off the north-east coast of Sakhalin have been discovered, no development has yet taken place. In June 1986 both the Soviet Union and Japan decided to reassess their plans to develop the discovered fields due to weak oil prices (Lawarne 1988, pp. 83–98).

At this point it seems some lessons can be drawn from the Soviet-Japanese joint exploration efforts off Sakhalin Island. First, despite her long history of offshore development of the shallow Caspian Sea oil fields, the Soviet Union at this time appears to lack the technology and equipment capable of developing offshore deposits, especially in turbulent, deep-water seas, in the Far East or elsewhere. While officially dealing primarily with Japan, the Soviets encountered major difficulties with both the Carter and Reagan administration embargoes on the numerous licences required for the Japanese rigs and equipment. As a result, in their efforts to circumvent (both directly and indirectly) the necessity of obtaining American licences, the Soviets have sought to engage Finnish companies with their shipbuilding experience and access to Norwegian North Sea oil-development technology to aid the Soviet offshore oil industry. This strategy, of course, has its own potential problems and limitations for the Soviets because Norway is a very strong supporter of the United States within NATO's Co-ordinating Committee on Multilateral Export Control (COCOM). Second, although the Soviets consciously retreated from the development phase of offshore Sakhalin oil while the Japanese appeared eager to move forward, one cannot simply assess the joint venture as a failure from the Soviet perspective. This seems true because, although the majority of the exploratory work had been completed, immediate production was not a Soviet priority despite the Far East being an oil-deficit region of the USSR (Lawarne 1988, pp. 97–8).

The North Sakhalin basin extending over a combined offshore and onshore area of 24,000 sq. km contains some fifty oil and gas fields. About 0.5 million tons of Sakhalin's 3 million tons of annual oil production is extracted from offshore wells (Danilov 1987, p. 102). Sakhalin's total production thus amounts to only about 20 per cent of the oil demands of the Far East. Sakhalin's oil is shipped by pipelines across the narrow strait separating the island from the mainland to the refinery at Komsomol'sk-na-amure. The regional deficit of about 12 million metric tons is shipped in from the west by tanker along the Trans-Siberian Railway (Shabad 1987, p. 72). These interregional oil imports supply the oil refinery at Khabarovsk.

The only other prospects for Soviet Far East petroleum production are located in Yakutia in the so-called Siberian Platform between the Yenisey and Lena rivers. In south-western Yakutia lies the north-east limit of the Lena-Tunguska oil and gas region. A series of anticlinal structures in the Sredneboturinskoye area appear to be highly promising oil and gas traps (Robinson 1985, pp. 91–7). In fact during the summer of 1987 at least two small nuclear explosions (yields of less than 20 kilotons) were detonated in this region south of the mining town of Mirnyy. Reportedly with 'good results', these explosions were designed to assist potential oil production (Sagers 1988a, p. 429).

The other promising area is the Khatanga-Vilyuy oil and gas province located essentially north-west of the city of Yakutsk. The province already is the centre of Yakutian natural gas exploitation, but new condensate fields have also been identified (Robinson 1985, p. 92).

The scope, tempo and timing of any potential new oil extraction operations within the Siberian Platform are circumscribed by several factors. First, the oil reserves tend to be spatially dispersed. Second, even within oil-bearing regions the oil tends to be trapped in a series of smaller anticlinal structures. Third, the physical and climatic conditions are harsh and the oil reservoirs tend to be very deep and expensive both to explore and exploit. On the positive side, the quality of Siberian Platform oils is high with low sulphur, solid paraffin, tars and asphaltene contents. Despite some current active development in the Irkutsk Oblast portion of the Lena-Tunguska oil and natural gas province (Sagers 1988a, p. 428), it appears very doubtful that Yakutian oil production will materialize, except for minor condensate, until well into the 1990s, if then.

Finally, while the vast majority of the Far East has not been subject to either oil or gas exploration, some very modest preliminary exploration has taken place in the north east. For example, many wells have been drilled on the Chukotsk Peninsula (Meyerhoff 1983, p. 341; *Selskaya zhizn'*, 26 October 1984, p. 1) and more than 100 wells (mainly natural gas) have been sunk along the east coast of Kamchatka (Meyerhoff 1983, p. 341; and *Gazovaya promyshlennost'*, vol. 1, no. 7, July 1985, p. 11). Only references to several minor oil shows and seeps have been found and no large potential appears to exist because of the volcanic origin of so many of the potential reservoirs. Meyerhoff (1983, pp. 338–42) provides a discussion of the geology of the essentially unexplored offshore basins in the seas surrounding the Soviet Far East. It is possible that these regions could contain vast reserves of petroleum, but exploration and possible development planning for these distant and environmentally difficult regions seems far off.

Gas

Overall Soviet natural gas development is continuing with far less difficulties than her oil development. For example, total 1988 gas production of 770,000 million cu. m increased by 5.9 per cent over 1987 production and was thus 18,000 million cu. m above the plan target. Similar to oil production, West Siberia dominates natural gas production, accounting for 64.5 per cent of the Soviet total in 1988. Other Siberian gas production in 1988 amounted to only 8,000 million cu. m. This production is centred on three regions: Noril'sk in Krasnoyarsk Oblast in East Siberia, and Yakutia and Sakhalin Island in the Far East (Sagers 1989a, pp. 317–27).

In 1987 total Yakutian gas production amounted to a little over 1,000 million cu. m (*Summary of World Broadcasts*, 27 November 1987, p. 5). Until very recently Yakutian gas production originated solely from the Mastakh gasfield centred on the settlement of Kysyl-Syr and the Srednevilyuskoye (Middle Vilyuy) gasfield, both situated some 200 km to 300 km north-west of the city of Yakutsk. The Ust'-Vilyuy gasfield which originally served Yakutsk is now depleted. The Mastakh and Srednevilyuskoye fields are linked by several pipeline strings to Yakutsk and the nearby industrial towns of Bestyakh, Pokrovsk and Mokhsogollokh (*Atlas SSSR*, 1984, p. 187). In 1987 the mining town of Mirnyy in the southwestern region of Yakutia was linked with the new Srednebotuobinskoye (Middle Botuobinskoye) gasfield via a new 180 km gas pipeline (Sagers 1988a, p. 440).

Gas reserves may be substantial in the Vilyuy basin. Currently there are eight known gasfields in the area. In the early 1980s 273,500 million cu. m had been proven and an additional 190,900 million cu. m were probable or potential (C_1 reserve category) and 161,400 million cu. m were speculative reserves (C_2). The Soviets claim that the proven natural gas reserves in Yakutia, including those of the Irkutsk amphitheatre fields such as the newly producing Srednebotuobinskoye field, are upwards of 792,900 million cu. m (Meyerhoff 1983, p. 338). Stern (1983, pp. 364–5) lists what appear to be slightly older reserve figures of 316,200 million cu. m for Yakutia and 68,200 million cu. m for the Sakhalin basin.

The two primary gas consumers of the first of the two above-mentioned gasfields appear to be the Yakutsk regional heat and power plant and the cement plant at Mokhsogollokh (Shabad 1987, p. 75). *Izvestiya* (21 November 1987, p. 6) disclosed that an explosion in the distribution pipeline occurred on 20 November 1987, forcing the Yakutsk power plant to be switched over to reserve fuel oil. Sagers (1988a, p. 440) noted that no back-up pipeline had yet been completed by early spring of 1988.

As discussed previously, the Vilyuy river gasfields are reported to contain large gas reserves and were the subject of intensive Japanese and

United States export-oriented liquefied-natural-gas (LNG) development interest in the early 1970s (Meyerhoff 1983, p. 338; Stern 1983, pp. 375–6). Discussions and negotiations surrounding this central Yakutian gas-based LNG project first ran into difficulties after the enactment of the US Trade Act of 1974, which incorporated the Jackson-Vanik and Stevenson Amendments. Other difficulties ensued with regard to the size and rate of proving up of gas reserves as called for in the agreement with the US and Japan (Stern 1983, pp. 375–6). Finally, the Yakutian LNG project was terminated by the western interests after the Soviet invasion of Afghanistan in 1979 (Shabad 1987, p. 75). Again, the relatively depressed world oil prices probably will prevent rekindled western interest in Yakutian natural gas for some time even though the Soviet Union has withdrawn from Afghanistan.

The largest producer and final Far East gas region to be discussed is Sakhalin Island. While Sakhalin oil has been flowing to the mainland via two pipelines for some time, Sakhalin natural gas did not reach the mainland until 1987 when the Okha–Komsomol'sk gas pipeline, completed in 1986, was put into service. The discovery in the 1970s of the Dagi-Mongi gas deposits on the east coast of Sakhalin Island allowed for sufficient expansion of Sakhalin gas production to warrant this new gas distribution pipeline to the mainland (Shabad 1985, p. 772). Initially this pipeline was scheduled to transmit about 1,500 million cu. m annually increasing to three times that amount by 1990 (*Sovetskaya Rossiya*, 4 June 1987). Part of this augmented supply, some 500 million cu. m annually, will come from the newly installed gas collection system at the Mirzoyeva oil field where the associated gas was previously flared (*Sotsialisticheskaya industriya*, 4 March 1988, p. 1). On the mainland Sakhalin gas is now being used to fuel the boilers at the two Komsomol'sk-na-Amure heat-and-power plants and fuel its small open-hearth Amurstal' steel mill furnaces and rolling mill. This gas is also being consumed for apartment space-heating (Sagers, 1988a, p. 440). Shabad (1985, pp. 771–2) reported that Sakhalin gas will also be used as the feedstock in a projected nitrogen fertilizer plant to be located at Nizhnetambovskoye on the Amur river along the trace of the pipeline some 100 km north-east of Komsomol'sk-na-Amure. If built, this plant would be the only nitrogen fertilizer plant in the Far East and given the limited agricultural potential of the Far East, its output could partly serve the Japanese market (Dienes 1985, p. 517; Shabad 1985, pp. 771–2).

Despite the regional oil deficit, it does not appear that the Soviets will rapidly expand either Yakutian or Sakhalin gas production without a strong export market and joint venture co-operation.

Hydro-power resources

Geographically, of course, the European Soviet Union is well known for its series of large hydro-electric cascades constructed within the Dnepr and Volga basins and for smaller cascades, especially in Karelia, the Kola Peninsula and the Transcaucasus. In the post-war era the construction of gigantic hydro-electric dams on the Yenisey-Angara river system in East Siberia has been an emotional rallying cry for the building of communism. Hydro developments in the Far East are much more modest in scale and are concentrated in three broad regions.

The Amur river basin has the two largest hydro developments in the Far East, both on left-bank tributaries of the Amur. The oldest is the 1,290 megawatt Zeya hydro station located about midway between the headwaters and mouth of the Zeya river in central Amur Oblast. This station is the source of power for a 500 kilovolt transmission line travelling south-eastwards along the old Trans-Siberian Railway through Svobodnyy and on to Khabarovsk and Komsomol'sk-na-Amure. This high voltage line is scheduled to be extended southwards to link up with power from the coal-fired Luchegorsk plant and on to Vladivostok during the 12th Five Year Plan, thus joining up with the Maritime Kray power grid. Farther to the south-east just above the confluence of the Bureya river with the Amur an even larger hydro-dam is under construction at Talakan, 180 km east of Blagoveshchensk. With an ultimate designed capacity of 2,000 megawatts, this Bureya hydro-electric station will significantly strengthen the Far East power grid. The first of its six 335 megawatt generating units is scheduled to go on line by 1990 (Shabad 1986a, p. 4). Heading west from the Zeya station an important 220 kv intertie line between Chita Oblast (Siberian grid) and Amur Oblast (Far East grid) was completed in January 1988 (Sagers 1988a, p. 455).

The second centre of Far East hydro-power development is associated with diamond mining in the upper reaches of the Vilyuy river basin in Yakutia. The first station at Chernyshevskiy was begun in 1960 and achieved its 308 megawatt designed capacity in 1969. This station was expanded during the so-called No. 2 Project in the 1970s to a total of 648 megawatts. A third project was started in 1979 downstream at Svetlyy located at the confluence of the Ochchugui (or Little) Botouobuya river, a right-bank tributary of the Vilyuy river. When fully completed the Svetlyy station will have four generating units each with a 90 megawatt capacity (Shabad 1987, p. 78). The first unit is planned to begin generating electricity by 1990 (Shabad 1986a, p. 4).

The need to supply the electricity demand of the important gold-mining industry in north-east Siberia has stimulated the third and final hydro-electric system in the Far East, namely, the Kolyma river system in Magadan Oblast. Begun in 1970 the first station at Sinegor'ye was shrouded in controversy over poor management and improper operating conditions at the station in 1981. By 1984, however, three of the five planned 180 megawatt generating units were in place. The current or 12th Five-Year Plan provides for the start of construction on the second Kolyma station, the Ust'-Srednekan hydro-electric dam, downstream from Debin at the mouth of the Srednekan river (Shabad 1986a, p. 4).

It seems unlikely that the Soviets will build any hydro stations in the Far East having capacity larger than either the Zeya or Bureya stations. It seems more likely that they would continue to focus more of their investment resources on trying to reinforce the reliability and spatial connectivity of the Far East power grid.

Geothermal energy

It is quite probable that the Soviet Union possesses the largest undeveloped geothermal resources in the world, but it lags behind such western nations as Italy, Iceland, Japan, New Zealand and the United States in research and development work. Almost one-half of the Soviet Union has 100°C or greater heat energy situated within 3,000 m to 4,000 m of the land surface. Exploration work has identified over fifty possible large geothermal development sites. The Soviet policy of development seems predicated on cost-effectiveness concerns. Except in remote areas and especially suitable areas where fossil fuels are lacking, geothermal resources apparently will not be developed. Even in these cases the main uses will be for municipal and industrial heat and hot water. In many Far East and Siberian mining operations the use of hot water is being considered to help ease mining operations and lengthen the mining season (Central Intelligence Agency 1985b, p. 64).

The Kamchatka Peninsula is composed of a series of both dormant and active coalesced volcanic islands. It is not surprising then that it is the location of the Soviet Union's only geothermal steam-driven power generating station. Rated at a mere 5 megawatts, it has been plagued with operating problems and has only been intermittently operational at a sub-design level of 3.5 megawatts. In the early 1980s the Soviets reported that a newly drilled steam borehole would be used to expand the capacity to eleven megawatts (Central Intelligence Agency 1985b, p. 64).

In terms of the Far East's geothermal potential and development plans four items may be noted. First, feasibility studies on the use of the energy from magma and hot rock to generate electricity have been undertaken at two volcanoes in the south-east part of the Kamchatka Peninsula, Avachinskaya and Mutnovskaya. Second, Soviet estimates suggest that geothermal steam and hot water deposits on the Kamchatka Peninsula would be sufficient to produce as much as 600 megawatts of electricity. Third, by the development and use of artificial circulation systems (created by underground explosions or hydraulic pressure) in hot rocks on Kamchatka this electrical generating capacity could be increased by 3,000 megawatts. And fourth, the only large-scale project under construction is a 200 megawatt installation at the Mutnovskaya volcano ('Power generation plans for Far East outlined', 1988, p. 68).

Other fuel and energy resources

The Aldan district contains uranium, thorium and other rare earths associated with gold mining (Central Intelligence Agency 1985b, pp. 42–3). Yakutia contains the Olenyok oil shale deposits lying to the west of the Lena river some 800 km north-west of the city of Yakutsk, but they are not now being exploited and it seems unlikely that they would be. (Only the adjacent Estonian and Leningrad oil shale fields and the middle Volga Syzran' oil shale deposits are being commercially used in the Soviet Union today.) On the other hand, this same geographically remote region of Yakutia contains a large fraction of the estimated 30,000 million metric tons of potential oil reserves which could be extracted from tar sands. Only in the European north at Ukhta are the Soviets making use of tar sand deposits. While the USSR possesses about 60 per cent of the world's supply of peat reserves, and peat reserves are extensive in the Far East; this low-grade organic fuel is used very little both in relative and absolute terms in the Far East. Fuelwood, by way of contrast, does appear to be quite important for winter space heating and year-round cooking in many of the more remote parts of the Far East. Finally, although the Soviets have two excellent tidal power sites in the Far East at Penzhinskaya Guba (maximum tidal range 11.0 m, average range 5.8 m) and Tugurskiy Zaliv (maximum tidal range 7.3 m, average range 4.0 m); the costs involved in the hydro-technical structures required to convert the tidal kinetic energy at these two sites into electricity will remain absolutely prohibitive for a very long time (Central Intelligence Agency 1985b, pp. 42–5, 63).

Mineral resources

Introduction

On the one hand, the Far East's mineral resources appear to be quite vast and varied, and in many instances not well known or explored due to the economic remoteness and harshness of the region. On the other hand, because of the high value and acute need for some of the Far East's mineral resources, especially diamonds, tin and gold, the Soviets have exploited some of these materials despite their severe locational

97

handicaps. The recent Soviet Nedra (Mineral Wealth) Programme is designed to overcome some of these development handicaps. For example, one of the main tasks of the Nedra Programme is the *perestroyka* or restructuring of the Far East's 'mining industry so as to extract all valuable ore components, to avoid waste, and to conserve existing ore deposits' (Chichkanov and Minakir 1986, pp. 106–7). One claimed success of the Nedra Programme has been the recycling of mine wastes to serve as inputs to auxiliary industries (Chichkanov and Minakir 1986, pp. 106–7).

Recently, the Far Eastern Geology Institute of the Academy of Sciences published a book which includes descriptions of eighty-seven different minerals and their deposits in the Far East (Shcheka 1987). Many of these listings include recently discovered deposits and many of them represent either industrial minerals and/or deposits of rare minerals. Unfortunately, more information is given about the chemical, mineralogical and geological associations of these deposits than about their commercial quality and size. As such it would be difficult to present a comprehensive overview of them, even if space allowed.

If this investigation's geographical coverage extended westwards to include East Siberia then the valuable nickel, cobalt and platinum-group metals centred in the north at Noril'sk would be a prominent topic as would the varied mineral deposits, such as the Udokan copper in Chita Oblast, along the BAM service area in southern East Siberia. However, the purpose here will be merely to highlight some of the more important minerals in the Far East region under the three categories of iron-ore resources, non-ferrous metallic ore resources, and non-metallic minerals (see Figure 5.2). Readers seeking additional detailed information about the mineral resources of the Far East are urged to refer to the volume *Soviet Natural Resources in the World Economy* especially Chapters 7, 8, 9, 19 and 24 (Jensen, Shabad and Wright 1983, pp. 138–231, 464–91, 556–96).

Far East iron-ore resources

The Soviet Union possesses more iron-ore reserves than any nation on earth. In 1987 total Soviet iron-ore production reached 251 million metric tons (Sagers 1988b, p. 535) but declined slightly to 250 million metric tons in 1988 (Sagers 1989b, p. 400). Soviet iron ore production and reserves are abundant in the European heartland (unlike the geographical distribution of many of the Soviet Union's industrial raw materials and energy resources. These latter are located from the Urals eastwards or are essentially inversely proportional to the current spatial pattern of demands which are concentrated in the industrially developed European part of the country). For example, as of 1975 the Kursk Magnetic Anomaly (KMA) in southern European Russia contained 16,700 million metric tons and Ukrainian deposits 21,400 million metric tons of the *proven*, *explored* or *probable* reserve categories, A + B + C₁. Combined, these two European ferrous metallurgy industrial areas held 59.8 per cent of the country's 63,700 million metric tons of proven and probable ore reserves. In the speculative category (C₂) the KMA alone accounted for 21,700 million metric tons or one-half of the 43,400 million metric tons of C₂ reserves. By contrast in 1975 the Far East's reserves of 1,800 million metric tons represented only 2.8 per cent of the proven or probable reserves and its 1,000 million metric tons of C₂ reserves only 2.3 per cent of the speculative reserves (Shiryayevo, Yarkho and Borts 1978, p. 9). As of 1 January 1981, the Soviet Union's total A + B + C₁ proven or probable reserves had grown to 77,600 million metric tons (*Razvedka i okhrana nedr*, 1981, p. 15). Then, too, more recent data on Far East ore deposits (see Table 5.2) indicate proven and probable iron-ore reserves ranging from 3,240 to 3,960 million metric tons, and speculative reserves ranging from 3,870 million to 4,590 million metric tons. For our purposes the Far East region contains four major iron-ore districts which will be discussed below.

Table 5.2 Selected Far East iron-ore deposits

Name or location of deposit	Status	Reserves (1,000 mill. m.t.) A+B+C₁	C₂	Ave. Fe content (in % Fe)	Main Fe minerals
Chara-Tokko	not used	2.00	3.00	30	FQ
Aldan:	not used	0.84	0.66	40–42	M
Tayezhnoye	not used	0.71	0.58	42.0	M
Desovskoye	not used	na	na	40.0	M
Pionerskoye	not used	na	na	40.0	M
Sivagli	not used	na	na	40.0	M
Other Yakutian:					
Olekma-Amga	not used	(not of industrial quality)			FQ
Khani	not used	(not of industrial quality)			FQ
Tarinakhskoye	not used	na (but large)		na	na
Imalyskoye	not used	na (but large)		na	na
Gorkitskoye	not used	na (but large)		na	na
Gar	not used	0.21	0.18	41.7	M
Malokhingan:	not used	(0.72 total)		30–36	FQ,M,H
Kimkan	not used	0.19	0.03	35.6	FQ,M,H
Udsko-Selemdzhinsk	not used	na	na	na	na

Sources: Misko, T. and ZumBrunnen, C. (1983) 'Soviet iron ore: its domestic and world implications', in Robert G. Jensen, Theodore Shabad and Arthur W. Wright (eds) *Soviet Natural Resources in the World Economy*, Chicago and London: The University of Chicago Press, p. 472; Bunich, P. G. (1977) 'Economic impact of the BAM', in Theodore Shabad and Victor Mote (authors and eds) *Gateway to Siberian Resources (The BAM)*, Washington, D.C.: Scripta Publishing Co., p. 138; Mote, V. (1983) 'The Baikal–Amur mainline and its implications for the Pacific basin', in Robert G. Jensen, Theodore Shabad and Arthur W. Wright (eds) *Soviet Natural Resources in the World Economy*, Chicago and London: The University of Chicago Press, pp. 152, 155–6; and Kuznetsov, V. A. (1985) 'Mineral'nyye resursy zony BAM i problem ikh osvoyeniya', in A. G. Aganbegyan and A. A. Kin (eds) *BAM: pervoye desyatiletiye*, Novosibirsk: Izdatel'stvo Nauka, pp. 126–8.

Notes: FQ: Ferruginous quartzites
M: Magnetite (Fe⁺²Fe₂⁺³O₄)
H: Hematite (Fe₂O₃)

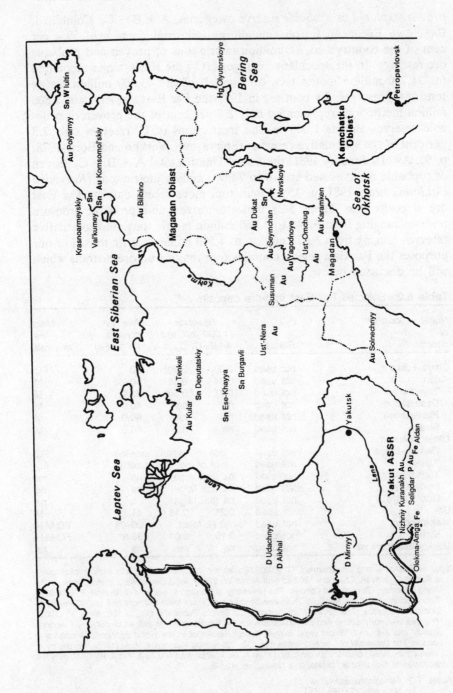

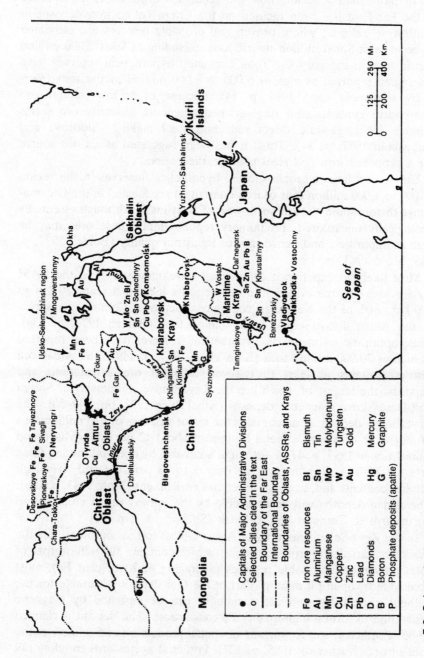

5.2 Selected mineral resources of the Soviet Far East

For more than a decade now the geological exploration for iron ore in the Far East has been focused on the Chara-Tokko iron deposits in south-west Yakutia, where proven and probable reserves are estimated to be at least 5,000 million metric tons, including at least 2,000 million metric tons in category C_1. Total estimated hypothetical reserves have even been reported as high as 6,000 to 8,000 million metric tons (*Razvedka i okhrana nedr*, 1981, p. 14). Because of its low sulphur and phosphorus contents, this large-grained magnetite quartzite ore seems suitable for large-scale direct reduction steel making (Sidorova and Vadyukhin 1977, p. 34). Thus, it has been suggested as an ore source for a proposed iron and steel plant in the region.

The second largest prognosticated hypothetical reserves in the region (2,000 to 4,000 million tons of iron quartzites) are located in the Olekma-Amga district some 400 km north-west of Neryungri. Although essentially geologically unexplored, preliminary reports imply these ores may be high in impurities, and hence, not of industrial quality (Bunich 1977, p. 138; Mote 1983, p. 155).

Most likely to experience commercial development first in the BAM service area are the Aldan iron-ore deposits located a scant 80 km to 100 km north of the Neryungri coking-coal deposits. Confirmed reserves in the Aldan district equal 1,500 million tons (Bunich 1977, p. 138), while optimistic estimates of 'hypothetical' reserves have been reported as high as 20,000 million tons (Kibal'chich 1977, p. 149). The Aldan ore district contains at least thirty-seven different ore associations and deposits, the largest of which are the Tayezhnoye, Pionerskoye, Sivagli and Desovskoye magnetite deposits (Kibal'chich 1977, pp. 149–50). The Tayezhnoye deposit alone contains the major reserves of 710 million tons of magnetite ores with a metal content of about 42 per cent (Misko and ZumBrunnen 1983, p. 472). Although accessible by open-pit mining these ores are high in silicates, ranging from 2 to 33 per cent and have high sulphur contents and, hence, will require enrichment (Mote 1983, p. 156). The Aldan deposits will be accessible by the railway under construction northwards to Tommot and Yakutsk (Shabad 1987, p. 81).

In the Zeya-Selemdzha area of Amur Oblast to the south-east lies the shallowly buried Gar iron-ore deposit which contains 390 million tons of total proven and probable reserves (averaging 41.7 per cent Fe), most accessible by open-pit mining. Part of the Gar deposit is suitable for use without enrichment and the remainder can be enriched by magnetic separation to form a high-quality Fe concentrate. The deposit is claimed to be of sufficient size to warrant an annual production of 5 million metric tons of ore (Kuznetsov 1985, p. 127). Potential northwards trending rail spurs from the completed BAM could potentially make available the reputed billions of tons of iron-ore reserves with associated manganese in the Udsko-Selemdzhinsk district about 400 km to 450 km north-east of Svobodnyy (Shabad 1980, pp. 54–5).

Finally, located in the Malyy (Lesser) Khingan mountains in the Yevreysk (Jewish) Autonomous Oblast of Khabarovsk Kray is the ferruginous quartzite Kimkan deposit containing a total of 190 million tons of A + B + C$_1$ class iron ores. It has a lower Fe content (35.6 per cent) than some of the other Far East iron-ore reserves and is in need of further exploration to determine its commercial significance (Misko and Zum-Brunnen 1983, p. 483). The total reserves of the Malokhingan iron district are claimed to be 722.6 million tons (Kuznetsov, 1985, p. 127).

In summary the Far East appears to have far more than adequate iron-ore reserves to supply any as yet unbuilt integrated regional iron and steel complex. Such facilities have been proposed for the Chul'man-Aldan district, Svobodnyy, or Komsomol'sk-na-Amure in the Far East, and Tayshet in East Siberia. This issue remains undecided; however, a new scrap-based, electric steel mini-mill commenced operation at Komsomol'sk-na-Amure in 1986 (Sagers 1987, p. 363). Interestingly enough, ZumBrunnen and Osleeb's SISEM optimization model of the USSR's iron and steel industry indicated that both Tayshet and the Chul'man-Aldan area would be optimal locations for steel complexes each producing over 2.5 million tons annually (ZumBrunnen and Osleeb 1986, pp. 183–6). At this point, however, the Far East's iron-ore reserves remain just that, reserves. It appears as though the Soviets have lost out to Australian and south-east Asian deposits as ore suppliers for the Japanese steel mills.

Non-ferrous metal ore resources

Manganese and aluminum

In 1988 total Soviet manganese production amounted to 9.1 million metric tons, but none of this production appears to have originated in the Far East (Sagers 1989b, p. 410). Manganese ores have been detected in the Udsko iron-ore region of Khabarovsk Kray where they are associated with iron ore and also phosphorite ores. They are not at present of industrial value, however, because of the remoteness and inaccessibility of this region (Kuznetsov 1985, p. 128). Another manganese district lies close to the Chinese border in the Yevreysk Autonomous Oblast (*Atlas SSSR*, 1984, p. 93). An aluminum-ore body has been discovered south of the mouth of the Amur river in Khabarovsk Kray (Kovrigin 1986, p. 5). No hard data on the size of reserves of any of these mineral deposits have been uncovered.

Copper, zinc, lead, bismuth

The Soviet Union smelts about 1.6 million tons of copper annually, which places her essentially on par with the US in terms of copper production. Only a small fraction of the current Soviet copper production comes from the Far East (Central Intelligence Agency 1987, p. 155). The main Far East copper producing site seems to be near Komsomol'sk-na-Amure where copper is a by-product of the Solnechnyy tin mine (Shabad 1987, p. 83). Given the technological complications which have now long plagued even the large-scale Udokan copper deposit to the west in Chita Oblast, it is not surprising that the lower-grade, copper-bearing sandstones of the Dzheltulakskiy flexure, located close to the route of the BAM south of Tynda, are also not currently of economic industrial value (Kuznetsov 1985, p. 128).

Current statistics are difficult to come by, but, none the less, it appears as though the USSR is the world's leading producer of both lead and zinc. The CIA estimated 1984 Soviet zinc production at 850,000 tons and lead production at 582,000 tons (Central Intelligence Agency 1987, pp. 157–8). New ore sites are being opened up at Ozernyy and Kholodnaya along the BAM east and north of Lake Baykal and west of the Far East Economic Region. The Far East proper is reported to contain the richest zinc deposits of the entire Soviet Union in the regions of the Sikhote-Alin' mountains of the Maritime Kray, the Chukchi Peninsula, the southern portions of Khabarovsk Kray, and in Yakutia. As is mentioned again below, tin-rich cassiterite-ore bodies in the Komsomol'sk-na-Amure region of Khabarovsk Kray have significant trace amounts of copper, lead, zinc, bismuth and wolfram (tungsten ore) (Kovrigin 1986, p. 5). Finally, the poly-metallic ores at Dal'negorsk in the Maritime Kray yield zinc and lead as well as tin and gold (Mote 1983, pp. 152–5).

Tin, molybdenum, tungsten

In 1986 Soviet tin production was reported at 26,000 tons, second only to Malyasia, but down from the USSR's peak tin production level of 34,000 tons in 1984 (Central Intelligence Agency 1987, p. 160). Despite the Soviet Union now being a significant world producer of tin, this has been a recent development since the early 1960s. Prior to that time the USSR was highly dependent on China for its supply of tin. The deteriorated political climate with China then led the Soviet Union to embark on a high priority crash programme to develop her modest tin resources. Although this costly effort has led to more than a tripling of her tin output since 1960, she is still not self-sufficient in tin. In fact, tin is one of the few valuable industrial minerals in which the USSR is not self-sufficient (Central Intelligence Agency 1985a, p. 158).

Most of the Soviet tin mines and deposits are located in the Far East.

The largest producing mines are located in the more easily accessible southern portions of the Far East in Khabarovsk Kray and Maritime Kray. Four tin-sulphide deposits are concentrated along the eastern regions of the BAM zone in the Komsomol'sk-na-Amure area, namely the Solnechnyy, Festival'nyy, Pereval'nyy and Sobolinyy ore bodies (Kuznetsov 1985, p. 129). The Solnechnyy ore-processing *kombinat*, situated 40 km north-west of Komsomol'sk-na-Amure, has been in operation since 1963 (Shabad 1987, p. 85). This mine and processing facility is of significant interest because it produces a wide range of by-product metals including copper as noted previously, lead, zinc and tungsten (Central Intelligence Agency 1977, pp. 7 and 14). Farther to the west in the neighbouring Badzhal range tin reserves are reputedly vast. This ore district contains a couple of other poly-metallic ores: the Verkhneurmiyskiy tin-wolfram-(tungsten ore)-molybdenum-ore body, and the Verkhnebadzhal'skiy tin-and-wolfram-ore body. Similar ore deposits are concentrated along the BAM's route, but at a greater distance, in particular in the Yan-Alinskiy tin district about 120 km from the railway (Kuznetsov 1985, p. 129).

An older tin-producing centre, the Khrustal'nyy mine and concentrator, lies to the south in Maritime Kray at Kavalerovo. Begun during the Second World War, this concentrator processes ores from several mines, presumably including tin ores from the Berezovskiy deposit located west of it (Mathieson 1980, pp. 107–10). A third tin centre in the southern Far East is the small operation at Khingansk near Obluch'ye in the Yevreysk Autonomous Oblast which was also developed during the Second World War (Shabad 1987, p. 85).

Because of the previously mentioned scarcity of tin in the Soviet economy, the Soviets have also pursued an active effort to develop other more remote and costly Far East tin resources in Magadan Oblast and the Yakut Autonomous Republic. In fact, Magadan was the Soviet Union's leading tin-producing region prior to the development of the Solnechnyy complex near Komsomol'sk-na-Amure. Three of the principal Magadan tin-mining areas are in the far north Chukotsk Autonomous Okrug. Two of these, the Krasnoarmeyskiy placer mine and the Val'kumey lode mine, are very near the town of Pevek on the Arctic Ocean coast. The third mine located to the east at Iultin is accessible by surface road from the port of Egvekinot on the southern Bering Sea coast. In addition to tin, the Iultin mine yields tungsten. Shabad (1987, p. 85) reports that the reserves at Iultin are dwindling and a new deposit called Svetlyy is under development. A fourth and older tin mine in Magadan Oblast at Galimyy reportedly became depleted in the 1970s after over thirty years of operation (Shabad 1987, p. 85). It is unclear whether or not the nearby Nevskoye tin deposit indicated on an *Atlas SSSR* (1984, p. 93 map) represents a separate new deposit.

The final foci of tin mining to be mentioned are situated in Yakut Autonomous Republic. Mining began at the Ese-Khayya deposit near Batagay in the Yana river basin (see Figure 5.2) in 1941. As this deposit gradually played out in the 1970s, the focus of tin-mining in Yakutia switched to the more inaccessible Deputatskiy deposit. Placer deposits were first worked here in 1951 and after lengthy and costly delays and construction difficulties the Deputatskiy lode mine and concentrator finally began test runs in 1986 and its first stage should be fully operational by the time this volume goes to press (Shabad 1986a, pp. 13–14). The *Atlas SSSR* (1984, p. 93) indicates a third tin deposit in the upper Yana river watershed called Burgavli.

Unfortunately, no data on reserves at any of these various tin and associated ore sites have been found. None the less, it seems clear that the Soviet domestic demand, production and reserves will probably not be sufficient for the Soviet Union ever to become a tin exporting nation.

In 1986 estimated Soviet tungsten production amounted to 9,200 metric tons and molybdenum to 11,400 metric tons (Central Intelligence Agency 1987, pp. 148 and 162). Unlike the case with tin where the Far East has a near monopoly on Soviet production, the Far East's contribution to tungsten and molybdenum production is marginal. In the Far East tungsten is usually associated with tin production except for the major tungsten mine opened in the late 1970s at Vostok in north-west Maritime Kray (Shabad 1987, p. 86).

Gold

Soviet gold production in 1987, estimated at 10.64 million troy ounces, equalled about half that of South Africa; but far more than that of any other country besides South Africa (Central Intelligence Agency 1988, p. 133). Although over the past twenty years gold production has rapidly developed in Uzbekistan and Armenia, Soviet gold production still comes overwhelmingly from the Far East and East Siberia. For example, in 1980 the Far East's estimated share of primary (i.e., non-by-product supplies) Soviet gold production was 59 per cent (derived from Kaser 1983, p. 585). Figure 5.2 shows selected gold mining centres in the Far East. Only general trends and geographical patterns in the Far East's gold industry will be included here; however, the interested reader is encouraged to refer to Kaser's (1983, pp. 556–77) detailed study of the Soviet gold industry in general and of the Far East in particular (Kaser 1983, pp. 559–72, 574–5).

The geographic focus of Soviet gold production has diffused across the Siberian landscape with a north-easterly trajectory. From early mining in the Transbaykal region the focus shifted to the Aldan district of the Far East in the 1920s, to the Kolyma district in the 1930s, and to the far north-eastern Chukchi Peninsula in the 1960s (Kaser 1983, pp. 556–77).

Technologically, Soviet gold mining has evolved from simple hand panning of placer deposits in streams and gravel beds to large dredges used for placer deposits. The former method used by forced-labourers was infamous in the Kolyma district during the Stalinist *gulag* era (for example, see: 'In depth look at the GULAG & Siberian corrective-labor camps', 1988, p. 1). While large placer-dredging continues on a large scale in the Far East, the pattern of production has shifted more towards large, deep lode mines requiring complex milling and recovery methods.

Because of the continuing lack of a convertible currency, chronic grain shortages and unstable oil prices, the Soviets use gold more for its international monetary value than for its metallic properties. The size of Soviet gold reserves and the quantities that they are able to place on the world market can and do significantly affect the world gold price. Because of the harsh environmental conditions and hence high exploitation costs of most of the Far East's gold mining districts, it is not clear just how profitable Soviet gold mining really is. In the past the need to obtain foreign currency weighed heavily in the willingness and size of Soviet gold disposals on the world market. In summary, it seems safe to argue that the Far East's gold production will remain significant nationally and internationally well into the next century.

Non-metallic minerals

The Far East currently functions as the monopoly supplier to the economy of some other important non-metallic minerals, especially diamonds from Yakutia and boron from the poly-metallic and poly-minerallic ore body at Dal'negorsk in Maritime Kray. Other important, but mostly undeveloped mineral resources include mercury, phosphate, apatite, graphite and building materials such as limestone for cement, clay for bricks and tile and gravel for roads and concrete. The latter three materials are more or less geographically ubiquitous and serve mainly local or regional needs. Briefly, then, some of the other non-metallic minerals will be discussed and evaluated.

Diamonds

Most important from the perspective of Soviet domestic and export markets, of course, are the diamonds from Yakutia. Combined with the Soviet Union's synthetic diamond production, the discovery of diamonds in western Yakutia in the mid–1950s and their subsequent mining and export have made the Soviet Union the largest player in the international diamond trade. Disregarding the annual fluctuations in Soviet gold sales, diamonds probably represent the Soviet Union's third-largest foreign currency earner behind crude oil and refined products and natural gas.

Shabad (1987, p. 87) claims that 70 to 80 per cent of the diamonds figure in trade with the London-based DeBeers corporation.

Yakutian diamonds are mined in three locations. The oldest diamond centre, the Mirnyy kimberlite pope, is located in western Yakutia south of the right bank of the Vilyuy river. The electricity needs of the Mirnyy diamond complex and settlement was the chief justification for the construction of the previously mentioned hydro-electric dam located at Chernyshevskiy some 120 km to the north-west of Mirnyy on the Vilyuy river. The other two more recent diamond centres, Aykhal and Udachnyy, are situated about 340 km and 380 km north of Chernyshevskiy, respectively (see Figure 5.2). The latter mine, an open-pit mine already excavated to a depth of over 200 m is the more important. Winter mining conditions are extremely difficult when the temperature commonly hovers around −40°C. Neither of these latter two mining complexes built during the 1960s has year-round surface transport access and materials, workers and much of the heavy equipment must be flown in by large Soviet transport planes (Shabad 1987, pp. 87–8). Estimated diamond reserves at these mines are not available, but presumably they are quite substantial. Furthermore, it seems likely that the lightly explored geologic formations in the regions contain other kimberlite pipes. Accordingly, it seems likely that the Soviet Union will be able to remain on centre stage in the international diamond theatre for a long time to come.

Boron

Many important applications have been found for boron in the modern aerospace and nuclear industries. Searles Lake (borax or sodium borate deposits) in the Mojave desert of southern California has long been the world's most important source of this exotic material. However, during the 1960s the Soviet Union began to develop its own domestic supply based on the datolite deposit at Dal'negorsk. Elsewhere in the world this calcium borosilicate mineral is not used commercially (Shabad 1987, p. 88).

Fertilizer minerals

Although the agricultural base and production of the Far East is inadequate to supply the region's needs, none the less, agricultural production is hampered by the necessity to long-haul fertilizers into the region as there has been no regional fertilizer production. The possibility and plans for natural gas-based nitrogen fertilizer production for both regional domestic and potential Japanese markets has previously been discussed. The mineral base for a Far East phosphate fertilizer industry remains a topic worthy of discussion.

The most commercially viable phosphate mineral base for development clearly appears to be the Seligdar apatite-dolomite deposit located in the

Aldan-Timpton interfluve about 30 km south-west of Aldan in the South Yakutian district. The 12th Five-Year Plan, in fact, calls for development of the Seligdar deposit (Shabad 1987, p. 89). Compared with the Kola Peninsula apatites which contain around 15 per cent P_2O_5, the Seligdar rock is of low grade, averaging only 6.3 per cent P_2O_5 (Kuznetskov 1985, pp. 130–31). Its commercial value is augmented, however, because the deposit seems amenable to enrichment to a 35 per cent P_2O_5 concentrate while allowing for the potential by-product recovery of fluorine and rare earths of the cerium group. The Yamal railroad being built north from Aldan to Yakutsk will allow for relatively easy access to potential Seligdar phosphate mines. Shabad (1987, p. 89) argues that the decision to go ahead with the Seligdar apatite project probably means that the plans to develop the apatite deposit at Oshurkovo in the Lake Baykal watershed has been shelved by concerns over water pollution problems.

It appears as though the Soviets have one other major potential mineral base for phosphate development. This is the Udsko-Selemdzhinsk phosphorite region of Khabarovsk Kray (see Figure 5.2). Deposits here are reported to be large, especially at the Lagapskoye and Nimiyskoye deposits among others. These ores contain 8 to 10 per cent P_2O_5 and laboratory tests indicate that it would be possible to achieve a concentrate having a 32 to 35 per cent P_2O_5 content from them. Unfortunately, from a commercial development perspective this phosphate district is located some 300 km north of the route of the BAM. Accordingly, near-term development prospects are at best very uncertain (Kuznetsov 1985, p. 130).

Mercury

The Far East has at least one known source of mercury ores, but of undisclosed size and quality. These ores are located near the Bering Sea coast at Olyutorskoye in the Koryaksk Autonomous Okrug (*Atlas SSSR*, 1984, p. 93).

Graphite

Finally, graphite is important in the Soviet nuclear reactor industry and to line the pencils of the millions of Soviet bureaucrats that Gorbachev is trying to cajole into becoming more efficient. Post-Chernobyl' reactor design changes could well reduce the Soviet demand for graphite and *perestroyka*, if successful, will clearly reduce the pencil pushers! None the less, the Far East has at least two graphite deposits, one at Syuznoye in the Yevreysk Autonomous Oblast and another at Tamginskoye in the Maritime Kray.

Conclusions and postscript on prospects for Asian–Pacific trade

This chapter has attempted to present an overview of selected non-biotic resources of the Soviet Far East. Many desirable data are lacking on detailed resource reserves, production and costs for many of these resources, and hence, much of this analysis is more descriptive and qualitative than one would wish. Accordingly, the always precarious art of forecasting economic development is even more seriously handicapped. None the less, some broadly sweeping generalizations are possible. First, as this chapter has hopefully pointed out, the energy and mineral resource endowment of the Far East is indeed vast. At the same time, throughout much of the region environmental factors pose very significant physical, transportation, human settlement and economic obstacles to rational economic resource development of the type called for in Gorbachev's economic reforms. Thus, it seems that Gorbachev's hopes for the economic future of the Far East, expressed in his various quotations which begin this chapter, face formidable obstacles before they will be realized. The size of the required regional investments and the general scarcity of Soviet investment capital do not augur well for such hopes.

A persuasive argument can be made that Gorbachev's dreams will require a major infusion of foreign investment and technology, something which the general consensus has found to be very unrealistic for a variety of reasons (for example, see Chapter 10 by Bradshaw and Chapter 11 by Dienes). Expressing much more optimism, however, are two recent papers dealing with Soviet foreign economic ties in the Far East (Baklanov 1988, pp. 1–9 and Ogawa 1988, pp. 1–8). Kazuo Ogawa, Director of the Japan Association for trade with the Soviet Union and Socialist Countries of Europe, ends his paper with a rather sanguine assessment of enhanced Soviet–Japanese trade noting a number of what he perceives to be significant developments and events. He first notes the March 1988 Soviet formation of the Asian Pacific Committee for Economic Co-operation, headed by Ye. Primakov, Director of the Soviet Union's Institute of World Economy and International Relations. Second, he considers Primakov's guest observer status at the sixth meeting of the Pacific Economic Council (PEC) in an effort to establish the Soviet Union's participation as positive and significant. Third, he claims real progress has been made in the expansion and utilization of the port of Nakhodka. The fourth and crucial new item Ogawa cites is the Soviet Union's new plans for the development of 'special economic zones' in the Far East. Ogawa's final note of optimism is based on the Soviet Union's landmark decision to allow domestic joint-enterprises with foreign capital. During a question-and-answer session he expressed even greater optimism for expanded Japanese–Soviet regional trade if the proposed new draft rules and regulations for joint-ventures (which would relax significantly some of the restrictions and difficulties in the initial Soviet joint-venture rules with regard to the repatriation of foreign profits and other important items) are adopted. He ended his presentation urging multi-lateral co-operation among the USSR, Japan, Korea, Taiwan and mainland China. At the same time, he remains critically aware of Soviet institutional, personnel and infrastructure difficulties and bottlenecks to enhanced Japanese–Soviet trade and co-operative resource development in Siberia and the Far East.

During the presentation of his paper and in response to questions afterwards, Baklanov, Director of the Institute of Economics of the Far Eastern Branch of the Academy of Sciences of the USSR, specifically mentioned plans for the formation of 'special economic zones' designed especially for joint-venture enterprises at Pos'yet and Nakhodka in the Maritime Kray, in Khabarovsk, at various places along the BAM, on the Kuril Islands and on Sakhalin Island. Among the economic activities envisaged by Baklanov for joint-venture development in the Far East are fishing, lumbering, mining, petroleum and natural gas development and ocean technology.

Only time will tell whether this optimism is simply old wine in new bottles which will turn to vinegar, or whether the 'special economic zones' and joint-venture legislation will really spawn significantly greater regional economic development co-operation and international trade between the Soviet Far East and her Pacific-Rim neighbours.

Bibliography

Atlas SSSR (1984) Moscow: Glavnoye upravleniye geodezii i kartografii pri Sovete Ministrov SSSR.

Baklanov, P. Ya. (1988) 'Formation of aqua-territorial complexes of national economy in the Soviet Far East and development of new forms of foreign economic ties', paper presented at session entitled: The USSR's Pacific Rim: New Initiatives in the Soviet Far East, American Association for the Advancement of Slavic Studies National Meeting, Honolulu, Hawaii, 21 November 1988.

Bunich, P. G. (1977) 'Economic impact of the BAM', in Theodore Shabad and Victor Mote (authors and eds) *Gateway to Siberian Resources (The BAM)*, Washington, DC: Scripta Publishing Co., pp. 135–44.

Central Intelligence Agency (1977) *Soviet Tin Industry: Recent Developments and Prospects through 1980*, Washington, DC: US Government Printing Office.

——(1985a) *Handbook of Economic Statistics, 1985*, Washington, DC: US Government Printing Office.

——(1985b) *USSR Energy Atlas*, Washington, DC: US Government Printing Office.

——(1987) *Handbook of Economic Statistics, 1987*, Washington, DC: US Government Printing Office.

_____(1988) *Handbook of Economic Statistics, 1988*, Washington, DC: US Government Printing Office.

Chichkanov, V. P. and Minakir, P. A. (1986) in John J. Stephan and V. P. Chichkanov (eds) *Soviet–American Horizons on the Pacific*, Honolulu: University of Hawaii Press, pp. 95–113.

'China–Soviet summit possible, Li says', *The Seattle Times/Post-Intelligencer*, 18 September 1988, p. A–18.

Danilov, V. A. (1987) *Neft' SSSR*, Moscow: Nedra, 1987.

Dienes, L. (1985) 'Soviet–Japanese economic relations: Are they beginning to fade?', *Soviet Geography*, vol. 26, no. 7, pp. 509–25.

_____(1988) 'A comment on the new development program for the Far East economic region', *Soviet Geography*, vol. 29, no. 4, pp. 420–22.

_____(1989) phone conversation 11 January 1989.

Gazovaya promyshlennost' (1985) vol. 1, no. 7, p. 11.

Gorbachev, M. (1988) 'Appendix: text of speech by Gorbachev in Vladivostok, 28 July 1986', in Ramesh Thakur and Carlyle A. Thayer (eds) *The Soviet Union as an Asian Pacific Power: Implications of Gorbachev's 1986 Vladivostok Initiative*, Boulder and London: Westview Press, 1988.

_____'In depth look at the GULAG & Siberian corrective-labor camps', (1988) *Current Digest of the Soviet Press*, vol. 40, no. 32, p. 1.

Jensen, R. G., Shabad, T. and Wright, A. W. (eds) (1983) *Soviet Natural Resources in the World Economy*, Chicago and London: The University of Chicago Press.

Kaser, M. (1983) 'The Soviet gold-mining industry', in Robert G. Jensen, Theodore Shabad and Arthur W. Wright (eds) *Soviet Natural Resources in the World Economy*, Chicago and London: The University of Chicago Press, pp. 556–96.

Kibal'chich, O. A. (1977) 'The BAM and its economic geography', in Theodore Shabad and Victor Mote (authors and eds) *Gateway to Siberian Resources (The BAM)*, Washington, D.C.: Scripta Publishing Co., pp. 145–54.

Kovrigin, E. B. (1986) in John J. Stephan and V. P. Chichkanov (eds) *Soviet–American Horizons on the Pacific*, Honolulu: University of Hawaii Press, pp. 1–16.

Kuznetsov, V. A. (1985) 'Mineral'nyye resursy zony BAM i problem ikh osvoyeniya', in A. G. Aganbegyan and A. A. Kin (eds) *BAM: pervoye desyatiletiye*, Novosibirsk: Izdatel'stvo Nauka, pp. 124–35.

Lawarne, S. (1988) *Soviet Oil: the Move Offshore*, Boulder and London: Westview Press.

Mathieson, R. S. (1980) 'Kavalerovo tin-mining district', *Soviet Geography*, vol. 21, no. 2, pp. 107–10.

Meyerhoff, A. A. (1983) 'Soviet petroleum: history, technology, geology, reserves, potential and policy', in Robert G. Jensen, Theodore Shabad and Arthur W. Wright (eds) *Soviet Natural Resources in the World Economy*, Chicago and London: The University of Chicago Press, pp. 306–2.

Misko, T. and Zumbrunnen, C. (1983) 'Soviet iron ore: its domestic and world implications', in Robert G. Jensen, Theodore Shabad and Arthur W. Wright (eds) *Soviet Natural Resources in the World Economy*, Chicago and London: The University of Chicago Press, pp. 464–91.

Mote, V. (1983) 'The Baikal–Amur mainline and its implications for the Pacific basin', in Robert G. Jensen, Theodore Shabad and Arthur W. Wright (eds) *Soviet Natural Resources in the World Economy*, Chicago and London: The University of Chicago Press, pp. 133–87.

Ogawa, K. (1988) 'Japan–Soviet trade and far eastern development in the Soviet Union', paper presented at session entitled: The USSR's Pacific Rim: New Initiatives in the Soviet Far East, American Association for the Advancement of Slavic Studies National Meeting, Honolulu, Hawaii, 21 November 1988.

Pomfret, J. (1988) 'Sino-Soviet relations at the border of a thaw', *Journal American*, 18 September 1988, p. B–6.

'Power generation plans for Far East outlined' (1988) *Foreign Broadcast Information Service SOV–88–111*, 9 June, pp. 67–8.

Razvedka i Okhrana Nedr (1981) no. 6, pp. 14–15.

Robinson, B. V. (1985) 'Economic-geographic assessment of oil resources in East Siberia and the Yakut ASSR', *Soviet Geography*, vol. 26, no. 2, pp. 91–7.

Rodgers, A. (1983) 'Commodity flows, resource potential and regional economic development: The example of the Soviet Far East', in Robert G. Jensen, Theodore Shabad and Arthur W. Wright (eds) *Soviet Natural Resources in the World Economy*, Chicago and London: The University of Chicago Press, pp. 188–213.

Sagers, M. J. (1987) 'News notes', *Soviet Geography*, vol. 28, no. 5, pp. 358–66.

———(1988a) 'News notes', *Soviet Geography*, vol. 29, no. 4, pp. 423–57.

———(1988b) 'News notes', *Soviet Geography*, vol. 29, no. 5, pp. 533–46.

———(1989a) 'News Notes', *Soviet Geography*, vol. 30, no. 4, pp. 306–52.

———(1989b) 'News Notes', *Soviet Geography*, vol. 30, no. 5, pp. 397–410.

Selskaya zhizn', 26 October 1984, p. 1.

Shabad, T. (1980) 'News notes', *Soviet Geography: Review & Translation*, vol. 21, no. 1, pp. 48–58.

———(1985) 'News notes', *Soviet Geography*, vol. 26, no. 1, pp. 762–79.

——— (1986a) 'Geographic aspects of the new Soviet five-year plan, 1986–1990', *Soviet Geography*, vol. 27, no. 1, pp. 1–16.

———(1986b) 'News notes', *Soviet Geography*, vol. 27, no. 4, pp. 248–79.

———(1987) 'Economic resources,' in Alan Wood (ed.) *Siberia: Problems and Prospects for Regional Development*, London: Croom Helm, pp. 62–95.

——— and Sagers, M. J. (1987) 'News notes,' *Soviet Geography*, vol. 28, no. 4, pp. 259–81.

Shcheka, S. A. (editor-in-chief) (1987) *Novyye i redkiye mineraly Dal'nego Vostoka*, Vladivostok: DVO AN SSR.

Shilo, N. and Chichkanov, V. (1981) 'Sotsial'no-ekonomicheskiye problemy razvitiya dal'nego vostoka', *Territorial'noye planirovaniye i ekonomika rayonov*, no. 11, November, pp. 74–8.

Shiryayevo, P. A., Yarkho, Ye. N. and Borts, Yu. M. (1978) *Metallurgicheskaya i ekonomicheskaya otsenka zhelezorudnyy bazy SSSR*, Moscow: Metallurgiya.

Sidorova, V. S. and Vadyukhin, A. A. (1977) 'New technology and the location of the iron and steel industry in the eastern portion of the USSR', *Soviet Geography: Review & Translation*, vol. 18, no. 1, pp. 33–8.

'Soviets who return to China report the welcome is warm', *Journal American*, 18 September 1988, p. B–6.

Stern, J. P. (1983) 'Soviet natural gas in the world economy', in Robert G. Jensen, Theodore Shabad and Arthur W. Wright (eds) *Soviet Natural Resources in the World Economy*, Chicago and London: The University of Chicago Press, pp. 363–84.

Stroitel'naya gazeta, 5 February 1987.

Ugol' (1988) no. 1, p. 22.

Zapasy ugley stran mira (1983) Moscow: Nedra.

ZumBrunnen, Craig and Osleeb, J. P. (1986) *The Soviet Iron and Steel Industry*, Totowa, NJ: Rowman & Allanheld.

Chapter six

Forest and fishing industries
Brenton M. Barr

Perspective

The Soviet Far East's role in the Soviet economy is distinguished by its national share of material supply for non-ferrous metallurgical industries (between 14 per cent and 100 per cent depending on the mineral), fish products (40 per cent) and forest products (9 per cent) (*CDSP*, 1984, vol. 36, no. 32, p. 12). In each case, these industrial sectors have a regional importance in the industrial output of the Far East which is twice as great as their role in total industrial output at the national level. Mining, fishing and forestry account for over half the Far East's industrial output (Bradshaw, 1988, p. 370) and approximately two-thirds of the region's industrial output (1975) when all sectors producing and processing raw materials and food (dominated by fishing) are taken together (Mozhin, 1980, p. 208). Fishing alone accounts for one-quarter of the region's industrial output (Singur, 1988, p. 95); the forest sector occupies 13 per cent of the region's labour force (Yakunin *et al.*, 1987, p. 5). Mining, fishing and forestry particularly reflect the region's peripheral national location, the spatial dispersion of its industrial nodes and settlements, and the somewhat vagarious role of foreign trade in its development.

Forestry, however, harvests a potentially renewable raw material over much of the accessible land area of the Far East, and fishing derives its catch from a wide expanse of the Pacific Ocean and adjacent seas; together they exert the dominant influence on the region's limited contiguous space economy and underdeveloped coastal settlement pattern. These two industries enhance the spatial importance of major transportation nodes on the region's two trunk rail lines and key harbours at strategic coastal locations. The extensive land or water areas needed to support forestry and fishing provide a rationale for the focus of this chapter on their contribution to the spatial structure of the Far East and the USSR.

Forest and fishing industries of the Soviet Far East are located far from national markets, are somewhat dispersed throughout the accessible areas of the region, and with the exception of log and woodchip exports to Japan or across the border to China, do not have close spatial proximity to international markets for their major existing or prospective industrial products. Both industries produce relatively low value and somewhat bulky products whose delivered price is adversely affected by lengthy land transportation. The existence of a fish and forest industry in the Soviet Far East is primarily caused by sheer size of resource stock, not locational advantage. Better quality forest growing stock exists much closer to national markets, and the locational advantages of forests west of Lake Baykal would obviate the need for major commercial utilization of the Soviet Far East if economics alone were taken into consideration. Fishing in the Far East is not so constrained by locational advantages of stocks in other major water bodies. Its development is necessitated by Soviet need to land fish harvested in the Pacific Ocean at the closest adjacent Soviet ports, and to satisfy nearly half the country's protein diet with fish wherever they are available. Distance to national markets is a problem for numerous resource-extractive industries in the Soviet Far East but appears to be more effective in reducing the size and complexity of harvesting and processing in forestry more than in fishing.

Thus forestry and fishing support relatively undeveloped communities with narrow occupational and industrial profiles, except in those few notable places such as Vladivostok, Komsomol'sk, Khabarovsk, Nakhodka and Soviet Harbour which clearly fulfil so many national functions associated with transportation and/or defence. Links of forestry and fishing activities with other industries and national markets extend out of this hinterland region many thousand kilometres to the west and help support complex regional economies and settlements elsewhere in the country. Notwithstanding the region's strategic military importance in the USSR's relations with China, Japan and the United States, the Far East's economy is basically contrived, dependent and peripheral. Much of the spatial economy reflects a political and social desire by central authorities to occupy an area which is badly mismatched to the population distribution of the USSR (Table 6.1), to the range of alternative physical and cultural environments available for habitation by the Soviet population, or to the location of raw materials, including timber, needed by the national economy. The region's major settlements and economic activities, including defence, appear as unconnected map overlays in that they appear independent of each other and have interconnections or commonalities in the region's space economy except for the export of primary commodities to the heartland, and the import of complex consumer and producer goods from it.

The regional spatial and economic impact of forestry and fishing reflects key differences in the location of their resource bases. Basic

Table 6.1 Regional distribution of forests and population

	Growing stock under central state forestry management		1986 population ('000s)	
	(millions of cu. m)	%		%
USSR	74,872	100	278,784	100
RSFSR	72,216	96.5	144,080	51.7
North-west	7,366	9.8	14,137	5.1
Volgo-Vyatka	1,209	1.6	8,362	3.0
Urals	3,014	4.0	16,110	5.8
RSFSR Europea–Uralia	14,670	19.6	112,347	40.3
Western Siberia	9,655	12.9	14,358	5.2
Eastern Siberia	27,191	36.3	8,875	3.2
Far East	20,700	27.6	7,651	2.7
Amur	1,814	2.4	1,041	0.4
Khabarovsk	4,693	6.3	1,760	0.6
Maritime	1,818	2.4	2,164	0.8
Sakhalin	623	0.8	700	0.3
(South)	8,948	12.0	5,665	2.0
Kamchatka	414	0.6	435	0.2
Magadan	661	0.9	542	0.2
Yakut	10,677	14.3	1,009	0.4
(North)	11,752	15.7	1,986	0.7

Sources: Population: *Narodnoye khozyaystvo SSSR v 1985 godu* pp. 12–17. Actual Growing Stock under Central State Forest Management: G. I. Vorob'yev *et al.*, (1979) *Ekonomicheskaya geografiya lesnykh resursov SSSR*, Moscow: Lesnaya Promyshlennost', various pages (Far Eastern section, pp. 312–34, includes various statistics for the oblasts and krays; W. Siberian data, p. 283; E. Siberian data, p. 295). Growing stock data are from the 1973 inventory, the last for which regional data are fully available.

forest activity, logging, has an intra-territorial focus which has been limited to date largely by access to a limited rail transportation system. Much of the spatially extensive growing stock, with a large potential calculated annual allowable cut (Table 6.2) is simply inaccessible for commercial exploitation and must await improvements in infrastructure and changes in value to transform it from resource to reserve. Fishing has an extra-territorial orientation extending through much of the Pacific basin and adjacent seas; it reflects both the industrial prowess of the Soviet Union in providing a modern fishing and processing fleet, and in the country's unwillingness to obtain protein from land-based sources in regions more coincident with the distribution of the USSR's population. The Soviet Far Eastern fish industry's size reflects a longstanding national priority in the supply of basic food which depends on vast resources outside the region but which enter the USSR via the region's ports. Forestry is quite underdeveloped in relation to the size of the region's growing stock, but in recent decades has developed partly on the basis of processing facilities developed when southern Sakhalin Island belonged to Japan, and partly in relation to extensive stands of mature and over-mature timber. Some of this timber is exported to Japan, a stagnant market, and increasingly to China. The remainder is consumed within the region or, in an expeditious manner, shipped westwards in the Soviet Union to offset the planners' reluctance to achieve appropriate levels of reforestation and rotation cutting in forests economically accessible to the Soviet European heartland.

Table 6.2 Calculated annual allowable cut in reserve and inaccessible forests[a] million cu. m

	Total	Conifers	SIHS[b]	STHS[c]
USSR	201.2	178.1	21.9	1.2
RSFSR	<201.2	<178.1	21.9	1.2
North-west	>4.5	4.1	>0.4	
North Caucasus	>-		>-	
RSFSR E-U	4.6	4.1	0.5	
Western Siberia	13.7	9.2	>4.5	
Eastern Siberia	100.1	87.5	12.6	
Far East	82.7	77.3	4.2	1.2
RSFSR Asia	<196.6	<174.0	21.4	1.2
USSR Europe–Uralia	4.6	4.1	0.5	
USSR–Asia	196.6	174	21.4	1.2

Source: G. I. Vorob'yev *et al.* (1979) *Ekonomicheskaya geografiya lesnykh resursov SSSR*, Moscow: Lesnaya Promyshlennost', p. 57.
Note: [a]Reserved for future or other (unspecified) uses
[b]SIHS = Shade-Intolerant Hardwood Species
[c]STHS = Shade-Tolerant Hardwood Species

Although construction of the Baykal–Amur Mainline (BAM) is generally believed to have been in part occasioned by the need for greater access to the timber resources of the Soviet Far East, in fact utilization of this growing stock is hampered by its great distance from national markets, the adverse domination of its species composition by larch, and the potential, but as yet only partially realized, economic advantages available to the national economy through effective silviculture in Soviet European–Uralian regions. Furthermore, Soviet planners themselves acknowledge that the costs of harvesting, industrial development, settlement construction and transportation adversely affect economic development east of the Urals not only in the Soviet Far East, but also in much of Eastern and Western Siberia. The BAM similarly will not affect the relative spatial utility of Far Eastern fishing although it will offer an alternative and slightly shorter route for transporting fish to national Soviet markets located primarily in Europe–Uralia, thereby relieving incremental pressure on the Trans-Siberian Railway and offering more direct access to fish originating in the Sea of Okhotsk and the North Pacific.

Focus

The two industries, however, vary greatly in their sensitivity to space or the friction of distance. Fishing is part of the world-wide Soviet search for food, and the priorities associated with it are far greater than the long land haul from the Pacific littoral to major centres of population. Processed fish with considerable value added to the primary commodity is shipped to western Soviet markets. Although the delivered real cost is inflated by the long rail journey, the political importance of ensuring adequate food supplies in the USSR makes the geographer's preoccupation with effects of space and distance appear relatively banal. Assessment of Far Eastern forests, however, must occur within the context of the Europe–Uralia versus Siberia debate widely reported in the recent literature. Real alternatives exist in the Soviet space economy for the derivation of timber and fibre, and the impact of long hauls on delivered costs is anything but banal. Forestry is ineluctably part of the continuing spatial dilemma facing Soviet planners and politicians. Should raw materials closer to the national heartland be used more efficiently and valued more appropriately to obviate the higher-cost exploitation of those in the east? Or should the apparently cheaper, more abundant and richer stocks in the east be utilized and the cost of their shipment to market and investment in related infrastructure be subsidized as in the past? This dichotomy underlies most analyses of Soviet eastern forests, and both perspectives have strong proponents among Soviet politicians, planners, administrators, economists and foresters.

Advocates of development in Europe–Uralia (widely discussed in Barr and Braden 1988) appear to have an irrefutable case because their arguments are based on economic considerations. The pragmatic realities of Soviet industrial management and international strategic considerations, however, cause strict factors of economic efficiency to be overlooked. The immediate need to meet various national political and industrial goals is paramount, and the rationale for use of eastern resources assumes an otherwise unwarranted legitimacy. Insufficient funds are allocated for appropriate reforestation, a long-term proposition, and expedient measures are taken by central forestry administrations and by the myriad non-forestry consumers of timber (the deleterious impact of commercial 'nomadic' loggers are cited by Barr and Braden, 1988, pp. 81, 91; Chelyshev (1986, cited in *Referativniy zhurnal*, 1988, no. 3, p. 40) notes that in the Yakut ASSR some thirty different ministries and administrations are engaged in logging) which led to *ad hoc* harvesting of eastern timber and absorption of extensive rail costs by managers desperate to meet numerous, often conflicting, performance criteria. In the mid–1980s, forty ministries and administrations were logging the Far Eastern forest, and the share harvested by Minlesbumprom (changed 10 March 1988 to Minlesprom) comprised 75 per cent (Yakunin *et al.* 1987, p. 40).

The Far East is also a pawn in relations between the USSR and major neighbours such as Japan and China. Dienes (1988, p. 422), perplexed by the somewhat whimsical and perhaps improbable brief announcements in 1987 that 198 billion roubles of centralized investment is proposed for the Far East through the year 2000, suggests that 'In part, perhaps, these figures may have been floated to gauge Japanese reaction and increase Japanese willingness towards economic participation' in joint Far Eastern investment. Dienes (1988, p. 421) meticulously points out that this investment in this time period 'would require an *annual* growth of over 9 percent, if a constant exponential rate is assumed', that the average annual growth of investment in the national economy in General Secretary Gorbachev's first three years (1985–7) did not exceed 5.3 per cent and for 1988 was predicted to rise at only 3.6 per cent. Dienes notes that 'As for the Far East, . . . capital allocation to the region rose by little more than a 4 per cent per annum average rate through the 1970s, much more slowly than that during the second half of that decade, and more slowly still during the first half of the 1980s'. This lack of coherence emanating from Moscow suggests that economic planning for the Far East is not in keeping with the clarity of Lenin's response to his challenge, 'Chto delat?' (What to do?). Moscow's recent proposal for the Far East seems designed instead to invoke consternation, often expressed by Soviet citizens when faced by confusion or inconsistency in official plans and promises as, 'Chto takoye?' (What is that?).

Revelations in the Soviet press under Gorbachev's current policy of *glasnost* strongly confirm that, whatever the direct economic or possible strategic benefits which might be deduced from the size and quality of Far Eastern resources for the region itself or the nation, day-to-day developments in the Far East may be largely guided by priorities which are anathema to ideology and official planning precepts. 'Freewheelingness and muddle' and a concern for 'Rush!', now attributed to construction of the BAM (*CDSP*, 1987, vol. 39, no. 34, p. 22) apparently lie at the heart of the project's cost overruns and the fact that by August 1987, it was carrying less than a million tons of freight instead of the 35 million tons envisaged for 1983. Furthermore, regular operation is not to commence until 1989 although the route opened for traffic in 1983. Continuing expediency and grievous degradation of the environment plague Far Eastern logging and drift-floating of timber (Mozhayev 1987), pious promises for remediation from Moscow mandarins notwithstanding (*Lesnoye Khozyaystvo*, 1988, p. 60). The need to overcome ineffective economic planning through improvements in project management and co-ordination, application of state of the art technology, scale of resource exploration and development, and utilization of raw materials underlie all major projects of Siberia and the Far East (Molodenkov 1987, cited in *Referativniy zhurnal*, 1988, no. 3, p. 35).

Forestry and fishing in the Soviet Far East are too extensive for comprehensive analysis here. Furthermore, other analysts have recently analysed the role of Soviet Asia in the national economy in depth (Dienes 1987), the Far East's economic structure (Dienes 1985), the role of Pacific trade to the Far East (Bradshaw 1988), and the re-evaluation of eastern investment under General Secretary Gorbachev, particularly that related to mega-resource and -transportation projects (Mote 1987b), and identified general factors and processes which directly pertain to the industrial sectors analysed here. Dienes (Chapter 11 of this volume) summarizes the essence of Far Eastern development by stating that 'Soviet policy towards the development of the Far East has always been influenced by the remote location of this huge region from the country's economic heartland and by its strategic position'. This chapter focusses, therefore, on the present relative regional and national significance of these industries, and examines some of the endogenous and exogenous factors which may affect them in the foreseeable future.

Analysis of Far Eastern forestry and fishing at this juncture also assumes that at least three great unknown factors are likely to affect future events in the Far East in an important, but unpredictable, manner. Gorbachev's economic reforms and his own antipathy towards 'frozen investment' in eastern regions makes the extent and function of future development there very uncertain (Mote 1987a, 1987b). Japanese investment, the most likely source of foreign capital for the Far East and apparently the most acceptable to Soviet leaders (Kol'tsov 1987, cited in *Referativniy zhurnal*, 1988, no. 1, p. 35), is problematic (Dienes 1985) and currently unpredictable because of numerous Soviet conditions on the operation of joint ventures, including the extent of profit repatriation (SUPAR, 1988, pp. 15, 18), snags in the movement of high technology goods from Japan to the USSR in return for Far Eastern raw materials (SUPAR, 1988, p. 11), and the lack of clarity among Japanese investors in the meaning of *perestroyka* and Soviet plans for the region (SUPAR, 1988, pp. 10 and 11). Finally, the strategic military and political role which the Far East currently plays in the USSR's international relations may be expected to affect every major development in the region for the foreseeable future. This role is likely to become even more acute if Japan begins to display a political and possibly military prowess in the Pacific basin and throughout the world commensurate with its existing economic importance. The future importance of the growing industrial might of the 'Four Tigers' of the western Pacific (Hong Kong, South Korea, Taiwan and Singapore) and China to the Far East cannot help but be profound.

Forestry

Two recent analyses (*CDSP*, 1984, vol. 36, no. 32, pp. 12–13; Yakunin *et al*. 1987, pp. 38–9) of the Far East's economy bluntly suggest that future development of forestry in the region requires many systemic changes. The quality of administration, management, production, mensuration and inventory techniques must be improved; greater investment is required in chemical wood processing and in the use of deciduous species, wood waste and low-quality timber. The quality and quantity of items produced in the Far East must be drastically improved. Exports to earn foreign currency should be broadened in variety and improved in quality; foreign expertise, capital and technology should be encouraged, particularly in joint ventures. Domestic shipments out of the region should focus on higher-value goods which can withstand increases in delivered value due to the addition of substantial transport costs, and shipments of roundwood need to be reduced to a minimum. To date, despite some expansion of logging in the BAM service area, and expansion of the Far East's output of wood chips, lumber, pulp, paperboard and furniture, the region's forest industries do not make comprehensive use of their raw materials. An amount equal to approximately one-half of the volume of harvested timber is left at the cutting site, and scraps from logging and timber processing are almost entirely unused. Singur (1988, p. 95) states that 10 million cu. m of low quality and poor marketability timber remains at the cutting site. Five million cu. m of fuelwood and logging waste and over 2 million cu. m of scraps from processing are unutilized. Furthermore, despite the widespread distribution of the region's forests, most logging occurs in southern and central Amur Oblast and Khabarovsk Kray, on Sakhalin Island, and in the Maritime Kray; the region's northern forests remain largely unused (*CDSP*, 1984, vol. 36, no. 32, p. 12).

Resource

The Far East is usually described as one of the richest forest regions of the USSR. It has 507 million ha of land in the state forest reserve, and 351 million ha of forest land, of which 270 million ha is forested. Of the forested land, however, 26 million ha is covered with brush, particularly Siberian dwarf-pine elfin wood (Sinitsyn *et al*. 1979, p. 193) suitable only for fuelwood. An additional 16 per cent is classified as Group 1 forest (protective) to protect major international spawning areas. The total volume of growing stock is 22 billion cu. m of which 15 billion is classified as mature and over-mature (13 billion of this comprises conifers). The relative importance of Far Eastern forests, however, varies considerably by forest category: the Far East has 40 per cent of the USSR's land area in the state forest reserve, 37 per cent of its forest land area, but only 33 per cent of its forested area. The region's general growing stock comprises 25.6 per cent of the national total, but its share of national mature and over-mature standing timber is 28.6 per cent, and its share of total coniferous volume in this category is 30 per cent (*Lesnaya entsiklopediya*, 1986, vol. 2, pp. 400–401). The relative importance of the region's national share of industrial or commercial stands, and the preponderance of mature and over-mature stands is summarized in Table 6.3. Like the commercial forests of Western and Eastern Siberia, the utilization of calculated allowable cut is low.

Table 6.3 Distribution of forested area: characteristics of industrial stands under central state forest management

	Forested area	Total growing stock	% of forested area by age group of stands				% of MOM comprising stands of:			MOM as % of GS vol.	Utilization of CAC for each species (%)			
	mill ha	mill cu. m	J	MA	AM	MOM	C	SIHS	STHS		C	SIHS	STHS	Total
USSR	675,206	74,872	15	18	10	57	85	13	2	69	n/a	n/a	n/a	n/a
RSFSR	646,102	72,216	14	17	10	59	86	(14-both)		71	n/a	n/a	n/a	n/a
North-west	68,806	7,366	20	17	5	58	89	11	0	72	104	41		87
RSFSR E–U	68,806	7,366	26	21	8	45	n/a	n/a	n/a	n/a	n/a	n/a	n/a	n/a
W. Siberia	75,716	9,655	7	15	15	63	68	32		72	44	16		30
E. Siberia	216,026	27,191	9	17	10	64	92	8		74	48	6		36
Far East	226,000	20,700	14	16	9	61	n/a	n/a	n/a	n/a	n/a	n/a	n/a	n/a
(South)	70,216	8,948					77	15	8	70	53	13	15	45
(North)	155,784	11,752					99	1		74	12			13
RSFSR Asia	517,742	57,546	11	16	10	63				74				

Source: Compiled and adapted from Vorob'yev *et al.* (1979) *Eknomischeskaya geografiya Lesnykh resursov SSSR*, Moscow: Lesnaya Promyshlennost', various pages; Timofeyev, N.V. (ed.) (1980) *Lesnaya industriya SSSR*, Moscow: Lesnaya Promyshlennost', pp. 17–18; *Lesnoye Khozyaystvo SSSR*, (1977) Moscow: Lesnaya Promyshlennost', p. 57.

Notes: SIHS = Shade-Intolerant Hardwood Species; STHS = Shade-Tolerant Hardwood Species; C = Conifers; J = Juvenile stands; MA = medium age stands; AM = stands approaching maturity; MOM = mature and over-mature stands.
CAC = calculated allowable cut.
GS = growing stock.
Values are net of forests administered by collective farms, non-forestry ministries and agencies, or held under long lease by sheep and reindeer herding enterprises.

Forestry in the Far East is greatly hampered, however, by the lack of infrastructure in the region (Sinitsyn *et al.* 1979, p. 194). Only 22 per cent of the commercially exploitable growing stock, or 12 per cent of the total growing stock of all species, is accessible to the existing road network; over half of the commercially exploitable forest is, nevertheless, accessible by water transport. Thus, approximately 12 billion cu. m are commercially exploitable, of which 71 per cent is mature or over-mature (Sinitsyn *et al.* 1979, pp. 193–4). Yakunin *et al.* (1987, p. 5), however, reflecting General Secretary Gorbachev's new realism towards eastern resources, state that only 6.5 billion cu. m should be seen as commercially exploitable in the near future; approximately 45 per cent of this volume is located in the most settled and climatically acceptable southern parts of Khabarovsk and Maritime Krays, and Amur and Sakhalin Oblasts. Difficulties in accessibility, plus the region's lack of a comprehensive wood-processing industry, however, limit the processing industry largely to the sawing of large-diameter, high-quality logs obtained from unsophisticated cutting methods, and constrain the possibilities for rapid expansion of forestry in this region. The high cost of producing logs, chips and lumber in the Far East, and the attendant high prices of these items, appear to confirm the spatial and environmental disadvantage of this region relative to the European north west and Siberia (Table 6.4), despite the case which some authors make that profitability of logging traditionally has been greater in the Far East than in other major forest regions (Table 6.5). In physical units, however, such as the relative share of forested area, and unutilized forest lands, the Far East is inferior to Eastern Siberia but superior to Western Siberia (Table 6.6). In broad terms, we may think of Eastern Siberia as a relatively rich Asian storehouse of Soviet forest resources flanked to the west and east by lower-quality forest environments.

With the completion of the BAM, the volume of accessible timber in the Far East could increase by 13 per cent from 12 billion cu. m to 13.6 (Vorob'yev *et al.*, 1979, p. 324), with half the increase derived from the Zeya-Selemdzhinsk region of Amur Oblast, and the other half from central Khabarovsk Kray. Although substantial by world standards, the incremental volume of timber made available by the BAM is quite out of keeping with the grandiose original projections for its impact on the region, and the large amount of forest which will remain inaccessible or unexploitable. Alternative sources for such incremental supplies of timber already exist or could be more efficaciously produced in economically developed regions closer to the Soviet heartland.

The annual allowable cut until at least 1990 for the Far East (107 million cu. m) represented a small share of the region's annual increment (191 million cu. m) but nevertheless was under-utilized (33 million cu. m) (Yakunin *et al.* 1987, p. 3). The rate of utilization of the allowable cut in coniferous timber exceeds that of deciduous species: in 1985, the utilization of conifers was 36 per cent, shade-tolerant species 20 per cent, and shade-intolerant species 15 per cent (Yakunin *et al.* 1987, p. 14). Utilization of allowable cut varies greatly among districts; in coniferous stands, for example, the four areas with major commercial timber harvesting are well above the Far Eastern average: Amur Oblast (69 per cent), Maritime Kray (55 per cent), Sakhalin Oblast (48 per cent) and Khabarovsk Kray (42 per cent). Kamchatka uses three-quarters of its annual allowable cut but its total output is small; Yakutia harvests 4.3 million cu. m per annum (more than Sakhalin Oblast) but only uses 13

Table 6.4 1973–4 regional costs and prices* for various forest products (roubles per unit of output**) (P=Price; C=Cost)

	USSR–Domestic	USSR–Export	North–west	West Siberia	East Siberia	Far East
Logs (P, 1964)	6.8	15.72	6.49		6.1	11.6
Logs (P, 1974)	16.6	21.8	12.0	13.2	11.5	18.0
Logs (C)	11.7		11.4	12.0	10.9	15.6
Lumber (P)***	32.6	57.3	48.2	45.2	47.1	48.5
Lumber (C)	39.6		41.6	41.1	46.6	50.9
Chips (P)	12.2		13.8	6.6	16.1	18.7
Chips (C)	11.5		11.1	5.7	15.9	19.5
Plywood (P)	170.2	128.0	158.2			
Plywood (C)	136.6		131.0			
Particleboard (P)	88.8	47.0	74.8	79.8		
Particleboard (C)	63.9		62.1	66.7		
Fibreboard (P)	380.0	197.3	362.0	400.0	392.0	
Fibreboard (C)	332.4		305.8	365.0	357.0	

Sources: Barr, B. M. (1970) *The Soviet Wood-Processing Industry: A Linear Programming Analysis of the Role of Transportation Costs in Location and Flow Patterns*, Toronto: University of Toronto Press, p. 108; Mugandin, S. I. (1977) *Povyshenniye effektivnosti lesopil'nogo proizvodstva*, Moscow: Lesnaya Promyshlennost', pp. 112, 120; Kozhin, V. M. and Styazhkin, V. P. (1976) *Sebestoimost', tseni i rentabelnost' na lesozagotovkakh*, Moscow: Lesnaya Promyshlennost', p. 152; Lobovikov, T. S. and Petrov, A. P. (1976) *Ekonomika kompleksnogo ispol'zovaniya dresvesiny*, Moscow: Lesnaya Promyshlennost', pp. 117, 121, 125; and USSR Ministry of Foreign Trade export prices 1973 (1969 log export price).

Notes: *Based on representative enterprises, averages weighted by volume of output, FOB prices.
**Roubles per cubic metre except fibreboard where price refers to roubles per thousand square metres.
***USSR-domestic (i.e. all-union) 1974 prices for logs, lumber and chips are obtained from Mugandin 1977, pp. 112, 120. All others are calculated by weighted regional averages, by profitability, or by data from Lobovikov and Petrov 1976, pp. 117, 121, 125. 1974 price for lumber may have been closer to 39–41 roubles. Mugandin (1977) lists 57.3 as the export price but Ministry of Foreign Trade Statistics for 1973 calculate it to be 45 roubles.

Table 6.5 Profitability of logging by region (1967–74)*

Region	1967	1968	1969	1970	1971	1972	1973	1974
North-west	20.0	21.2	19.8	17.6	14.7	12.3	8.1	5.1
Volgo-Vyatka	18.3	29.2	27.1	25.3	17.6	17.3	25.2	12.5
Urals	20.4	19.1	19.8	20.6	19.1	17.4	13.1	10.9
Western Siberia	20.4	21.5	11.2	11.3	12.3	12.1	13.3	10.1
Eastern Siberia	20.0	28.4	22.5	19.8	17.6	15.4	12.2	7.8
Far East	21.8	37.5	37.1	35.7	27.3	25.2	24.5	15.3
Average**	19.7	25.9	22.9	22.1	18.1	16.9	15.5	8.4

Source: Kozhin, V. M. and Styazhkin, V. P. (1976) *Sebestoimost', tseni i rentabelnost' na lesozagotovkakh*, Moscow: Lesnaya Promyshlennost', pp. 102–3.

Notes: *Profitability derived from ratio of price to cost, per cu. m, of logs.
Data in table represent per cent by which price exceeds cost per cu. m.
**Average of above regions.

Table 6.6 Land use composition of state industrial forests administered by central state forestry agencies (% of areal total)

	Forestry forest lands									Non-forestry forest lands						
		Forested lands			Non-forested lands						Miscellaneous usable lands					
Total forest lands	Total forestry lands	Total	Including plantations		Total	Unspecified	Under-stocked	B&DS	Cut-over	Total	Arable lands	Natural hay lands	Grazing lands	Water bodies	Rights of way	Unutilized lands
			Closed	Unclosed												
USSR	100	76	64	1	<1	12	1	6	4	1	24	·	<1	<1	<2	<1	21
RSFSR	100	76	64	<1	<1	12	1	7	4	1	24	·	<1	<1	<2	<1	21
North-west	100	76	71	>1	<1	5		<1	1	4	24	·	<1	<1	3	<1	21
RSFSR Europe	100	82	77	3	<2	5	<2	<1	1	<3	18	>1	<1	<1	<2	<1	14
W. Siberia	100	61	55	·	·	>6		4	2	<1	39	<1	<1	<1	<5		33
E. Siberia	100	83	76	·	·	7		>3	<4	>1	17	·	·	<1	>1		16
Far East	100	73	55	·	·	18		9	8	>1	27	·	<1	>1	1	>1	26
RSFSR Asia	100	74	62	>1	·	12		7	<5	<1	26	·	<1	<1	<2	<1	23

Source: Compiled and adapted from Vorob'yev, G. I. et al. (1979) *Ekonomicheskaya geografiya lesnykh resursov SSSR*, Moscow: Lesnaya Promyshlennost', pp. 31, 32 and 35; Sinitsyn, S. G. (1976) *Lesnoy fond i organizatsiya ispol'zovaniya lesnykh resursov SSSR*, Moscow: Lesnaya Promyshlennost', pp. 8–23.

Notes: 100 per cent of the total USSR comprises 1043.49 million ha of state forests administered by central state forestry agencies excluding long-lease grazing lands. (When including long-lease grazing lands, they comprise 1162 million ha.)

Percentages in the third column have been rounded. They represent values of forested area shown in Barr and Braden (1988) Table 3.3, column 1. Thus total USSR forested lands administered by central state forestry agencies excluding long-lease grazing comprise 64.7 per cent of total forestry forest land or 675.706 million ha.

Figures in the sixth column represent an area of regeneration backlog. Thus 12 per cent of total forestry forest land, or 124.653 million ha administered by central state forestry agencies represents an area of regeneration backlog or an area currently existing in a deforested condition.

B&DS: Burned and destroyed stands.

per cent of its coniferous allowable cut. The greatest existing potential for industrial production, however, is associated with forests in the Maritime and Khabarovsk Krays which have the major species desired by Soviet Wood processors and wood workers (Vorob'yev *et al.* 1979, p. 332).

The volume of wood left at the felling site ranges from 7 per cent in the clear-cut larch forests of Kamchatka and Magadan Oblasts, to 69 per cent in the obligatory selection felling in the stone pine-broadleaf forests of Khabarovsk and Maritime Krays. Consequently, the volume of wood remaining at the felling sites ranges from 15 to 250 cu. m of wood per ha, of which approximately 32 to 46 per cent is commercial roundwood (Sinitsyn *et al.* 1979, p. 194). Much of the allowable cut throughout the Far East is so poorly utilized that some areas such as Magadan Oblast and the Okhotsk littoral of Khabarovsk Kray depend heavily on timber felled elsewhere in the region. Furthermore, the maturation of spruce and fir forests even on well nurtured sites in southern districts of the Far East may take at least 100 to 150 years, although the aggregate extent of reforestation is also limited by environmental traumas such as fire (Sinitsyn *et al.* 1979, pp. 191–2) which appears to be a greater problem in the eastern RSFSR than elsewhere due to inadequate control services. This seems particularly acute in the Far East as evidenced by the relatively larger share of understocked forest lands, and burned and destroyed stands (Table 6.6).

Much of the potential of the large Far Eastern forest volume, therefore, cannot be realized without significant improvements in logging methods, which in turn are largely dependent on regional development of wood-processing industries to absorb wood waste and inferior grades of roundwood (Sinitsyn *et al.* 1979, p. 194; Yakunin *et al.* 1987, p. 14). New railways such as the BAM and AYAM will not by themselves overcome the serious deficiencies associated with Far Eastern forestry, although the development of railways, roads and ports throughout this century have greatly influenced logging in the region. Planned use of the Far East's vast forests must be differentiated according to their subregional physical and economic characteristics (Sinitsyn *et al.* 1979, p. 195), including the large volumes (Yakunin *et al.* 1987, p. 11) of larch in northern Amur Oblast and Khabarovsk Kray, and shade-tolerant hardwood species such as ash (two-thirds of the USSR reserve) and elm (one-half of the USSR reserve) in the Maritime Kray.

Most observers of the Far Eastern forest focus on the significance of the species composition and the proportion of the growing stock which is mature and over-mature. The influence of terrain, however, on the utility of the entire Soviet forest, particularly that located in the Far East and Siberia, is almost entirely neglected outside the USSR and is now a major research focus of this paper's author. A brief examination of the Far East's mountain forests prior to general assessment of the region's total timber resources helps place important qualitative limits on the extent to which the overall growing stock and specific sub-regions might be expected to undergo commercial exploitation in the near future.

Mountain forests

Approximately 40 per cent (474 million ha) of Soviet forest land comprises mountain forests (*gornye lesa*), half of which is forested land. The Asian RSFSR has 92.5 per cent of the general mountain forest land; most (65 per cent) of the country's forest land area is found in the Far East or in Eastern Siberia (25 per cent). Mountain forests comprise 36 per cent of the USSR's forested land, 39 per cent of the forested land of Eastern Siberia but 56 per cent of that of the Far East (Sinitsyn *et al.* 1979, pp. 8–9).

Three-fifths of the Far Eastern state forest reserve's general volume of standing timber, but 64 per cent of its volume of mature timber, is located in mountain forests. Comparable figures for the entire RSFSR are 36 per cent in both cases, but in the European RSFSR the respective figures are only 13 per cent and 14 per cent. Almost all of the general (89.3 per cent) or mature (91.5 per cent) volume of the USSR's basic timber species found in mountain forests are in the Asian RSFSR. The European RSFSR or USSR are thus relatively insignificant despite the sub-regional importance of mountain forests in the North Caucasus, the Carpathians and Georgia.

The regional concentration of mature timber in mountain forests in Asia, however, is even greater than that of the general forest volume (Tables 6.7 and 6.8). The Soviet Far East is the dominant mountain forest region of the USSR in all three groups of timber species. In particular, it has 80 per cent of the USSR's shade-tolerant hardwood species found in mountain forests, but 100 per cent of the country's Erman (stone) birch (Betula ermanii Cham.) and 100 per cent of the USSR's Asian stock of beech, ash and high-stem oak (Sinitsyn *et al.* 1979, p. 22). Over half of the USSR's volume of mature mountain coniferous forest (52 per cent) is found in the Far East, but although a large share (55 per cent) of this regional growing stock comprises larch, most of the remainder is made up of spruce and fir (30 per cent) or Siberian stone pine (10 per cent). (The species composition of the region's total forested area, however, although still dominated by larch, is considerably influenced by deciduous species, see Table 6.9.) The Far East and Eastern Siberia together have 91 per cent of the USSR's mature mountain coniferous forest, but that of Eastern Siberia has a greater share of larch (57 per cent) and Siberian stone pine (17 per cent). The mountain forests of these two regions are thus deemed (Sinitsyn *et al.* 1979, p. 23) to have the most likely potential for long-term development

Table 6.7 Species composition of mature mountain forests, million cu. m

	Total	All	Larch	Conifers Spruce and Fir	Siberian Stone Pine	All	SIHS[a] Birch	Aspen	All	STHS[b] Erman B,A, and O[c] Birch	
USSR	17,340	15,126	7,895	4,128	1,888	1,309	778	123	905	418	356
RSFSR	16,990	14,938	7,870	3,975	1,886	1,282	767	122	770	418	243
North-west	153	146	3	141		7	7				
Centre											
Volgo-Vyatka											
Black Earth											
Volga Littoral	245	45	2	33		160	73	28	40		1
North Caucasus	157	32		25		14	8	1	111		83
Urals	580	444	3	386	16	134	112	2	2		
RSFSR E-U	1,135	667	8	585	16	315	200	31	153		84
W. Siberia	603	468	129	262	76	135	59	1			
E. Siberia	6,266	5,869	3,338	788	1,007	397	277				
Far East	8,986	7,934	4,395	2,340	787	435	231	90	617	418	159
RSFSR Asia	15,855	14,271	7,862	3,390	1,870	967	567	91	617	418	159
USSR E-U	1,362	754	8	670	16	320	202	32	288		197
USSR–Asia	15,978	14,372	7,887	3,458	1,872	989	576	91	617	418	159

Source: Compiled from Sinitsyn, S. G. et al. (1979) Gornye lesa, Moscow: Lesnaya Promyshlennost', p. 22.

Notes: [a]Shade-Intolerant Hardwood Species
[b]Shade-Tolerant Hardwood Species
[c]Beech, Ash and High Stem Oak

Table 6.8 Mature timber

	Volume of mature timber in forests under central state forest management (million cu. m)				Volume of mature timber in mountain forests (million cu. m)				Mature mountain timber as % of total timber			
	Total	C[a]	SIHS	STHS	Total	C	SIHS	STHS	Total	C	SIHS	STHS
USSR	51,367	43,943	6,424	1,000	17,340	15,126	1,309	905	33.8	34.4	20.4	90.5
RSFSR	50,857	43,682	6,340	835	16,990	14,938	1,282	770	33.4	34.2	20.2	92.2
North-west	5,318	4,748	570		153	146	7		2.9	3.1	1.2	0.0
Centre	289	100	181	8	0				0.0	0.0	0.0	0.0
Volgo-Vyatka	527	302	218	7	0				0.0	0.0	0.0	0.0
Black Earth	16	3	8	5	0				0.0	0.0	0.0	0.0
Volga Littoral	387	78	238	71	245	45	160	40	63.3	57.7	67.2	56.3
N. Caucasus	166	32	18	116	157	32	14	111	94.6	100.0	77.8	95.7
Urals	1,831	1,415	413	3	580	444	134	2	31.7	31.4	32.4	66.7
RSFSR E-U	8,534	6,678	1,646	210	1,135	667	315	153	13.3	10.0	19.1	72.9
W. Siberia	7,001	4,760	2,241		603	468	135		8.6	9.8	6.0	
E. Siberia	20,414	18,456	1,958		6,266	5,869	397		30.7	31.8	20.3	
Far East	14,908	13,788	495	625	8,986	7,934	435	617	60.3	57.5	87.9	98.7
RSFSR Asia	42,323	37,004	4,694	625	15,855	14,271	967	617	37.5	38.6	20.6	98.7
USSR E-U	8,894	6,825	1,696	373	1,362	754	320	288	15.3	11.0	18.9	77.2
USSR-Asia	42,473	37,118	4,728	627	15,978	14,372	989	617	37.6	38.7	20.9	98.4

Sources: Compiled from Sinitsyn, S. G. et al. (1979) *Gornye lesa,* Moscow: Lesnaya Promyshlennost', p. 22; *Lesnoye khozyaystvo SSSR* (1977) Moscow: Lesnaya Promyshlennost', p. 39.

Notes: [a]Conifers

of logging, particularly because their age structure closely corresponds to that required for sustained yield harvesting on the basis of selective felling, a harvesting method favoured in the USSR for preservation of soils on slopes associated with mountain forests.

Table 6.9 Species composition of forested areas (% of area total)

| | Total | Conifers | | Larch | SIHS | STHS | Hardwoods | | | Scrub |
		Pine	Spruce and fir				High Forest	Of which: Oak	Beech	
USSR	75	17	14	38	17	3	2	1	<1	5
RSFSR	76	17	14	40	>16	2	<2	<1	>–	5
North-west	80	32	47	<1	20	–	–	–	–	–
Volgo-Vyatka	51	29	22	–	45	4	2	2	–	–
Urals	77	18	54	–	15	–	–	–	–	–
RSFSR E–U	65	29	35	<1	30	5	<2	1	<1	<1
Western Siberia	69	35	6	6	31	–	–	–	–	–
Eastern Siberia	85	16	6	48	15	–	–	–	–	–
Far East*	78	5	6	66	6	4	2	1	–	12
RSFSR Asian ST	79	13	9	50	13	2	2	<1	–	6

Source: Derived from Vorob'yev, G. I., et al. (1979) Ekonomicheskaya geografiya lesnykh resursov SSSR, Moscow: Lesnaya Promyshlennost', p. 78.

Notes: *Far East data from Yakunin, A. G., et al. (1987) Lesnaya industriya Dal'nego Vostoka, Moscow: Lesnaya Promyshlennost', p. 8.

The volume of growing stock and annual increment, per hectare, in Far Eastern mountain forests reflects the general decline from west to east in the USSR in the productivity of all forests. The volume per hectare and the annual increment of mature coniferous timber, however, is greater than in Eastern Siberia due to limited but still significant modifications to continental severity provided by Pacific air masses during the growing season. In keeping with generally greater levels of productivity in some European forest regions, however, the volumes of growing stock and annual increments per unit area in Far Eastern mountain forests are inferior to those of the North Caucasus, Georgia and particularly the Ukraine (Carpathian mountains) (Table 6.10).

Although long-term allocation of forest land in the USSR to other ministries and administrations for essentially non-forestry use is relatively small (4.5 per cent of the general area, 5.9 per cent of the forested area, and 6.3 per cent of the general growing stock) (Lesnaya entsiklopediya 1986, vol. 2, p. 402), it comprises a significant portion of the mountain forest area in some regions. In the Asian RSFSR, for example, such allocated land comprises 16 per cent of the general mountain forest area and 5 per cent of its volume. In the Far East, this allocated land takes up nearly one-quarter of the mountain forest's area but only 9 per cent

Table 6.10 Mature growing stock and mean annual increment/ha of mountain forest

	Conifers cu. m	MAI cu. m/ha	SIHS cu. m	MAI cu. m/ha	STHS cu. m	MAI cu. m/ha
USSR	156	1.2	133	2.1	132	1.4
RSFSR	148	1.2	133	2.1	123	1.3
North-west	144	0.9	66	0.8		
Volga Littoral	181	2.1	144	2.3	114	1.8
North Caucasus	346	2.1	104	2.4	215	2.3
Urals	197	1.8	156	2.4	121	1.6
RSFSR E–U	181	2	142	2.3	173	2.2
W. Siberia	171	1.7	130	2.5		
E. Siberia	150	1.3	128	2		
Far East	193	1.5	117	2.2	122	1.2
RSFSR Asia	146	1.2	131	2	116	1
Ukraine	445	5.3	172	3.2	263	3.2
Georgia	433	2.3	93	1.8	544	1.7
USSR E-U	194	2	140	2.3	193	2.1
Kazakhstan	176	1.6	105	2		
USSR Asia	146	1.2	130	2	116	1

Source: Sinitsyn, S. G. et al. (1979) Gornye lesa, Moscow: Lesnaya Promyshlennost', p. 25.

of its general growing stock (Sinitsyn *et al.* 1979, p. 11). Much of this land is associated with reindeer herding in northern districts.

Of the total national growing stock in mountain forests, 56 per cent of the volume is growing on slopes of over 20 degrees, and 30 per cent on slopes of 20 to 30 degrees. Each year approximately 50 million cu. m are harvested from mountain forests, but only 4 million or 8 per cent of this originates on slopes of over 15–20 degrees. The stocking density in Far Eastern mountain forests is 39 per cent, but that of Eastern Siberia is 48 per cent (*Lesnaya entsiklopediya* 1985, vol. 1, p. 214).

The regional forest base

The Far East has a much larger forest area than any other Soviet region regardless of which land category is considered, but the volume of the region's growing stock is smaller than that of European–Uralian USSR or of Eastern Siberia. The volume of mature and over-mature coniferous timber in the Far East, however, is 1.9 times greater than that of Europe-Uralia, but slightly less than three-quarters that of Eastern Siberia. The Far East's average stocking density is 44 per cent; that of Europe-Uralia is 35 per cent (almost equal to the USSR average), but the figure for Eastern Siberia is 59 per cent. Whatever the relative size and stocking density of the Far East's growing stock in relation to these two other major areas, however, the large volumes of timber in Europe-Uralia and Siberia represent a viable alternative to the forests of the Far East, especially as an intervening (and less remote) spatial opportunity between them and the industrial heartland which Soviet planners of either partisan geographical persuasion cannot ignore. Indeed, two major regions of the country have already developed a strong national profile related to pulp, paper and paperboard (the north west, with a strong fibre base) or lumber and plywood (Eastern Siberia, with large supplies of structural and peeler timber).

The quality of the forest varies extensively within the Far East. Average stocking densities, for example, range from 76 per cent in the Maritime territory, 64 per cent in Sakhalin Oblast, 60 per cent in Amur Oblast, to approximately 15 per cent in Magadan Oblast (*Lesnaya entsiklopediya* 1986, vol. 2, p. 401). The physiography becomes less rugged from south to north, but greater climatic extremes cause the range of commercial species to decline sharply. The southern Far East is mountainous with a more varied species composition than similar forest zones of Siberia or the European USSR. The greatest variety comprises Manchurian (Korean) flora found in the Ussuri river basin, the Japan Sea littoral and in the middle Amur river basin (*Lesnaya entsiklopediya*, 1986, vol. 2, p. 401). Relief, site and climate give these forests a significant protective role in control of run-off, erosion and fish habitat. The run-off control and shelter functions of forests are particularly pronounced in the Kuril Islands.

The major commercial forests of the Far East are found in the four south-western districts – Amur, Khabarovsk, Maritime and Sakhalin (Table 6.11). The bulk of their growing stock is mature and over-mature. Coniferous species occupy 73 per cent of the forested area, and 84 per cent of the growing stock. The most valuable commerical species, however, occupy a relatively small but nevertheless important area: Korean and Scotch pine (6 per cent of the forested area), spruce and fir (21 per cent). Larch dominates the growing stock and comprises 56 per cent of the forested area. Scrub is relatively minor – 6 per cent. The principal stands of larch are concentrated in Khabarovsk Kray and Amur and Sakhalin Oblasts. Korean pine favours Maritime and Khabarovsk Krays, and spruce and fir are important in all regions except Amur Oblast.

Unlike the other eastern forests of the RSFSR, those of parts of the southern Far East have important stands of deciduous species. Most important in terms of area is birch which occupies half of the forested deciduous area and comprises approximately twenty species. Commercial birch species throughout the region include flatleaf birch (Betula platyphylla), and in Khabarovsk and Maritime Kray – yellow birch (Betula lutea) and river birch (Betula nigra), and in southern Maritime Kray – schmidt birch (Betula schmidtii Regel.) which has a very limited growing stock and is protected from cutting. The southern Far East is widely touted as having significant commercial stands of shade-tolerant hardwood species such as ash, oak, elm and maple. These species are most prevalent in the moderate climatic zones of Maritime and Khabarovsk

Table 6.11 Forests of the Soviet Far East

Region	Forested Area thousand ha					Growing stock million cu. m				
	Total	Conifers	SIHS	STHS	Other	Total	Conifers	SIHS	STHS	Other
Maritime	11,132	6,257	1,765	3,074	37	1,818	1,287	185	346	1
Khabarovsk	34,789	27,385	3,553	1,690	2,161	4,693	4,081	342	203	67
Amur	19,948	14,479	4,162	512	795	1,814	1,557	216	18	23
Sakhalin	4,347	2,982	222	929	214	623	545	12	55	11
Total-South	70,216	51,104	9,701	6,204	3,207	8,948	7,470	754	621	103
Kamchatka	6,092	549	582	2,804	2,157	414	171	70	128	45
Magadan	19,038	7,605	348		11,086	661	402	28	0	231
Yakut	131,021	121,891	1,856		7,274	10,677	10,529	55	0	93
Total-North	156,151	130,045	2,785	2,804	20,517	11,752	11,102	153	128	369
Far East Total	226,367	181,149	12,487	9,008	23,723	20,700	18,572	907	749	472

Sources: Vorob'yev, G. I. et al. (1979) *Ekonomicheskaya geografiya lesnykh resursov SSSR*, Moscow, Lesnaya Promyshlennost', pp. 316, 328–30; Gladyshev, A. N. et al. (1974) *Prolemy razvitiya i razmeshcheniya proizvoditel'nykh sil Dal'nego Vostoka*, Moscow: Mysl', p. 36; Antsyshkin, S. P. et al. (1965) *Spravochnik lesnichego* (2nd edn) Moscow: Lesnaya Promyshlennost', pp. 647–8; some data for Kamchatka are estimated from Fenton, R. T. and Maplesden, F. M. (1986) *The Eastern USSR: Forest Resources and Forest Products Exports to Japan*, Rotorua, New Zealand: New Zealand Forest Service, pp. 75–7.

Krays. In general, the climatic characteristics of the Maritime Kray are the most favourable for timber growth of all species, followedby Sakhalin Khabarovsk and finally Amur Oblast, which like northern Khabarovsk Kray, has the least favourable environmental conditions for tree growth (harsh climate, extensive permafrost). The most productive forests in southern Far East are deemed to be those of the mixed stone pine-broadleaf deciduous type, especially when stone pine or other conifers are available in relatively large amounts, and also the spruce-fir forests of the lower Amur river and on the island of Sakhalin. Larch forests in Amur and Sakhalin Oblasts and in Khabarovsk Kray display the lowest productivity of the region's major forests species (Vorob'yev *et al.* 1979, p. 317).

The north-eastern Far East is characterized by three zones, commencing in the north with the tundra, followed by the forest-tundra and then the taiga. The forests of Kamchatka largely comprise Erman birch (*kamennaya*), Korean Dahurian larch (Larix olgensis Henry), Yezo spruce and aspen. In Magadan Oblast, the scattered distribution and low productivity of the forest make logging problematic. In Yakutia, the plains are basically covered by taiga, with the prevailing species being Dahurian larch with various mixtures of Erman birch, pine and spruce. A large area is covered with pine elfin wood formations. Most logging in Yakutia occurs in the south-western districts, and its accelerated future development will be facilitated by the BAM and its major offshoot into Yakutia, the Amur–Yakutsk Mainline (AYAM). Nevertheless, forests in the northern Far East tend to have low productivity and be dominated by scrub formations and lightly forested areas.

Throughout the Far East strong environmental limitations adversely affect the possibilities for economic utilization of the forest. Definition of commercial (exploitable) forests is limited not only by an inadequate data base, but also by the extensive areas of scrub formations and sparse stands of birch on Sakhalin Island, northern Khabarovsk Kray and Amur Oblast, steep-sloped mountain forests, birch and larch bog forests, and forests which have been subjected to repeated cutting and fire, and stands displaying sparse stocking of valuable species.

Output

The Far East is not a major producer of forest products, particularly given the large relative size of its growing stock. The region produces approximately 10 per cent (1982) of the national timber harvest, 6 per cent of the lumber, 4.8 per cent of the paperboard, 4 per cent of the paper and 1.6 per cent of the plywood (Tables 6.12a-d) and 4 per cent of the fibreboard. The most comprehensive output (lumber, pulp, paperboard, fibreboard and plywood) is concentrated in Khabarovsk territory; paper is mainly produced in Sakhalin Oblast, and particleboard in the Maritime territory. Yakunin *et al.* (1987, p. 22) state that 85 per cent of the gross value of output in the Far Eastern forest sector originates in four districts: Khabarovsk Kray (33 per cent), Sakhalin Oblast (24 per cent), Maritime Kray (17 per cent) and Amur Oblast (11 per cent).

Table 6.12a) Lumber output, 1960, 1975, 1985 (thousand cu. m and % change by period)

	1960	%	1975	%	1985	%	1960–75	1960–85	1975–85
USSR	105,556	100.0	114,511	100.0	96,831	100.0	8.5	−8.3	−15.4
RSFSR	83,568	79.2	93,513	81.7	79,549	82.2	11.9	−4.8	−14.9
North-west	16,622	15.7	18,530	16.2	15,661	16.2	11.5	−5.8	−15.5
Urals	12,758	12.1	13,450	11.7	10,784	11.1	5.4	−15.5	−19.8
RSFSR E–U	59,449	56.3	60,580	52.9	47,202	48.7	1.9	−20.6	−22.1
Western Siberia	7,237	6.9	8,790	7.7	9,157	9.5	21.5	26.5	4.2
Eastern Siberia	11,585	11.0	17,339	15.1	16,862	17.4	49.7	45.6	−2.8
Far East	5,089	4.8	6,580	5.7	6,179	6.4	29.3	21.4	−6.1
Amur	585	0.6	807	0.7	756	0.8	37.9	29.2	−6.3
Khabarovsk	1,634	1.5	2,389	2.1	2,075	2.1	46.2	27.0	−13.1
Maritime	1,394	1.3	1,703	1.5	1,495	1.5	22.2	7.2	−12.2
Sakhalin	693	0.7	660	0.6	585	0.6	−4.8	−15.6	−11.4
(South)	4,306	4.1	5,559	4.9	4,911	5.1	29.1	14.1	−11.7
Kamchatka	153	0.1	212	0.2	270	0.3	38.6	76.5	27.4
Magadan	187	0.2	227	0.2	188	0.2	21.4	0.5	−17.2
Yakut	443	0.4	582	0.5	810	0.8	31.4	82.8	39.2
(North)	783	0.7	1,021	0.9	1,268	1.3	30.4	61.9	24.2
RSFSR Asia	23,911	22.7	32,709	28.6	32,198	33.3	36.8	34.7	−1.6

Sources: Tables 12a-d: 1960 and 1985 – *Nar. Khoz. SSSR v 1985g*, pp. 150–51; *Nar. Khoz. RSFSR za 70 let*, pp. 26–130; 1975 – *Nar. Khoz. SSSR v 1975g*, pp. 274–6; *Nar. Khoz. RSFSR za 60 let*, pp. 78–84.

Table 6.12b) Plywood output, 1960, 1975, 1985 (thousand cu. m and % change by period)

	1960	%	1975	%	1985	%	1960–75	1960–85	1975–85
USSR	1,354	100.0	2,199	100.0	2,187	100.0	62.4	61.5	−0.6
RSFSR	900	66.5	1,559	70.9	1,573	71.9	73.2	74.8	0.9
North-west	236	17.4	417	19.0	391	17.9	77.1	65.9	−6.3
Urals	108	8.0	233	10.6	231	10.5	115.0	112.9	−0.9
RSFSR E–U	814	60.1	1,394	63.4	1,296	59.2	71.3	59.3	−7.1
Western Siberia	38	2.8	68	3.1	52	2.4	79.3	37.3	−23.4
Eastern Siberia	17	1.3	50	2.3	190	8.7	190.2	1,000.0	279.1
Far East	31	2.3	47	2.1	35	1.6	50.3	11.9	−25.5
Amur	−	0.0	−	0.0	−	0.0			
Khabarovsk	6	0.4	16	0.7	11	0.5	178.9	96.5	−29.6
Maritime	25	1.9	31	1.4	24	1.1	21.3	−7.1	−23.5
Sakhalin	−	0.0	−	0.0	−	0.0	−	−	−
(South)	31	2.3	47	2.1	35	1.6	50.3	11.9	−25.5
Kamchatka	−	0.0	−	0.0	−	0.0	−	−	−
Magadan	−	0.0	−	0.0	−	0.0	−	−	−
Yakut	−	0.0	−	0.0	−	0.0	−	−	−
(North)	−	0.0	−	0.0	−	0.0	−	−	−
RSFSR Asia	86	6.4	165	7.5	277	12.7	91.1	220.9	68.0

Table 6.12c) Paper output, 1960, 1975, 1985 (thousand tons and % change by period)

	1960	%	1975	%	1985	%	1960–75	1960–85	1975–85
USSR	2,421	100.0	5,215	100.0	5,986	100.0	115.4	147.3	14.8
RSFSR	1,941	80.2	4,318	82.8	5,030	84.0	122.5	159.2	16.5
North-west	672	27.8	1,875	35.9	2,575	43.0	179.0	283.3	37.4
Urals	507	20.9	1,028	19.7	1,045	17.5	102.7	106.1	1.7
RSFSR E–U	1,625	67.1	3,814	73.1	4,532	75.7	134.8	178.9	18.8
Western Siberia	1	0.1	2	0.0	2	0.0	28.6	57.1	22.2
Eastern Siberia	10	0.4	133	2.6	121	2.0	1,246.5	1,123.2	−9.2
Far East	197	8.1	229	4.4	228	3.8	16.2	15.9	−0.3
Amur	–	0.0	4	0.1	4	0.1	–	–	−18.6
Khabarovsk	8	0.3	9	0.2	9	0.2	20.0	22.7	2.2
Maritime	–	0.0	–	0.0	–	0.0	–	–	–
Sakhalin	190	7.8	216	4.1	216	3.6	13.8	13.8	0.0
(South)	197	8.1	229	4.4	228	3.8	16.2	15.9	−0.3
Kamchatka	–	0.0	–	0.0	–	0.0	–	–	–
Magadan	–	0.0	–	0.0	–	0.0	–	–	–
Yakut	–	0.0	–	0.0	–	0.0	–	–	–
(North)	–	0.0	–	0.0	–	0.0	–	–	–
RSFSR Asia	208	8.6	364	7.0	352	5.9	74.8	68.8	−3.4

Table 6.12d) Paperboard output 1960, 1975, 1985 (thousand tons and % change by period)

	1960	%	1975	%	1985	%	1960–75	1960–85	1975–85
USSR	806	100.0	3,370	100.0	4,034	100.0	318.0	400.4	19.7
RSFSR	491	60.8	2,514	74.6	2,877	71.3	412.5	486.4	14.4
North-west	148	18.3	790	23.4	981	24.3	434.6	564.0	24.2
Urals	18	2.2	137	4.1	140	3.5	654.1	674.0	2.6
RSFSR E–U	427	53.0	1,886	56.0	2,079	51.5	341.5	386.8	10.3
Western Siberia	–	0.0	56	1.7	85	2.1	–	–	50.6
Eastern Siberia	3	0.4	394	11.7	487	12.1	11,485.3	14,226.5	23.7
Far East	30	3.7	134	4.0	192	4.8	353.2	550.8	43.6
Amur	–	0.0	–	0.0	–	0.0	–	–	–
Khabarovsk	10	1.3	43	1.3	120	3.0	312.5	1,056.7	180.4
Maritime	–	0.0	–	0.0	–	0.0	–	–	–
Sakhalin	19	2.4	91	2.7	72	1.8	375.4	275.4	−21.0
(South)	30	3.7	134	4.0	192	4.8	353.2	550.8	43.6
Kamchatka	–	0.0	–	0.0	–	0.0	–	–	–
Magadan	–	0.0	–	0.0	–	0.0	–	–	–
Yakut	–	0.0	–	0.0	–	0.0	–	–	–
(North)	–	0.0	–	0.0	–	0.0	–	–	–
RSFSR Asia	33	4.1	584	17.3	764	18.9	1,674.8	2,221.9	30.8

National perspective

When the annual output of lumber, plywood, pulp and paperboard, the four industries for which sub-regional data are published annually, is converted into roundwood equivalent units, the relative change in the national importance of the Far East to the USSR's wood-processing can be assessed (Table 6.13). From 1960 to 1985, a period of considerable growth for major regions such as the north west and Eastern Siberia, the Far East's relative share of national production of these commodities increased by only 1 per cent, although the growth of the region from 1960 to 1985 (29.3 per cent) surpassed that of the north west (26 per cent) (Table 6.14). For this period, the forest regions of the RSFSR European-Uralian zone together declined by 2.5 per cent. Overall, the Far East remains the least significant wood processor among the USSR's major forest regions. In the decade 1975–85, however, the Far East, like all major regions except Western and Eastern Siberia, declined in total output of these four commodities, largely due to the decline in lumber production. Lumber's relative importance far surpasses that of the other three commodities in the FSFSR's major forest regions (Tables 6.15–17) for each of the time periods utilized here, although in particular administrative areas like Sakhalin Oblast, other commodities are relatively strong.

Roundwood equivalent units also offer the possibility of comparing regional and national change. One technique, mix-and-share analysis (m & s), considers regional change relative to national change to be the net result of three so-called 'effects': (1) the regional impact of total national change (can be growth or decline but is usually termed growth (+ or −), (2) the regional industry mix compared with the national industrial structure, and (3) the changing regional shares of total national activity in each industry. M&s has been applied to the Soviet forest industry

Table 6.13 Total roundwood equivalents, million cu m(r) 1960, 1975, 1985

	1960	% of total	1975	% of total	1985	% of total
USSR	175,871	100	205,376	100	187,942	100
RSFSR	138,534	78.8	166,775	81.2	153,499	81.7
North-west	28,729	16.3	37,117	18.1	36,158	19.2
Volgo-Vyatka	14,090	8	15,145	7.4	12,397	6.6
Urals	21,641	12.3	24,956	12.2	21,614	11.5
RSFSR E–U	100,005	56.9	112,921	55	97,536	51.9
Western Siberia	11,270	6.4	13,848	6.7	15,108	8
Eastern Siberia	17,933	10.2	27,917	13.6	28,959	15.4
Far East	8,613	4.9	11,227	5.5	11,134	5.9
Amur	901	0.5	1,257	0.6	1,236	0.7
Khabarovsk	2,575	1.5	3,827	1.9	3,611	1.9
Maritime	2,227	1.3	2,720	1.3	2,481	1.3
Sakhalin	1,704	1	1,852	0.9	1,752	0.9
(South)	7,407	4.2	9,655	4.7	9,080	4.8
Kamchatka	236	0.1	326	0.2	437	0.2
Magadan	288	0.2	350	0.2	305	0.2
Yakut	682	0.4	896	0.4	1,312	0.7
(North)	1,206	0.7	1,572	0.8	2,054	1.1
RSFSR Asia	37,815	21.5	52,993	25.8	55,201	29.4

Source: Calculated by author. Methodology and coefficients are described in Barr and Braden (1988), pp. 96–7.

Table 6.14 Change in regional distribution of roundwood equivalents, million cu. m(r)

	1960–75	% change	1960–85	% change	1975–85	% change
USSR	29,505	16.8	12,071	6.9	−17.434	−8.5
RSFSR	28,241	20.4	14,965	10.8	−13,276	−8
North-west	8,388	29.2	7,429	25.9	−960	−2.6
Volgo-Vyatka	1,055	7.5	−1,693	−12	−2,748	−18.1
Urals	3,315	15.3	−27	−0.1	−3,342	−13.4
RSFSR E–U	12,916	12.9	−2,469	−2.5	−15,386	−13.6
Western Siberia	2,578	22.9	3,838	34.1	1,260	9.1
Eastern Siberia	9,985	55.7	11,026	61.5	1,042	3.7
Far East	2,615	30.4	2,522	29.3	−93	−0.8
Amur	356	39.5	335	37.2	−21	−1.6
Khabarovsk	1,252	48.6	1,036	40.2	−215	−5.6
Maritime	493	22.1	254	11.4	−239	−8.8
Sakhalin	148	8.7	48	2.8	−100	−5.4
(South)	2,248	30.4	1,674	22.6	−575	−6
Kamchatka	91	38.6	202	85.6	111	34
Magadan	62	21.4	17	5.8	−45	−12.9
Yakut	214	31.4	630	92.3	416	46.4
(North)	367	30.4	848	70.4	482	30.6
RSFSR Asia	15,178	40.1	17,386	46	2,208	4.2

Source: Calculated by author.

Table 6.15 Regional composition of structure of major forest industries, 1960 (% of regional total)

	Lumber	Plywood	Paper	Paperboard	Total
USSR	92.4	2.4	4.4	0.7	100
RSFSR	92.9	2.1	4.5	0.6	100
North-west	89.1	2.6	7.5	0.8	100
Volgo-Vyatka	91.7	1.4	6	0.9	100
Urals	90.8	1.6	7.5	0.1	100
RSFSR E–U	91.5	2.6	5.2	0.7	100
Western Siberia	98.9	1.1	0	0	100
Eastern Siberia	99.5	0.3	0.2	0	100
Far East	91	1.1	7.3	0.5	100
Amur	100	0	0	0	100
Khabarovsk	97.7	0.7	0.9	0.6	100
Maritime	96.4	3.6	0	0	100
Sakhalin	62.6	0	35.6	1.8	100
(South)	89.5	1.3	8.5	0.6	100
Kamchatka	100	0	0	0	100
Magadan	100	0	0	0	100
Yakut	100	0	0	0	100
(North)	100	0	0	0	100
RSFSR Asia	97.4	0.7	1.8	0.1	100

Source: Calculated by author from roundwood equivalents.

Table 6.16 Regional composition structure of major forest industries, 1975 (% of regional total)

	Lumber	Plywood	Paper	Paperboard	Total
USSR	85.9	3.4	8.1	2.6	100
RSFSR	86.4	3	8.3	2.4	100
North-west	76.9	3.6	16.2	3.4	100
Volgo-Vyatka	83.3	1.9	12.9	1.8	100
Urals	83	2.9	13.2	0.9	100
RSFSR E–U	82.6	3.9	10.8	2.7	100
Western Siberia	97.7	1.6	0	0.7	100
Eastern Siberia	95.6	0.6	1.5	2.3	100
Far East	90.3	1.3	6.5	1.9	100
Amur	98.9	0	1.1	0	100
Khabarovsk	96.1	1.3	0.8	1.8	100
Maritime	96.4	3.6	0	0	100
Sakhalin	54.9	0	37.3	7.8	100
(South)	88.7	1.5	7.6	2.2	100
Kamchatka	100	0	0	0	100
Magadan	100	0	0	0	100
Yakut	100	0	0	0	100
(North)	100	0	0	0	100
RSFSR Asia	95.1	1	2.2	1.8	100

Source: Calculated by author from roundwood equivalents.

Table 6.17 Regional composition structure of major forest industries, 1985 (% of regional total)

	Lumber	Plywood	Paper	Paperboard	Total
USSR	83.5	2.9	10.2	3.4	100
RSFSR	84	2.6	10.5	3	100
North-west	70.2	2.7	22.8	4.3	100
Volgo-Vyatka	79.9	1.9	16.1	2.1	100
Urals	80.8	2.7	15.5	1	100
RSFSR E–U	78.4	3.3	14.9	3.4	100
Western Siberia	98.2	0.9	0	0.9	100
Eastern Siberia	94.3	1.6	1.3	2.7	100
Far East	89.9	0.8	6.6	2.8	100
Amur	99.1	0	0.9	0	100
Khabarovsk	93.1	0.8	0.8	5.3	100
Maritime	97.6	2.4	0	0	100
Sakhalin	54.1	0	39.4	6.5	100
(South)	87.6	1	8	3.4	100
Kamchatka	100	0	0	0	100
Magadan	100	0	0	0	100
Yakut	100	0	0	0	100
(North)	100	0	0	0	100
RSFSR Asia	94.5	1.3	2	2.2	100

Source: Calculated by author from roundwood equivalents.

elsewhere (Barr and Braden 1988) to evaluate regional shift of production between 1960 and 1984. RSFSR Asia as a whole, and Eastern Siberia, among the regions analysed by m&s, suggest, particularly for the time periods 1960–75, and 1975–84, 'a growing polarity of wood-processing in major regions of the RSFSR, and the essential differences between Europe and Asia in the sectors of importance' (Barr and Braden, p. 130). Asia has more structural and peeler timber to support sawmilling and plywood manufacture; Europe has a large supply of fibre to support existing investment in pulp, paper and paperboard, and offers important access to related secondary manufacturers. Overall, however, the wood-processing industry is shifting eastwards; the regional shares effect in lumber, the most important processing industry, continues to exert a strong influence on the Asian RSFSR although, as this study demonstrates, this effect is somewhat more problematic in the Far East than in Eastern Siberia.

M&s has been repeated for this study to include the Far East, and to lengthen the period of analysis from 1960 to 1985. The results are summarized for seven regions in Table 6.18. If Far Eastern lumber output, for example, had changed at the national total rate of change (national growth effect) in wood-processing, output in the region would have declined by 860,000 cu. m. Instead, regional lumber output decreased by only 123,000 cu. m. The difference between the two figures, net relative change, of 737,000 cu. m can be accounted for by two factors. First, the national lumber sector declined more than the decline in total wood-processing. Hence lumber output decreased as a proportion of total national wood-processing. A smaller share of Far Eastern production (91 per cent) was accounted for by lumber, however, than in the national total (92.4 per cent). The smaller share of regional than national employment in lumber (industry mix effect) means that the Far East was less strongly affected by a drop in lumber output than the nation but was still responsible for a decline of 259,000 cu. m. During this period, however, the Far East's share (regional shares effect) of total lumber output increased (Table 6.12a). This regional shares effect of +996,000 cu. m helped to offset the impact of the industry mix effect (−259,000) and the national growth effect (−860,000). Consequently the actual decline was only −123,000 cu. m (−860,000 −259,000 +996,000). In the case of paperboard, a positive regional shares effect and a positive industry mix effect offset a negative national growth effect to ensure modest growth in paperboard for the period. Negative regional shares effects for paper and plywood contributed to a net decline of these industries in the Far East; overall, the strong regional shares effect of lumber could not prevent a minor decline in the total output of the four industries in the Far East between 1975 and 1985.

Over the period of major change in wood-processing, 1960–75, the Far

Table 6.18 Mix-and-share: summary of three effects, selected regions, 1960–85, thousand cu. m(r)

Shift/effect	Total	Lumber	Plywood	Paper	Paperboard
RSFSR					
1960–75					
R	28,241	15,315	2,083	7,605	3,237
N	23,274	21,621	478	1,043	132
M	−883	−10,672	1,298	6,127	2,364
S	5,851	4,367	307	435	742
1960–85					
R	14,965	175	1,089	9,884	3,818
N	9,508	8,833	195	426	54
M	−935	−13,338	595	8,719	3,089
S	6,392	4,679	298	739	675
1975–85					
R	−13,276	−15,141	−994	2,279	580
N	−14,157	−12,225	−418	−1,173	−341
M	33	−3,684	−633	3,215	1,135
S	849	768	57	237	−213
RSFSR Europe-Uralia					
1960–75					
R	12,916	1,742	1,834	7,007	2,334
N	16,801	15,381	432	873	115
M	769	−7,592	1,174	5,129	2,058
S	−4,653	−6,047	228	1,004	161
1960–85					
R	−2,469	−15,084	668	9,303	2,644
N	6,864	6,284	176	357	47
M	1,039	−9,488	538	7,300	2,690
S	−10,372	−11,880	−46	1,646	−93
1975–85					
R	−15,386	−16,826	−1,166	2,296	310
N	−9,586	−7,919	−374	−1,036	−256
M	739	−2,386	−566	2,840	851
S	−6,539	−6,520	−226	492	−285
North-west					
1960–75					
R	8,388	2,938	574	3,849	1,027
N	4,826	4,300	125	361	40
M	1,050	−2,123	340	2,121	712
S	2,512	761	109	1,367	275
1960–85					
R	7,429	−227	233	6,090	1,333
N	1,972	1,757	51	148	16
M	1,451	−2,653	156	3,018	930
S	4,006	669	26	2,924	387
1975–85					
R	−960	−3,165	−341	2,241	306
N	−3,151	−2,422	−112	−509	−107
M	853	−730	−169	1,396	356
S	1,338	−13	−60	1,354	57
Urals					
1960–75					
R	3,315	1,066	393	1,667	189
N	3,636	3,301	57	273	5
M	215	−1,629	156	1,601	87
S	−536	−606	180	−207	97
1960–85					
R	−27	−2,177	234	1,721	195
N	1,485	1,349	23	111	2
M	428	−2,036	72	2,278	114
S	−1,940	−1,490	139	−669	79

Table 6.18 (*continued*)

Shift/effect	Total	Lumber	Plywood	Paper	Paperboard
1975–85					
R	−3,342	−3,243	−159	54	6
N	−2,118	−1,758	−62	−279	−19
M	203	−530	−94	765	62
S	−1,426	−955	−2	−432	−37
RSFSR Asia					
1960–75					
R	15,178	13,549	249	499	882
N	6,353	6,186	46	112	9
M	−2,113	−3,054	125	658	159
S	10,938	10,416	78	−271	714
1960–85					
R	17,386	15,338	420	459	1,170
N	2,595	2,527	19	46	4
M	−2,616	−3,816	57	936	207
S	17,407	16,627	344	−523	959
1975–85					
R	2,208	1,789	172	−40	288
N	−4,498	−4,276	−44	−99	−79
M	−821	−1,289	−67	271	264
S	7,528	7,353	283	−212	104
Eastern Siberia					
1960–75					
R	9,985	8,861	104	395	625
N	3,013	2,997	9	5	1
M	−1,407	−1,480	25	31	16
S	8,379	7,343	70	358	607
1960–85					
R	11,026	9,476	421	356	774
N	1,231	1,225	4	2	0
M	−1,772	−1,849	11	44	21
S	11,567	10,100	406	309	752
1975–85					
R	1,042	614	317	−39	149
N	−2,370	−2,267	−13	−36	−53
M	−426	−683	−20	99	178
S	3,838	3,564	351	−102	25
Far East					
1960–75					
R	2,615	2,296	49	103	167
N	1,445	1,315	16	106	8
M	158	−650	45	621	142
S	1,012	1,631	−12	−624	18
1960–85					
R	2,522	2,173	−11	101	260
N	591	538	7	43	3
M	278	−812	21	885	185
S	1,653	2,447	−38	−827	72
1975–85					
R	−93	−123	−60	−2	93
N	−953	−860	−12	−62	−18
M	−47	−259	−19	171	60
S	908	996	−29	−110	51

Notes: R=net actual change in regional activity; N=national growth effect; M=industry mix effect; S=regional shares effect.

East displays characteristics not entirely similar to those reported for Eastern Siberia and RSFSR Asia by Barr and Braden (1988) and repeated in Table 6.18. The regional shares effect is not the strongest contributor to the overall net positive regional change in wood-processing, although it is the strongest effect for lumber production, and lumber throughout the eastern RSFSR is the most important component of wood-processing. In the Far East, however, unlike Eastern Siberia and RSFSR Asia, all three effects for wood-processing were positive for the period 1960–75, and 1960–85, suggesting that overall the wood-processing industry in the region was a positive phenomenon regardless of which processes or effects were operative nationally. By the period 1975–85, however, the Far East, like the other two comparative units registered negative impacts from the national growth effect and the industry mix effect, but unlike the other two regions had a negative overall actual change for the period. In short, the recent decade has not favoured wood-processing in the Soviet Far East and seems to confirm that, even for sawmilling, the region is recognized in Moscow as having comparative disadvantages relative to those closer to the USSR's industrial and population concentrations. Three of the four sectors analysed here registered a decline in output during the decade 1975–85, including sawmilling, the impact of whose net actual negative regional change was mitigated only by the regional increase in paperboard output.

If the Soviet lumber industry continues to stagnate nationally but to focus on a few regions to sustain itself between now and the end of this century, the relative position of the Far East nevertheless is likely to continue to attenuate in the national space economy because other sectors such as paper and plywood are showing much stronger regional shares effects in other regions and thus may not be able to offset lumber's overall decline in the Far East. For the remainder of this century at least, the Far East is likely to continue to support the USSR's export of roundwood and fibre (chips) to Asian markets, but is unlikely to become a strong contributor of finished products to those markets. Surplus output from wood-processing in the Far East is not available for export, and new capacity to bolster regional, national and export needs is accruing to other regions better situated in the national space economy and with more acceptable levels of construction and production costs.

Nevertheless, these observations stand in rather stark contrast to statements made by A. Reut, the First Deputy Chairman of Gosplan (*Pravda*, 26 August 1987, p. 2; discussed by Dienes 1988, pp. 420–22), and to extensive proposals for expanded output identified by Yakunin *et al.* (1987). According to Reut, the Far East by the year 2000 is to increase the output of pulp by 2.1 times, of paperboard by 3.2 times, of particleboard by 5 times, of plywood by almost 6 times, and of wood chips by 4.7 times (Singur 1988, p. 97), but the output of commercial roundwood is to increase by only 11 per cent. The level of output of wood-products, except lumber, in the Far East is so low that any sizeable increment could produce apparently large percentage increases in output and still not significantly alter the region's relative national importance. Reut's figures for these sectors also do not reflect recent trends in the evolution of the region's profile, and they exclude lumber, the major sector in Far Eastern wood-processing. They must be interpreted within the national goals and priorities of the current plan or those of the guidelines adopted through the year 2000 (discussed in Barr and Braden 1988, pp. 71, 138). In the 12th Five-Year Plan, for the entire USSR, the output of timber is to increase by 10–12 per cent (*CDSP* 1985, vol. 37, no. 46, p. 13), pulp by 15–18 per cent, paper by 11–15 per cent, fibreboard by 17–20 per cent, and paperboard and particleboard each by about 30 per cent (*CDSP* 1985, vol. 37, no. 47, p. 24). A small portion of the increment in any of these industries in the current plan alone could have a major impact on the existing volume of Far Eastern output. Nevertheless, the long-range programme for development of the Far East (including Chita Oblast and Buryat ASSR) by the year 2000 (Sobolev 1988, cited in *Referativniy zhurnal*, 1988, vol. 5, p. 38) envisions the region's share (including the two provinces of Eastern Siberia) of RSFSR pulp output only increasing to 11.4 per cent from 7 per cent, and paperboard's share to 8.4 per cent from 5.4 per cent on the basis of comprehensive resource utilization – the consumption of 5 million cu. m of additional wood waste per year. Singur foresees by the year 2000 the Far East's share of the USSR's pulp output reaching 7.8 per cent from 6.9 per cent, and paperboard accounting for 9.1 per cent instead of 4.8 per cent as in 1985 (Yakunin *et al.* 1987, p. 19).

Some of Sobolev's figures, however, may qualify as vapour-data because the Far East plus the Buryat ASSR in 1986 already comprised 12 per cent of the RSFSR's paperboard output and 8.5 per cent of the USSR's production (*Narodnoye khozyaystvo RSFSR za 70 let* 1987, p. 130; *Narodnoye khozyaystvo SSSR za 70 let* 1987, p. 183); the Buryat pulp figure cannot be verified because appropriate regional data are not available, but in 1985 the Far East alone produced 7.2 per cent of the RSFSR's pulp (Yakunin *et al.* 1987, p. 19). Singur's data are similar to those of Yakunin *et al.* for 1985.

Although Yakunin *et al.* (1987, p. 39) do not identify the expected levels of forest-product output for the RSFSR or the USSR in 1990, they suggest that some branches in the Far East will show sizeable increases in 1990 compared to 1985: wood chips (137 per cent), particleboard (172 per cent), paperboard (61 per cent), chemical pulp (39 per cent), plywood (20 per cent), and fibreboard (18 per cent); timber harvested will increase by only 8 per cent, but output of lumber and paper will hardly change.

Thus, construction of one or two new plants or the expansion of existing facilities could produce the increases for each commodity in the Far East envisaged by Reut and Yakunin *et al.* without altering the basic pattern of polarization of wood-processing in the USSR if new capacity also came on stream in other regions (current and possible expansion projects in each of the major branches of the Far Eastern forest sector, and expected 1990 sub-regional levels of output for some planning groups (inconsistent with the present analysis) are identified in Yakunin *et al.* 1987, pp. 38–49). For example, commissioning of a new paperboard mill at Ussuriysk (Eronen 1985, pp. 77–9) and expansion (Yakunin *et al.* 1987, p. 47) of pulp capacity by 200,000 tons and paperboard production by 95,000 tons at the Amur pulp and paper mill will account for most of the growth in these branches in the Far East by 1990 in addition to modest increases at the old Japanese plants on Sakhalin Island (22,000 tons of pulp, 7,000 tons of paper and 21,000 tons of paperboard). Japanese equipment deliveries to Amursk, Khabarovsk Novomikhailovka (Maritime Kray), Poronaysk and Soviet Harbour (described by Bradshaw 1987, pp. 185–290) and major changes to some of the old Japanese mills on Sakhalin may be part of the expansion described by Reut because of the lengthy periods between delivery and complete installation. The Far East could increase its output of pulp by 4.4 times and paperboard by 3.4 times by 2000–2003 (Yakunin *et al.* 1987, p. 48), however, if the following major projects are undertaken in the 1990s: construction of the first market pulp section and a start on construction of the second paperboard line of the Svobodnensk pulp and paperboard mill in Amur Oblast; the start of construction on a market-pulp mill on the lower Amur and one at Tommot in Yakutia; continuation of reconstruction of four Sakhalin pulp and paper mills (Dolinsk, Makarov, Chekhov and Uglegorsk) plus major changes at three others on the island – installation of new equipment permitting the Poronaysk pulp and paper mill to orient its entire output towards sulphate pulp from continuous digesters, the Tomarinsk pulp and paper mill towards wallpaper, and the Kholmsk pulp and paper mill towards both wallpaper and personal care products. The long periods normally associated with Soviet industrial construction, however, also suggest that Reut and others at Gosplan may be predicating some of the envisaged growth in the Far East on more efficient construction methods and processing technology available if joint ventures with Japanese companies, and not simply equipment purchases, are soon undertaken. Joint ventures are a key element in the Far Eastern forest-sector's future envisaged by Yakunin *et al.* (1987, p. 54).

The first major joint forest venture between the USSR and Japan, for example, under the reforms instituted by General Secretary Gorbachev since his elevation to the country's leadership in 1985, has been a sawmill in Irkutsk Oblast (SUPAR Report 1988, vol. 5, p. 38) which held its opening ceremonies on 28 March 1988. Situated in Novaya Igirma, nearly

1,000 km north of Irkutsk, it is owned 51 per cent by the Irkutsk Forestry Association and 49 per cent by Tairiku Boeki of Japan. Of its expected annual output of 90,000 cu. m of Siberian red pine lumber, 70,000 cu. m, meeting Japanese home construction standards, will be exported to Japan via the BAM and Vanino. The Soviet Union has collaborative logging agreements with China and Cuba in Amur Oblast and Khabarovsk Kray (Yakunin *et al.* 1987, p. 6).

Timber

Although regional logging data were not published in official statistical handbooks between 1976 and 1986, Yakunin *et al.* (1987, p. 7) state that the region produced 34.5 million cu. m in 1985 (apparently excluding timber outside the principal cut); 40 per cent originated in Khabarovsk Kray, 18 per cent in Maritime Kray and 17 per cent in Amur Oblast which has shown the largest increase of all since 1970–68 per cent compared to an average over all seven districts of the Far East of 17 per cent for the same period. The Forestry Encyclopedia (*Lesnaya entsiklopediya* 1986, vol. 2, p. 404) shows that, in 1982, the principal cut by central state forestry organs (i.e. excluding collective farms) in the Far East was 33.7 million cu. m, of which 31.4 million cu. m comprised conifers. Improvement felling provided an additional 1.3 million cu. m, and 1.9 million cu. m was obtained from 'other forms of harvesting' (defined in *Lesnaya entsiklopediya* 1985, vol. 1, p. 518) such as those preceding expansion of industrial, urban and transportation facilities. The total harvest for 1985 thus could be at least 38 million cu. m. The greatest influence on the expansion of logging in Khabarovsk Kray and Amur Oblast is the BAM, and although Yakunin *et al.* (1987, p. 25) do not state how much is currently harvested in the Far Eastern BAM area, they note that 13.6 million cu. m should be harvested there by 1990. We might expect, therefore, that the BAM forests will account for 40 per cent of the Far Eastern timber harvest by the early 1990s, and that the share from newly-accessible forests in the region will increase even further when the AYAM forests of southern Yakutia become accessible in the mid–1990s (Yakunin *et al.* 1987, p. 14).

As a whole, the Far East produces approximately 6.7 million cu. m net of surplus timber (Tables 6.19–21). Singur (1988, p. 95) states that 10 million cu. m of roundwood and 1 million cu. m of lumber annually is shipped out of the Far East. (Yakunin *et al.* (1987, p. 35) report that 24.8 million cu. m of wood, including waste, is consumed within the Far East (1985), out of a total harvest of 34.5 million cu. m.) The mix of timber supply and demand in the eastern RSFSR is dominated by saw logs, unprocessed timber and fuelwood. Given that the combined demand of Japan and China for roundwood today is approximately equal to that of Japan alone in the 1970s, and that the level of wood-processing in

the region has only slightly declined, then the pattern of timber surplus
and deficit in the Far East is probably much the same today as in Tables
6.19–21. Surplus timber in the saw-log category is destined for export to
Japan and China, and for domestic shipment mainly to Kazakhstan and
Soviet Central Asia (Yakunin *et al.* 1987, p. 34). In 1985, 1.5 million
cu. m of roundwood and 0.9 million cu. m of lumber were shipped out
of the region to domestic markets; 56 per cent of the roundwood originated in Amur Oblast, 33 per cent in Khabarovsk Kray and the remainder in Maritime Kray. All of the domestic lumber shipments out of the
region originated from these three districts, but 55 per cent was shipped
from Khabarovsk Kray, and 17 per cent from Maritime Kray. Thus, after
the various deficits shown in Table 6.21 are met by substitution of timber
of different categories, the Far East produces a sizeable timber surplus
similar to that of other relatively unimportant timber surplus regions.
Without the adjacent Pacific markets of Japan and China, however, it is
not likely that the Far East would be encouraged to produce much
surplus roundwood. Forest industry planners would like to see domestic
shipments of forest products out of the region dominated by higher value-
added items which can better withstand long-distance transportation costs
(Yakunin *et al.* 1987, pp. 34–5).

Table 6.19 Volume and mix of consumed timber, 1973 (thousand cu. m)

	Total	Saw logs	Peeler logs	Pulp-wood	Pit props	Constrn timber	Other c. timber	Fuelwood
USSR	289,270	116,423	7,397	18,814	15,964	21,207	49,815	59,650
RSFSR	230,910	99,239	5,123	17,180	6,899	11,152	43,038	48,279
North-west	56,830	23,274	1,177	6,964	840	1,778	13,008	9,789
Volgo-Vyatka	19,352	8,944	284	1,852	14	731	3,083	4,444
Urals	35,115	14,904	695	3,326	1,204	1,270	5,926	7,790
RSFSR E–U	154,281	66,269	4,414	12,364	3,532	7,278	29,902	31,332
W. Siberia	17,995	7,796	268		2,088	1,130	3,015	3,698
E. Siberia	36,066	16,351	197	2,743	302	959	9,369	6,145
Far East	22,568	8,823	244	2,073	977	1,785	1,562	7,104
RSFSR Asia	76,629	32,970	709	4,816	3,367	3,874	13,946	16,947

Source: Vorob'yev, G. I. *et al.* (1979) *Ekonomicheskaya geografiya lesnykh resursov SSSR*, Moscow:
Lesnaya Promyshlennost', various pages.

Note: Refers to industrial timber administered by Soyuzglavlesa (a division of Gossnab) for 1973 (See
Vorob'yev *et al.* 1979, p. 193).

The total volume of timber harvested in the Far East by central organs,
approximately 34.4 million cu. m, represents 9 per cent of the national
harvest. This share is greater than that of 1975 (8.4 per cent of national
cut, *see* Table 6.22); between 1975 and 1985 the volume harvested
increased by 1.2 million cu. m while in Eastern Siberia, felling increased
by approximately 0.1 million cu. m although that region's share of

Table 6.20 Volume and mix of harvested timber, 1973 (thousand cu. m)

	Total	Saw logs	Peeler logs	Pulp-wood	Pit props	Constrn timber	Other c. timber	Fuelwood
USSR	302,564	125,175	7,033	23,824	16,486	20,064	55,061	54,921
RSFSR	283,982	118,561	6,093	22,832	16,328	17,972	51,769	50,427
North-west	81,325	29,461	1,405	11,623	6,018	6,213	15,576	11,029
Volgo-Vyatka	24,978	10,248	832	1,484	1,198	2,272	4,192	4,752
Urals	48,678	19,628	1,679	4,049	3,287	3,064	8,472	8,499
RSFSR E–U	184,499	71,220	5,223	18,742	10,696	13,992	32,247	32,379
W. Siberia	22,463	10,200	432		2,001	1,574	3,545	4,711
E. Siberia	47,720	24,140	282	2,663	2,360	1,559	10,379	6,337
Far East	29,300	13,001	156	1,427	1,271	847	5,598	7,000
RSFSR Asia	99,483	47,341	870	4,090	5,632	3,980	19,522	18,048

Source: Vorob'yev, G. I. et al. (1979) Ekonomicheskaya geografiya lesnykh resursov SSSR, Moscow: Lesnaya Promyshlennost', various pages.

Note: Refers to industrial timber administered by Soyuzglavlesa (a division of Gossnab) for 1973 (See Vorob'yev et al. 1979, p. 193).

Table 6.21 Regional surplus/deficit (−) in mix of harvested timber, 1973 (thousand cu. m)

	Total	Saw logs	Peeler logs	Pulp-wood	Pit props	Constrn timber	Other c. timber	Fuelwood
USSR	13,294	8,752	−364	5,010	522	−1,143	5,246	−4,729
RSFSR	53,072	19,322	970	5,652	9,429	6,820	8,731	2,148
North-west	24,495	6,187	228	4,659	5,178	4,435	2,568	1,240
Volgo-Vyatka	5,626	1,304	548	−368	1,184	1,541	1,109	308
Urals	13,563	4,724	984	723	2,083	1,794	2,546	709
RSFSR E–U	30,218	4,951	809	6,378	7,164	6,714	3,155	1,047
W. Siberia	4,468	2,404	164	0	−87	444	530	1,013
E. Siberia	11,654	7,789	85	−80	2,058	600	1,010	192
Far East	6,732	4,178	−88	−646	294	−938	4,036	−104
RSFSR Asia	22,854	14,371	161	−726	2,265	106	5,576	1,101

Source: Vorob'yev, G. I. et al. (1979) Ekonomicheskaya geografiya lesnykh resursov SSSR, Moscow: Lesnaya Promyshlennost', various pages.

Note: Refers to industrial timber administered by Soyuzglavlesa (a division of Gossnab) for 1973 (See Vorob'yev et al. 1979, p. 193).

national output increased by 1.3 per cent. Western Siberia, on the other hand, produced 9.1 per cent of national timber output in 1985, an increase of 0.7 per cent over 1975, and an absolute increase of 0.5 million cu. m. Between 1960 and 1975, the Asian RSFSR increased its annual timber harvest by 44.2 million cu. m, and its relative share of the USSR's timber output from 24.7 per cent to 34.3 per cent. Between 1975 and 1985, however, the Asian RSFSR's output increased by 1.8 million cu. m and the region's share comprised 37.3 per cent of the national output. These figures suggest that the much-touted move of logging towards Siberia has been relatively stagnant since 1975, and that much attention

Table 6.22 Regional shift in industrial removals of timber, 1960–1985 (thousand cu. m)

	1960	% of total	1975	% of total	1985	% of total
USSR	369,550	100.0	395,039	100.0	367,962	100.0
RSFSR	336,365	91.0	366,900	92.9	337,275	91.7
North-west	89,829	24.3	97,300	24.6	87,141	23.7
Volgo-Vyatka	38,242	10.3	30,300	7.7	24,001	6.5
Urals	65,086	17.6	58,400	14.8	46,911	12.7
RSFSR E–U	244,769	66.2	231,100	58.5	199,622	54.3
Western Siberia	25,135	6.8	33,000	8.4	33,530	9.1
Eastern Siberia	45,416	12.3	69,200	17.5	69,308	18.8
Far East	20,640	5.6	33,200	8.4	34,451	9.4
(South)	15,637	4.2	28,600	7.2	29,251	7.9
(North)	5,003	1.4	4,600	1.2	5,200	1.4
RSFSR Asia	91,191	24.7	135,400	34.3	137,289	37.3

	Change 1960–75	% Change 1960–75	Change 1975–85	% Change 1975–85	Change 1960–85	% Change 1960–85
USSR	25,489	6.9	−27,077	−6.9	−1,588	−0.4
RSFSR	30,535	9.1	−29,625	−8.1	910	0.3
North-west	7,471	8.3	−10,159	−10.4	−2,688	−3.0
Volgo-Vyatka	−7,942	−20.8	−6,299	−20.8	−14,241	−37.2
Urals	−6,686	−10.3	−11,489	−19.7	−18,175	−27.9
RSFSR E–U	−13,669	−5.6	−31,478	−13.6	−45,147	−18.4
Western Siberia	7,865	31.3	530	1.6	8,395	33.4
Eastern Siberia	23,784	52.4	108	0.2	23,892	52.6
Far East	12,560	60.9	1,251	3.8	13,811	66.9
(South)	12,963	82.9	651	2.3	13,614	87.1
(North)	−403	−8.1	600	13.0	197	3.9
RSFSR Asia	44,209	48.5	1,889	1.4	46,098	50.6

Source: Narodnoye khozyaystvo SSSR v 1970 g, p. 225; Narodnoye khozyaystvo RSFSR v 1970 g, pp. 97–9; Narodnoye khozyaystvo SSSR v 1975 g, p. 273; Narodnoye khozyaystvo RSFSR v 1975 g, pp. 87–9; Narodnoye khozyaystvo SSSR v 1987 g, pp. 142–3; Narodnoye khozyaystvo RSFSR v 1987 g, pp. 102–7 (first handbook publication of regional timber removals since 1975).

is being paid to more effective use of accessible forests in Europe–Uralia, including those outside the RSFSR. Within the Asian RSFSR, the forests more accessible to Europe–Uralia or to Pacific export markets (especially Japan and China – Yakunin *et al.* 1987, pp. 49–54) appear to be undergoing a slight increase in utilization. Expansion of harvesting in the heart of Siberia, however, may have been curtailed as the economic advantages of logging more accessible forests elsewhere have come to be accepted by central administrators.

Wood chip production in the Soviet Far East grew rapidly in the 1970s primarily to serve the Japanese market (Yakunin *et al.* 1987, p. 28) but output has stagnated in the 1980s at approximately 777,000 cu. m (1985) and opportunities for expansion of foreign sales have been missed. In 1985, 227,000 cu. m of chips were consumed in the region's pulp and paper industry, and the remainder were available for export (Yakunin *et al.* 1987, p. 30). Sixty per cent of the region's chips are produced in Khabarovsk Kray, and 26 per cent in the Maritime Kray. Almost all the remainder originates in Amur and Sakhalin Oblasts. Adequate funding to convert wood waste and low-quality timber into wood chips for use within the region and for export has not been forthcoming, and represents a major form of potential growth for the Far Eastern forest sector (discussed in Yakunin *et al.* 1987, pp. 33–8).

Trade

The Soviet Union exported 18.1 million cu. m of roundwood in 1986 comprised mainly of sawlogs (9.2 million) and pulpwood (7.3 million) (*Vneshnyaya Torgovlya SSSR v 1986 g*, 1987, p. 27). The dominant share of the USSR's sawlog exports originates in the Far East and is mainly destined for Japan (5.3 million). Annual sales of sawlogs to China, however, suddenly increased by approximately 1 million cu. m per annum in 1984 and jumped again by nearly 600 cu. m in 1986 to reach a current level of 2.5 million cu. m, or approximately half of that annually purchased by Japan (*Vneshnyaya torgovlya SSSR v 1986, g*, 1987, p. 69). In 1986, China was the USSR's second largest customer for sawlogs, and third after Japan and Finland for all types of roundwood. Nearly three-quarters of Finland's total roundwood purchases of 3.6 million cu. m from the USSR, however, are pulpwood; in 1986, in addition to sawlogs, Japan purchased 1.2 million cu. m of pulpwood. Japan thus accounts for approximately one-third of Soviet roundwood exports.

China and Japan are served by Far Eastern timber supplies, supplemented by non-larch coniferous species from Eastern Siberia (Eronen 1983, p. 209); Finland's purchases emanate primarily from Europe-Uralia and in recent years, surprisingly also from Siberia. Although recent data on the regional origin of exported forest commodities are not available, the pattern prevailing in the 1970s (Table 6.23) suggests that, except for roundwood, the Far East has traditionally been a small contributor to Soviet forest exports. Recent changes appear to confirm that a stronger export role for the Far East is dependent on the strength of Japanese and Chinese markets for roundwood and chips (Japan). Export markets do not appear to be a factor in the development of comprehensive wood-processing facilities in the Far East.

Most Soviet timber is exported through a few ports in the Far East (Kanevskiy *et al.* 1975, p. 251). Of 8.9 million cu. m exported by water (accounting for approximately half of all roundwood exports) in 1972, for example, over half moved through three Far Eastern ports: Nakhodka (20 per cent), Vladivostok (17 per cent), and Vanino (16 per cent). Mago, Pos'yet and Sakhalin Island (as a whole) each handled 5–6 per

Table 6.23 Regional origin of exported commodities (% of total commodity exported)

	Round wood	Lumber	Chemical pulp	Paper/ paper- board	Plywood	Particle- board	Fibre- board
USSR	100.0	100.0	100.0	100.0	100.0	100.0	100.0
RSFSR	89.6	86.2	80.4	71.0	41.9	25.5	62.4
North-west	29.5	55.9	52.2	52.8	21.1	8.8	37.2
Urals	4.5	10.2	4.2	13.6	20.8	16.7	19.6
Eastern Siberia	10.00	16.8	12.6				5.6
Far East	45.6	3.3	11.4	4.6			

Source: Kanevskiy, M. V., et al. (1975) Lesnoy eksport SSSR, Moscow: Lesnaya Promyshlennost', pp. 110–11.

cent of Soviet log exports. Five other mainland Far Eastern ports accounted for a total of 15 per cent; Ust'-Kamchatka handled approximately 1 per cent. The Chinese market for Soviet logs has traditionally been served by rail shipments (Eronen 1983, p. 209) but the recent expansion in volume of Chinese purchases suggests that Soviet logs will move from Far Eastern ports into the Chinese eastern seaboard. Thus, log export ports which have served Japan in recent decades are now likely to redirect an increasing portion of their flows towards China. Japan, the Pacific market for Far Eastern chips, is served by Vostochniy and Vanino.

The Fourth Forest Resources Agreement, 1988–95 (earlier compensatory forest trade agreements, referred to as the K-S agreements, are discussed in Barr and Braden 1988, pp. 205–7) is indicative of the role envisaged by the USSR and Japan for the Far East's timber resources through the remainder of most of this century (SUPAR Report 1988, p. 15). Japan will obtain 12.3 million cu. m of timber (half to be delivered in the first five years), about 100 million tons of South Yakutian coal (by 1998), 8.2 million cu. m of chips and 3 million cu. m of hardwood pulp. The USSR will purchase US$ 200 million worth of equipment (Japan Lumber Journal, 31 January 1986, p. 15 cited in Barr and Braden 1988, p. 207). This latest agreement thus continues the patterns of forest resources-for-manufactured-goods compensation trade established between the USSR and Japan (discussed in Bradshaw 1987, pp. 212–14, 285–90) in the late 1960s which clearly reveal not only the USSR's dependency role vis-à-vis major western industrial nations, but the key role of the Far East in that pattern of dependency and peripherality. The current Soviet study of the establishment of special economic zones in the Far East (SUPAR Report 1988, p. 41) could lead to a more diversified export base for the Far East, but any significant impact of such a scheme appears unlikely to be felt before the first decade of the twenty-first century.

Discussion

When viewed in terms of utilization of allowable cut, harvesting in the Far East or remote areas of Eastern Siberia comprises only about one-third of the annual potential and large incremental supplies of roundwood that could be obtained from the region's existing growing stock. Greater harvesting in less-peripheral eastern regions with shorter transportation distances to European Soviet markets, however, is likely to occur long before many of the trees in the Far East are felled. Of even greater significance, however, is the observation that harvesting in Europe–Uralia comprised only three-quarters of the allowable cut; this means that 57.7 million cu. m more could have originated from this well-situated region and thereby obviated the need for a significant portion of the Far Eastern harvest not destined for Japan, China, or the diverse processing activities of Khabarovsk Kray and Sakhalin Oblast. Admittedly the portion of the European–Uralian coniferous allowable cut used, 88.5 per cent, was higher than that of deciduous species (63–9 per cent), but the volume of coniferous timber unutilized, 15.5 million cu. m, was nevertheless equal to approximately half the coniferous harvest of the Far East. Thus, although some of the Far Eastern harvest is consumed by local and regional users, including manufacturing plants inherited from the Japanese on Sakhalin Island, and by adjacent foreign markets, incremental timber needs could be satisfied in forests much closer to national markets and shipped over railroads other than the chronically over-used Trans-Siberian.

Acceptance of the idea of alternative regional sources for some of the timber currently harvested in the Far East, or likely to be harvested to meet the incremental needs of the national economy, seems to conform to the demands for intensification currently being promoted by General Secretary Gorbachev, and to the call by the 12th Five-Year Plan and its predecessor for greater production from plantations and more effective silviculture in Europe–Uralia, particularly in the provision of fibre for the pulp and paper industry. Furthermore, the incremental supplies of Soviet timber available in accessible regions closer to the market could more than offset the need to supply domestic markets from forests adjacent to the BAM. Although the BAM in the Far East theoretically can provide an additional 15–20 million cu. m per annum to the domestic timber supply, that amount of material or the 11 million cu. m actually originating there each year could easily be provided by coniferous and deciduous stands under-harvested in the principal cut in European-Uralian commercial forests (Group 3 and part of Group 2), or available from greater use of improvement felling in those same forests, or in the various European protection forests (Group 1). Investment in advanced equipment capable of producing structural lumber from small pieces of wood or of producing lumber from waste, i.e. technology developed in the western industrial economies to utilize timber from cut-over deciduous and coniferous forests, would be quite suitable for the accessible heartland forests of the European-Uralian USSR. Economic factors alone do not favour unrestrained expansion of Far Eastern forests, and actually support actions to the contrary.

Of the 3,000 million cu. m of timber available in the Far Eastern service area of the BAM, 20 million cu. m are designated as allowable cut but only 11 million cu. m are harvested annually (approximately one-third of the Far East's annual volume) (Chumin and Sheyngauz 1983, no. 147, p. 2). The annual rate of growth of harvesting in the Far Eastern service area of the BAM, however, has diminished from 17.2 per cent during the second half of the 1960s to 2.5 per cent a decade later (Chumin and Sheyngauz 1983, p. 2). At least in the Far East, the BAM is not entirely providing access to virgin stands of timber, but also to forests which were cut over in the 1950s (Komsomol'sk-Postyshevo), the 1960s (Izvestkovaya-Urgal) or even during the gold rush at the turn of the last century (Selemdzha river basin). Forests in the Tynda region have been extensively logged since construction of the BAM commenced. Furthermore, significant amounts of fuelwood and larch are not felled, and hence the full allowable cut is under-utilized in the immediate railway service area. The same problem will occur if forests far from the BAM are subjected to intensive use (Chumin and Sheyngauz 1983, p. 2). The key issue in the BAM forests, as throughout the Far East (Yakunin *et al.* 1987) is said to lie with development of comprehensive utilization of the growing stock – but this factor is cited in remedies for all other Soviet forests as well.

Although unwilling to conclude that harvesting of the BAM forests has repeated all the 'mistakes' which occurred when southern forests in the Far East were harvested, Chumin and Sheyngauz reveal that similar types of inappropriate practices are being repeated along the BAM. Silviculture in the Far East seems to be planned in relation to traditional labour-intensive techniques, whereas shortages of labour and policies to encourage labour-saving prevent the effective replenishment of felled timber. Forest-fire control has traditionally had low priority in the eastern USSR, but with development in new areas such as the BAM service zone, the occurrence of fires has increased despite investment in new suppression facilities. Chumin and Sheyngauz suggest that each rouble in forest-fire suppression in the Far Eastern taiga will facilitate timber regeneration 10–11 times more than the same rouble spent directly on silviculture. While recognizing that spending priorities in Far Eastern forestry cannot be altered without the involvement of central forestry agencies, Chumin and Sheyngauz point out that forest-fire expenditures are not included in the economic and social performance indices and are almost entirely absent from work plans. Hence they are absent from productivity indices, wages, etc. Hence, economic incentives promote activities other than forest-fire control, whereas in the Far East at least, better fire control would promote better forestry.

To the extent that some domestic timber consumers outside the Far East currently receiving Far Eastern timber and processed items could be more effectively supplied from forests in other regions, then utilization of some of the Far East's forests must be subject to an agenda unrelated to resource use *per se*. The core of the problem probably lies in the ability of extra-regional consumers not to be affected by the real cost of consuming timber or wood products from the Far East. Although the entire economy suffers from inflated costs in any major industry, the planned cost and profit indicators for enterprises traditionally insulate them from excessive transport charges or higher material costs incurred from consumption of timber harvested in peripheral, difficult environments. Furthermore, the Far Eastern forest is so heavily mature and over-mature that it must appear to planners as a 'free' good which will continue to deteriorate on the stem if not harvested. Many foresters ask why capital should be invested in plantations and effective silviculture in Europe-Uralia when large amounts of wood are waiting to be felled in the Far East. Some analysts argue that exploitation of the eastern forests will buy time for implementation of better forest management in Europe-Uralia. These arguments, however, ignore the enormous annual expenditures associated with such expediency, and suggest that a 'quick and direct' approach will enhance the well-being of those currently in authority and leave difficult solutions to the problem to be faced by their successors. As shown above, accessible forests of the Far East, including many 'new' areas served by the BAM are depleted or cut over and now require replenishment. From this perspective, forestry in the Far East reflects arcane political and administrative processes not usually evaluated in assessment of regional development.

Fishing

The Soviet Union accounted for approximately 10.5 million tons or 12 per cent of the world's harvest of ocean products (Sysoyev 1988, p. 92) in 1985, a share which has been growing steadily. The USSR, like the world fishery, has increasingly been expanding its fishing activities away from internal waterbodies and into the distant reaches of the world's oceans. The world now obtains some 88 per cent of its fish products from ocean fishing (a proportion probably only slightly higher than that of the USSR – *Soviet Geography* 1978, vol. 19, no. 6, p. 426). Prior to 1939, over 83 per cent of the world catch came from oceans north of 30°N latitude; today, 54 per cent is derived from the northern zone, 27 per cent from the centre, and 19 per cent from the south. In 1938, 55 per cent of the world's catch originated in the Atlantic Ocean, 40 per cent in the Pacific, and 5 per cent in the Indian; the remainder was obtained from freshwater fisheries. In 1985, the Pacific Ocean accounted for 55.3 per cent, the Atlantic 28.1 per cent, but the Indian for only 5 per cent. The remainder from fresh water was approximately equal to that of 1938 although the overall world harvest had increased more than fourfold (Sysoyev 1988, p. 93).

Output

The USSR has participated in the expansion of the world's fishing industry into the far reaches of the globe. The Soviet Far East now provides approximately 40 per cent of the country's fish catch (SUPAR 1988, p. 8), up from 33 per cent reported for 1975 (*Soviet Geography* 1978, vol. 19, no. 6, p. 427). Singur (1988, p. 98) suggests that the region's output of fish and marine products will increase 28 per cent by the year 2000 and will account for 43 per cent of the nation's output. In 1975, the North Pacific fishery yielded 3.4 million tons, of which 1.4 originated in the Sea of Okhotsk, 0.6 million in the Bering Sea and 0.42 in the Northern Kuril Islands. The additional 1 million tons is reported to have originated mainly off the coast of North America (*Soviet Geography* 1978, vol. 19, no. 6, p. 427). The SUPAR data support the estimate that the Far East now provides the USSR with 4.6 million tons of fish and marine products (*Narodnoye khozyaystvo SSSR za 70 let*, 1987, p. 6) out of a total in 1986 of 11.4 million tons, a figure which was expected to decline to 10.8 million in 1987 probably as pollution and other forms of environmental degradation continue to destroy large inland and offshore areas of the European USSR including highly publicized water bodies in Asia such as Lake Baykal and the Aral Sea (Solecki 1979, pp. 114–16).

The growth in the Soviet Far East has been facilitated by the expansion of the USSR's Pacific fleet and the development of related comprehensive fishing, processing, storage, maintenance and service facilities which focus on some of the major mainland ports of the Far East, particularly on the Seas of Okhotsk and Japan, and on Sakhalin and the Kuril Islands and Kamchatka. In his 1979 review of the USSR fishing industry based extensively on the writings of Sysoyev, Solecki (1979, pp. 97–123) reported twenty-one sea ports sustaining the two major fishing fleets of the Soviet fishing industry. Seven were located in the Far East: Valdivostok, Nakhodka, Nevel'sk, Petropavlovsk-Kamchatskiy, Okhotsk, Mago and Magadan. While these may be major administrative centres for the fishing industry, an older volume by Kustov (1968, pp. 92–6), identifies additional fishing ports in the Far East at Soviet Harbour (near Vanino), southern Sakhalin Island (including Korsakov, Kholmsk and Aleksandrovsk-Sakhalinskiy), the southern and northern-most Kuril Islands, southern Kamchatka and a few relatively minor fishing ports or harbours at occasional intervals along the extensive Far Eastern Sea coast. Dienes (1987, p. 217) claims that, of the sixty-nine urban-type settlements lacking city status out of a total of ninety-four urban places on the Pacific coast in 1983, 'the great bulk . . . must be fishing settlements'. All maritime activities occupy 8 per cent of the Far East's employed population, but 20 per cent of the coastal areas whose employment opportunities are dominated by fishing and marine activities (Revaykin and Olenicheva 1987 cited in *Referativniy zhurnal*, 1988, vol. 3, p. 40).

The main foci of the post Second World War growth in fish processing were ocean-going processing facilities whereas future growth, although likely to incorporate agreements with foreign nations for access to the increasing number of 200-mile economic zones, will place more emphasis on the country's coastal waters and those within its own 200-mile limit despite Singur's (1988, p. 95) claim that only 60 per cent of the Pacific Ocean's potential harvest is utilized. Sobolev (1988, cited in *Referativniy zhurnal*, 1988, p. 39) claims the waters within this limit could provide 90 per cent of the Far Eastern catch. Improved rail–seaport interfaces, new onshore fish-processing facilities, expanded ship repair installations, and better navigation and fish detection equipment are planned for the region and its fishing fleet. Changes envisaged for the Far Eastern fishing industry (Chichkanov and Orlov 1983, pp. 67–9) thus involve to a large degree a major reorientation towards coastal waters and the inshore fishery. The industry in the future will harvest species hitherto largely ignored and is likely to develop economic activities across a wider spectrum of the settlement system.

Future changes are also to stress a comprehensive, i.e. integrated, approach to growth of the fishing industry through expanded investment in fish breeding and aquaculture at coastal installations involving the culture of relatively valuable commodities and species (Pistun 1984, p. 189; Chichkanov and Orlov 1983, pp. 67–8). At the moment, aquaculture is significant in the USSR for assisting in the catch of half (10–15,000 tons) of all sturgeon, and 6–8,000 tons of Far Eastern salmon (Sysoyev 1988, p. 94). Such activity is expected to increase sharply in the future, particularly in the Far East, and to enhance the catch of Pacific salmon, Okhotsk herring, marine invertebrates and plants (Sysoyev 1988, p. 95).

The growth of the Pacific fishing fleet and attendant processing facilities, however, should not be seen as evidence that the Soviet Far East has an ineluctable role to play in the future supply of fish to the Soviet domestic market. As with the slow, but nevertheless important, growth of specialized high performance forest plantations in European-Uralian USSR to supply specific fibre requirements to the wood-processing industry, the Soviet Union in the 1970s and 1980s has worked to develop its inland waters (Solecki 1979, pp. 106–9) in major populated regions. Although the harvest from inland waters has not been an unequivocal success due to the depletion of stocks in polluted and over-fished areas, the USSR's ability to recognize that major amounts of fish could be obtained nearer to market areas suggests that fishing, like forestry, is not an uncontested element of the Far East's economic base.

Trade

One of the four tasks reported by Solecki (1979, pp. 104, 117) as guiding the recent development of the Soviet fishing industry has been to help the USSR achieve a positive trade balance. The main customers for canned fish are within COMECON, although higher-value products like crab earn a disproportionately high income in western countries (Solecki 1979, pp. 117–19). The USSR is a net exporter of fish and in 1986 exports exceeded imports by more than 4:1. Fish and fish products comprise 0.6 per cent of the USSR's exports, whereas timber and major wood products make up 3.1 per cent (*Vneshnyaya torgovlya SSSR v 1986 g.* 1987, p. 19). Approximately half the value of imported fish accrues to the United States, presumably from fishing in American North Pacific waters, and probably is processed by the Far Eastern fleet (USSR fish exports to the USA are approximately one-tenth the size of imports). In this sense, the Far East is a major region of entry for imported fish prior to its dissemination throughout the USSR or to export markets. In fish exports as in timber, however, the Far East appears to be most closely linked with Japan which is the USSR's number one purchaser of fish by weight, and prime international purchaser of canned crab by number of standard cans (France, which in 1975 took over half of Soviet crab exports according to Solecki (1979, p. 118) is now in second place) (*Vneshnyaya torgovlya SSSR v 1986 g.* 1987, p. 74). Mongolia is the only Asian consumer of Soviet canned fish but buys less than 2 per cent of the USSR export of that commodity. The United States is a small importer of Soviet fish but its imports are still smaller than those of Singapore. Japan and Singapore are thus the major Asian Pacific consumers of Soviet fish by weight. The role of the Far East in Soviet fish exports thus not only is small in all cases except canned crab, it is also relatively insignificant in terms of the value of national exports and pales in comparison to that of forest products, particularly timber. The 1986 value of crab exports to Japan was almost equal to that of wood waste (chips) (*Vneshnyaya torgovlya SSSR v 1986 g.* 1987, p. 238).

Discussion

The Far Eastern fishing industry is relatively much more important to the USSR than forestry and wood-processing. The fishing industry is undergoing reorientation towards the greater use of the Far East's coastal waters and resources of the continental shelf, although pond fisheries are to expand everywhere in the USSR during the current five-year plan (*CDSP*, 1985, vol. 37, no. 48, p. 18), and fish breeding and fishing are supposed to increase in inland waters. Nevertheless, the Far Eastern fishing industry has fewer spatial competitors for investment capital than the region's forests. Despite reorientation in technique, processing and species harvested, the Far Eastern fishing industry has a well-developed scientific, technical, industrial, personnel and settlement base to draw upon, and has a clearly defined profile in the Soviet food industry. The country already has accepted and accommodated the distant location of the Far Eastern fishery, but is still not prepared to do the same in forestry. Forestry exhibits distance-decay characteristics away from the heartland characterized by poor reforestation and unsophisticated sawmilling.

Fishing is relatively independent of the large landmass intervening between the Pacific coast and Moscow because of the nation's constant quest for sustained and incremental supplies of protein. The Far Eastern fishing industry is largely intended to supply the domestic market although delicacy items such as crab are a small source of foreign exchange. In sum, however, the fishing industry scarcely impacts upon the vast landmass of the Far East comprising 27.7 per cent of the USSR and is really independent from it. Despite its relatively tenuous hold on fewer than 100 coastal urban settlements, however, the Far Eastern fishing industry is indispensable to the Soviet Union's food supply and thus comprises an important element in the region's strategic significance to the USSR.

Associated settlements are also a visible reminder to neighbouring countries of the USSR's determination to retain control over the area, even if refusing to return the southern Kuril Islands to Japan continues to hinder economic relations between the two countries. Ironically, labour shortages in the forest industry enhance the region's strategic vulnerability by weakening the settlement system and leading the Soviet Union to increase the presence of 'guest workers'. Timber sold to China may eventually be harvested by temporary Chinese labour operating in Far Eastern forests (Backman 1988).

Conclusions

The forest and fishing industries of the Soviet Far East reflect Moscow's selective approach to investment in the region. Where the region can provide materials like food unavailable elsewhere, a modern capital-intensive industry has emerged with a national profile able to overcome the great distances to domestic markets. Where materials like timber for heartland consumption are more accessible from other Soviet regions, central administrators have made token investments in the Far East to sustain regional demands for timber and some processed items, and to meet export opportunities in important Pacific markets. Major processing investments seem to focus on a few nodes in the zone of intensive development adjacent to north-eastern China (Bradshaw 1988, p. 371), but investment opportunities along the new trans-regional railway have been foregone in favour of locations closer to the heartland. The overall costs of development including infrastructure and a stable labour force are too great to warrant massive national subsidies. Genial relations with China probably have diminished the justification for priority economic development of the Far East. Growing national shortages of labour, or of labour willing to relocate towards the harsh eastern environment, seem further to encourage central administrators to seek alternative locations. Bradshaw's (1988, p. 371) assessment of the region is confirmed by forestry and fishing: 'The Soviet Far East can be characterized as being a highly truncated regional economy, dependent on external sources of capital and equipment, specializing in the production of a limited number of natural resources, with the bulk of economic activity concentrated in urban settlements in the southern parts of the region.'

Bond (1987, p. 507) notes that 'the region is justifiably considered by Soviet planners to be a graveyard for investment capital'. He adds that transport costs comprise a larger proportion of commodities delivered to the region than elsewhere, capital and labour are less productive and returns on investment are slower than in other major regions, and capital productivity has declined since the 1970s. In both forestry and fishing, future harvesting will occur in areas hitherto relatively unimportant. This is the most important element common to both sectors in the Far East. Incremental supplies of timber in the future will be drawn increasingly from mountain forests; larger proportions of the fish catch are expected to be derived from the continental shelf and coastal waters. In both industries, these relatively unfamiliar sources will necessitate new techniques and equipment and will likely produce lower returns on capital investment in the initial periods of activity. These expected developments in both industries are likely to exacerbate the existing problems of low productivity in eastern regions described by Bond.

The present Far Eastern forest industry seems strongly to support both

Bond and Bradshaw, whereas fishing now has a national importance and momentum which to date has overcome the regional malaise common to other sectors. If the present Soviet leadership is able to restructure the country's economy, then the region's fishing industry should continue to prosper. The Far Eastern forest industry, however, is unlikely to deviate significantly from its present narrow structural and spatial pattern even if its modest goals for the year 2000 are achieved. On the other hand, if environmental conditions in the European-Uralian forests become irrevocably degraded, harvesting in all eastern forests, including those of the Far East, may significantly increase. Environmental degradation of Soviet inland waters outside of the Far East could similarly affect the region's fishing industry and cause its national importance to become even more pronounced. The processes and trends identified in this chapter, however, support the conclusion that, at the present time, the Far Eastern fishing industry seems to have a guaranteed strong future, whereas the region's forestry and wood-processing industries, except for roundwood exports, are likely to remain relatively undeveloped and insignificant.

Bibliography

Antsyshkin, S. P., *et al.* (1965) *Spravochnik lesnichego* (2nd edn) Moscow: Lesnaya Promyshlennost'.

Backman, C., Personal Communication, 19 August 1988.

Barr, B. M. (1970) *The Soviet Wood-Processing Industry: A Linear Programming Analysis of the Role of Transportation Costs in Location and Flow Patterns*, Toronto: University of Toronto Press.

——, and Braden, K. (1988) *The Disappearing Russian Forest: A Dilemma in Soviet Resource Management*, Totowa, N.J.: Rowman & Littlefield.

Bond, A. (1987) 'Spatial Dimensions of Gorbachev's Economic Strategy', *Soviet Geography*, vol. 28, no. 7, pp. 490–523.

Bradshaw, M. J. (1987) 'East-West Trade and the Regional Development of Siberia and the Soviet Far East', Unpublished doctoral dissertation, Vancouver, B.C.: University of British Columbia, Department of Geography.

—— (1988) 'Soviet Asian-Pacific Trade and the Regional Development of the Soviet Far East', *Soviet Geography*, vol. 29, no. 4, pp. 367–93.

Chelyshev, V. A. (1986) 'Razvitiye lesokhozyaystvennogo proizvodstva Yakutskoy ASSR v 1965–1983 gg. i voprosy uluchsheniya lesopol'zovaniy', *Tr. Dal'nevost. NII les x-va*, no. 28, pp. 146–51. Quoted in *Referativniy zhurnal. Geografiya. 07E. Geografiya SSSR. Vypusk svodnogo Toma*, 3 E276, 1988, p. 40.

Chichkanov, V., and Orlov, V. (1983) 'Biologicheskiye Resursy Okeana', *Planovoye khozyaystvo*, no. 1, pp. 62–9.

Chumin, V., and Sheyngauz, A. (1983) 'S Prirodoy ne Sporyat. Tak li nado Osvaivat' Lesa Zony BAMa?' *Lesnaya Promyshlennost'* no. 147, 10 December 1983, p. 2.

CDSP (Current Digest of the Soviet Press). (Various dates and years), weekly publication. Columbus, Ohio.

Dienes, L. (1985) 'Economic and Strategic Position of the Soviet Far East', *Soviet Economy*, vol. 1, no. 2, pp. 146–76.

_____ (1987) *Soviet Asia: Economic Development and National Policy Choices*, Boulder, Co: Westview Press.

_____ (1988) 'A Comment on the New Development Program for the Far East Economic Region', *Soviet Geography*, vol. 29, no. 4, pp. 420–22.

Eronen, J. (1983) 'Routes of Soviet Timber to World Markets', *Geoforum*, vol. 14, no. 2, pp. 205–10.

_____ (1985) *Future Prospects of Pulp and Paper Industry in the Soviet Union*, Helsinki: Report for INDUFOR, mimeo.

Fenton, R. T. and Maplesden, F. M. (1986) *The Eastern USSR: Forest Resources and Forest Products Exports to Japan*, Forest Research Institute Bulletin no. 23. Rotorua, New Zealand: New Zealand Forest Service.

Gladyshev, A. N., *et al.* (1974) *Problemy razvitiya i razmeshcheniya proizvoditel'nykh sil Dal'nego Vostoka*, Moscow: Mysl'.

Japan Lumber Journal (Various issues and years), Tokyo, Japan: Central PO Box 1945, Tokyo 100–91.

Kanevskiy, M. V., *et al.* (1975) *Lesnoy eksport SSSR*, Moscow: Lesnaya Promyshlennost'.

Kol'tsov, V. V. (1987) 'Ispol'zovaniye vneshneekonomicheskikh svyazey dlya povysheniya effektivnosti ekonomiki Dal'nego Vostoka', *Probl. soversh. khoz. mekhanizma region. ekon.* Vladivostok 1987, pp. 126–9. Quoted in *Referativniy zhurnal. Geografiya. 07E. Geografiya SSSR. Vypusk svodnogo toma. 1 E229*, 1988, p. 35.

Kozhin, V. M., and Styazhkin, V. P. (1976) *Sebestoimost', tseni i rentabelnost' na lesozagotovkakh*, Moscow: Lesnaya Promyshlennost'.

Kustov, E. d. (1968) *Geografiya rybnoy promyshlennosti*, Moscow: Pishchevaya Promyshlennost'.

Lesnaya entsiklopediya (1985), vol. 1. Moscow: Sovetskaya Entsiklopediya.

Lesnaya entsiklopediya (1986), vol. 2. Moscow: Sovetskaya Entsiklopediya.

Lesnoye khozyaystvo (1988), no. 1, p. 60.

Lesnoye khozyaystvo SSSR (1977) Moscow: Lesnaya Promyshlennost'.

Lobovikov, T. S., and Petrov, A. P. (1976) *Ekonomika kompleksnogo ispol'zovaniya drevesiny*, Moscow: Lesnaya Promyshlennost'.

Molodenkov, L. V. (1987) 'Problemy osvoeniya novykh khozyaystvennykh territoriy v Sibiri i na Dal'nem Vostoke', *Osvoeniye nov. khoz. territoriy v Vost. R-nakh RSFSR*, Novosibirsk, 1987, pp. 5–10. Quoted in *Referativniy zhurnal. Geografiya. 07E. Geografiya SSSR. Vypusk svodnogo Toma, 3 E237*, 1988, p. 35.

Mote, V. L. (1987a) 'The Amur–Yakutsk Mainline: A Soviet Concept or Reality' *The Professional Geographer*, vol. 39, no. 1, pp. 13–23.

_____ (1987b) 'Regional Planning: The BAM and the Pyramids of Power'. U.S. Congress, Joint Economic Committee, *Gorbachev's Economic Plans*, vol. 2, pp. 365–81. Washington, D.C.: U.S. Government Printing Office.

Mozhayev, B. (1987) 'Puteshestviye v lesnoye golovotyapstvo', *Literaturnaya gazeta*, 10 June 1987, p. 12.

Mozhin, V. P. (1980) *Ekonomicheskoye razvitiye Sibiri i Dal'nego Vostoka*, Moscow: Mysl'.

Mugandin, S. I. (1977) *Povysheniye effektivnosti lesopil'nogo proizvodstva*, Moscow: Lesnaya Promyshlennost'.

Narodnoye Khozyaystvo RSFSR za 70 let (1987) Moscow: Finansy i Statistika.

Narodnoye khozyaystvo RSFSR. Statisticheskiy yezhegodnik (Various years), Moscow: Statistika (after 1980: Finansy i Statistika).

Narodnoye khozyaystvo SSSR za 70 let (1987) Moscow: Finansy i Statistika.

Narodnoye khozyaystvo SSSR. Statisticheskiy yezhegodnik (Various years), Moscow: Statistika (after 1980: Finansy i Statistika).

Pistun, N. D., *et al.* (1984) *Ekonomicheskaya geografiya SSSR. Rayonnaya chast'*, Kiev: Golovnoye Izdatelstvo Izdatel'skogo Ob'edineniya 'Vishcha Shkola'.

Pravda (1987) 26 August, Moscow.

Revaykin, A. S. and Olenicheva, M. R. (1987) 'Problemy trudoobespecheniya otrasley primorskoy spetsializatsii na Dal'nem Vostoke', *3 S'ezd sov. okeanol. Leningrad, 14–19 dek., 1987. Tez. Dokl. Sekts*, Leningrad. Quoted in *Referativniy zhurnal. Geografiya. 07E. Geografiya SSSR. Vypusk svodnogo Toma*, 3 E270, 1988, p. 40.

Singur, N. (1988) 'Dal'niy Vostok: kompleksnoye razvitiye proizvoditel'nykh sil', *Planovoye khozyaystvo*, no. 3, pp. 94–8.

Sinitsyn, S. G. (1976) *Lesnoy fond i organizatsiya ispol'zovaniya lesnykh resursov SSSR*, Moscow: Lesnaya Promyshlennost'.

_____ *et al.* (1979) *Gornye Lesa*. Moscow: Lesnaya Promyshlennost'.

Sobolev, Yu. A. (1988) 'Problemy ekonomicheskogo i sotsial'nogo razvitiya Dal'nego Vostoka. (O dolgovremennoy gosudarstvennoy programme kompleksnogo razvitiya proizvoditel'nykh sil Dal'nevostovchnogo ekonomicheskogo rayona, Buryatskoy ASSR i Chitinskoy Oblasti na Period do 2000 g.)', *Geografiya v Shkole*, no. 1, pp. 9–14. Quoted in *Referativniy zhurnal. Geografiya. 07E. Geografiya SSSR. Vypusk Svodnogo Toma*, 5 E261, 1988, p. 38.

SUPAR (Center for Soviet Union in the Pacific–Asian Region. University of Hawaii at Manoa), (1988), *Report-Chronicle*, no. 5, 42pp.

Solecki, J. J. (1979), 'A Review of the USSR Fishing Industry', *Ocean Management*, no. 5, pp. 97–123.

Soviet Geography (Published ten times per year), Washington, D.C.: V. H. Winston & Son, Inc.

Sysoyev, N. (1988), 'Osvoeniye i ispol'zovaniye bioresursov Mirovogo Okeana', *Planovoye khozyaystvo*, no. 2, pp. 92–6.

Timofeyev, N. V. (ed.) (1980) *Lesnaya industriya SSSR*, Moscow: Lesnaya Promyshlennost'.

Vneshnyaya torgovlya SSSR v 1986 g. (1987) Moscow: Finansy i Statistika.

Vorob'yev, G. I., *et al.* (1979) *Ekonomicheskaya geografiya lesnykh resursov SSSR*, Moscow: Lesnaya Promyshlennost'.

Yakunin, A. G., *et al.* (1987) *Lesnaya industriya Dal'nego Vostoka*, Moscow: Lesnaya Promyshlennost'.

Chapter seven

The South Yakutian Territorial Production Complex

Victor L. Mote

During the latter half of the Brezhnev era, the Baykal–Amur Mainline (BAM) railway construction project was a permanent feature in the Soviet press. Since the completion of track-laying in late 1984 and the installation of the Gorbachev administration in early 1985, however, the railway and its ancillary regional development projects have received far less attention.[1]

Originally envisioned as a catalyst for economic expansion in the Soviet Far East as well as a magnet attracting hard currency for the benefit of the entire country, BAM suffered from the hardships of a changing world economy. As the railway was built, both domestic and international marketing potentials and priorities shifted. In part an indirect offshoot of the OPEC oil embargo and the world-wide materials' crises that characterized the 1970s, BAM in the 1980s found itself basically built but without markets for the potential commodities of its service area. Intervening resource opportunities had arisen that all but obviated much of BAM's materials' potential. The production of West Siberian oil, originally projected to be the main cargo hauled by the BAM, became so uneven that it is now doubtful that much of it will ever be exported to customers in the Pacific basin.

Even if petroleum production in West Siberia had continued to expand, it would have had to cope with an international glut of the resource. Conservation and substitution of energy and other resources (steel, for instance) in the United States, Japan and other developed countries contributed to an entirely new economic orientation within the Organization for Economic Co-operation and Development (OECD), including all of the world's developed market economies except Finland. In many of these countries, high-technology industries superseded smoke-stack industries, which the native raw materials of the BAM require. Indeed, as Alvin and Heidi Toffler have noted in their internationally acclaimed book, *The Third Wave* (New York, 1980), a work well known to Gorbachev, the very nature of the Industrial Revolution may have changed, leaving the USSR languishing with a smoke-stack economy in the trough of the 'second wave', surrounded by a sea of technically advanced countries riding the crest of the 'third wave'.

There were gluts of other resources as well. An over-supply of copper at least temporarily nullified the need for mining that metal in the vicinity of Udokan. At one time estimated to be one of the most promising 'profit-making' resources in the BAM zone[2], the extraction of asbestos at Molodezhnoye was faced with an international movement to ban the mining of the mineral because of its cancer-causing properties; moreover, as late as 1987 a thirteen-mile (twenty-km) rail spur was needed between the main BAM track and the Molodezhnoye minehead.[3] In fact, among the eight territorial-production complexes envisioned by planners in the 1970s,[4] with the exception of Komsomol'sk, only one, the South Yakutian, was genuinely under way, and it was in trouble.[5]

Yakutia and the BAM

The Yakut Autonomous Soviet Socialist Republic (ASSR) is the second-largest administrative-territorial unit in the USSR after the RSFSR (even larger than the Kazakh Republic), comprising around 2 per cent of the earth's landmass. Such a gargantuan area is bound to include a rich bounty of resources, and Yakutia does. But the Yakut Republic suffers from inaccessibility and physical harshness that can be equalled only on the continent of Antarctica. (Its extensive thermokarst landscapes are reputedly the only terrestrial analogs of the planet Mars.[6]) For these and other justifiable reasons, with 3.1 million sq. km (1.2 million sq. miles) and only 984,000 people, Yakutia is one of the most sparsely populated regions on earth, averaging only one inhabitant for every 3 sq. km. The actual densities are even less than that because 68 per cent of the population lives in cities (180,000 in Yakutsk alone).[7]

During the 1970s, immigration contributed over half the population growth of the Yakut ASSR. The titular ethnic group, the Yakuts, a Turkic-speaking, Mongoloid people, who themselves migrated into the region some 700 years ago, witnessed their share of republican population fall from 46.4 per cent in 1959 to 36.9 per cent in 1979. The proportion of Russians and Ukrainians, by contrast, increased to almost 56.4 per cent. Three-quarters of the Yakuts were rural whereas most of the Slavs were urban.[8] The vast proportion of the new Slavic arrivals were BAM workers and their families, and they resided in South Yakutia.

Throughout its history, the Yakut ASSR has been extremely isolated. Supplying the region can be extraordinarily complex and, always, expensive. 'The transport network is characterized by very long links joining small nodes.'[9] Dienes discovered that in terms of the average level of supply of consumer goods and services, this ASSR is in last place among provinces of the RSFSR. 'Per rouble of wages paid out, the Yakut ASSR produces seven times fewer consumer goods than the RSFSR average and six times less than Kamchatka Oblast.'[10] Yakutian shops may lack even the most essential items, such as soap and toothpaste for months and 'even years'.

Part of the supply problem rests with the fact that regional imports and exports use the same transport facilities, with imports far outweighing exports. Three-fifths of the republic's incoming freight is shipped over the Lena river by way of Ust'-Kut's river port of Osetrovo. Navigation of the Lena's main channel lasts an absolute maximum of six months (May to October), but on some of its larger tributaries (the Aldan, for example), because of shoaling and irregular flow, the season can be curtailed to as few as 10 to 12 days.[11] Such irregularities encourage excessive breaks-in-bulk that are 2.7 times the average for the country.[12] Recently, the Lena has been able to cope with only 80 per cent of its potential traffic, in part because of incapacities at Osetrovo. Robert North has shown why: (1) the river suffers from low-water problems; (2) moorage access in Osetrovo requires consistent dredging; (3) poor co-ordination of work between river and railway officials; (4) traffic congestion due to BAM service area construction.[13]

The rest of the Yakutian incoming traffic comes via the so-called Little BAM railway (18 per cent); by motor roads (13 per cent); by ship over the Northern Sea Route to Tiksi (8 per cent), and by aircraft (1 per cent).[14] Even here there are problems. In service for only a decade, the single track of the Little BAM, by 1985, was already used in excess of its capacity and required capital maintenance.[15] Little BAM railway officials demanded a second track, pointing out that freight was projected to increase by 40 per cent by 1990.[16] The two-lane asphalt highway from Never to Yakutsk, with only ten filling stations and one repair garage (in Yakutsk) over its entire length (740 miles or 1,177 km), employed 9,000 new drivers during the 1980s in order to meet the excessive demand.[17] The opening of the navigation season on the Northern Sea Route often coincides with low water on the Lena, Yana and Kolyma rivers, requiring precise co-ordination. Lastly, aviation obviously suffers from severe weather and fog problems.

Given these difficulties, together with record lengths of haul, transport costs in Yakutia range from 50 to 70 per cent of the cost of retail value versus a Soviet average of only 10 to 12 per cent.[18] Because of deficiencies at Osetrovo and other freight-handling facilities, 2,000 million roubles ($2,400 million 1985 US dollars) are spent each year on the storage of surplus supplies within the region.[19]

South Yakutia and the AYAM

The fastest growing sub-region of Yakutia is in the extreme south of the republic. This is an area that is clustered about the termini of the Little

BAM railway and the coalfields of Neryungri and Chul'man, which in the mid–1980s contained around 105,000 people.[20] The population was concentrated primarily in seven settlements of urban type, including Neryungri (1987 population 68,000), Chul'man (12,000), the coal-mining headquarters at Ugol'naya (estimated 1988 population under 500 permanent residents), the railhead of Berkakit (estimated 1988 population over 5,000); Serebyannyy Bor, site of the Neryungri central electric station (with perhaps 1,500 people in 1988); and two distant stations, Zolotinka and Nagornyy (combined population estimated at 5,000). Since 1978, when the first permanent high-rise apartments were built, the conurbation has grown by 10,000 people per year, the birth rate supposedly accounting for 5,000 to 6,000 of these. The population that is projected for the zone in the year 2000 is 200,000 (Figure 7.1).[21]

Along with the existence of the coalfields and the Little BAM, South Yakutia's recent explosive growth has been augmented by the arrival of new construction workers bent on extending the railway northwards to Yakutsk. The Little BAM is the first leg of a long-projected meridional trunkline running between the Trans-Siberian rail station of Bamovskaya and the Yakutian capital. This expanded version has been called AYAM for Amur–Yakutsk Mainline, a 1,230-km (760-mile) railway with important implications for the future development of the republic as a whole. The Little BAM is 402 km (250 miles) long, leaving an 830-km (515-mile) gap between Neryungri and Yakutsk. The original plan was to build roughly half of the AYAM to the Aldan river crossing of Tommot by 1990 and to finish the project by 1995. As of the spring of 1988, only 40 km of rails had been laid, and construction on the line had been constantly interrupted for want of sufficient funds.[22]

The railway is critical not only to the development of Yakutia but also to the fortunes of the burgeoning South Yakutian Territorial-Production Complex (TPC). Regional party officials, whose opinions undoubtedly are not entirely unbiased, have emphasized that the AYAM is virtually the only solution to Yakutia's problems of inaccessibility. Transport economists have reckoned that just completing the railway will: (1) save the ASSR 100 million roubles ($120 million 1985 US dollars) per year; (2) increase transport capacity and real use; (3) reduce traffic on the overworked Lena route; (4) reduce throughflow at Magadan's port of Nagayevo; and (5) divert some of the traffic now moving expensively over the Never-Yakutsk highway to the railway.[23] In addition to its basic supply function, passing along the spine of the South Yakutian TPC, the AYAM will serve as a locus for the regional development of a myriad of new (and old) resources all the way to Tommot. In fact, without the existence of the AYAM at least as far as the Aldan, it is difficult to envision a thriving TPC in South Yakutia (Figures 7.2 and 7.3).

Unfortunately, the construction of the AYAM has been faced with

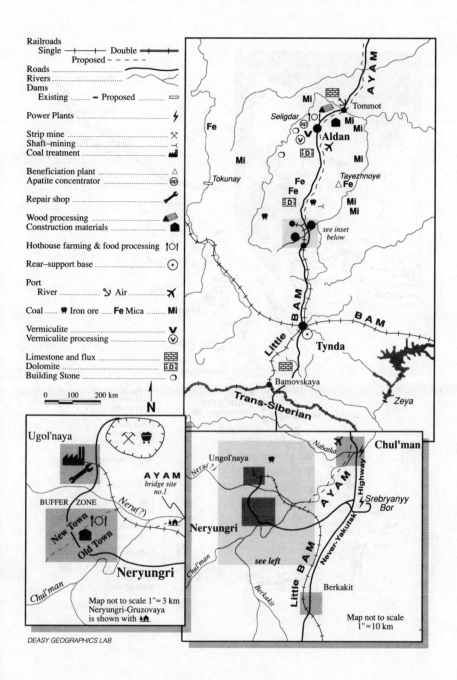

7.1 South Yakutian territorial production complex

7.2 The Amur–Yakutsk mainline

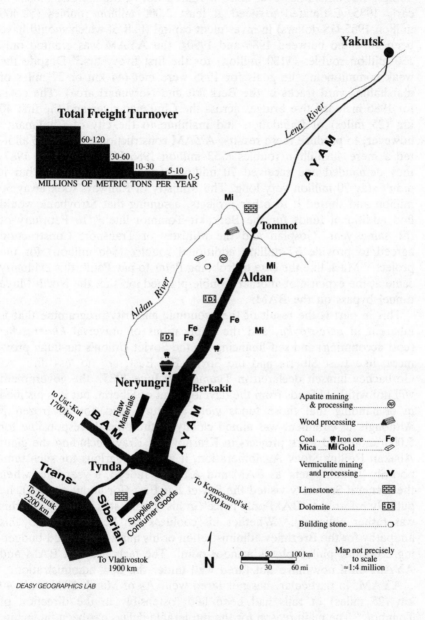

Total Freight Turnover

60-120

30-60

10-30

5-10

0-5

MILLIONS OF METRIC TONS PER YEAR

Yakutsk

Lena River

A Y A M

Mi

Tommot

Aldan River

Mi

ap

V   Aldan

D

Fe

Fe   Fe

Mi

Neryungri

Berkakit

to Ust'-Kut
1700 km

B A M

Raw
Materials

A Y A M

Trans-

Tynda

To Komosomol'sk
1500 km

Siberian

Supplies and
Consumer Goods

To Irkutsk
2200 km

To Vladivostok
1900 km

N

Apatite mining
& processing............ap

Wood processing ....

Coal ......   Iron ore ...... Fe
Mica ..... Mi Gold ....

Vermiculite mining
and processing............V

Limestone ....

Dolomite .... D

Building stone ....

0   50   100 km

0   30   60 mi

Map not precisely
to scale
≈1:4 million

DEASY GEOGRAPHICS LAB

7.3 Amur–Yakutsk mainline projected freight flows (c. 2000)
Source: *Izvestiya*, 20 March 1985

nothing but difficulties ever since it roared out of the starting blocks in early 1985. Estimated to need at least 2,000 million roubles ($2,400 million 1985 US dollars) in investment capital (half of which should have been allocated between 1986 and 1990), the AYAM was granted only 150 million roubles ($180 million) for the first five years.[24] Despite the weak commitment, the goals for 1985 were met (44 km or 27 miles of marshalling-yard tracks in the Berkakit and Neryungri area). The goals for 1986 included five bridges across the Chul'man river and the first 40 km (25 miles) of foundation and mainline to the city of Chul'man;[25] however, to produce these results, AYAM construction crews were allotted a mere 40 million roubles ($53 million 1986 US dollars). In 1987, they demanded and received 70 million roubles ($93 million), 'but it didn't stay 70 million very long. The Ministry of Railways took away 30 million and shifted it to other projects, assuming that Stroybank would find additional funds for the Berkakit–Tommot line'.[26] In February of the same year, Gosplan and the Ministry of Transport Construction agreed to provide 37 million additional roubles ($46 million) for the project.[27] Much like the case of 'robbing Peter to pay Paul', this evidently came at the expense of another trouble-plagued job site, the North Muya tunnel bypass on the BAM.

This in part is the result of the economic austerity programme that is inherent in *perestroyka*, and the implications of universal *khozraschet* (cost accounting) and self-financing for the Soviet Union's far-flung provinces, like East Siberia and the Soviet Far East, are no less dire. As Gorbachev himself declared in Tyuman' in June 1985, 'the government will not withhold funds from the development of Siberia, but it is justified in demanding that those funds yield a return, and not be frozen'.[28] Although his comment was aimed chiefly at the officials responsible for 5,000 unfinished work projects in Krasnoyarsk Kray, including the giant Abakan Rolling Stock Agglomeration, it was also parlous for such temporarily 'frozen assets' as BAM and AYAM. Indeed, a year later when the General Secretary visited the Soviet Far East (for about a week), he publicly referred to BAM only twice (in his Vladivostok speech) and this was rather obliquely.[29] Whether his coolness can be explained by his antipathy for the Brezhnev administration or his quasi-zero-based-budgeting economic philosophy is a moot point. The facts are that BAM and AYAM until now have not fared well under the new administration.

AYAM, in particular, has not fared well. As of March 1988, only 40 km (25 miles) of rails had been laid, ostensibly in the direction of Tommot.[30] The main reason for the intolerable delays has been financing, of course, but another is the flagrant 'departmentalism' that has prevailed over the issues of planning and design. To date (summer 1988), the official design of the new railway, which was produced by Mosgiprotrans (the Moscow State Agency for Transport Design) prior to 1985 (!) had not been approved by the Ministry of Railways (MPS). (This controversy had been reported earlier in *Izvestiya*, 11 May 1987). Without MPS approval, construction of AYAM cannot be carried out legally.

For two years (1987 and 1988), AYAM construction workers toiled at half-speed. Their lives, and those of their families, were literally in limbo. Many had already departed for other construction sites in the Kuzbas and Kansk-Achinsk basin.[31] Supervisors acknowledged that *khozraschet* and self-financing on the much-subsidized BAM and AYAM (as well as the South Yakutian TPC) were taking their toll.

This gloomy picture brightened somewhat in May 1988, when work began on the railway bridge over the Aldan river. The bridge is the second-longest on the AYAM route and will be composed of seven spans.[32] This manoeuvre as well as preparatory work on the giant 10-km (6-mile) Lena river bridge near Yakutsk was approved in the previous year by Gosplan, but under the existing economic constraints, they were in doubt. Since both bridges were inherent aspects of the Mosgiprotrans design for AYAM, this work, too, was illegal.

The South Yakutian Territorial-Production Complex (TPC)

Coal is the key

The key to the South Yakutian Territorial-Production Complex is the mining of coking coal. British geographer Tony French once said that when all the earth's fossil fuels are used up, the last to go will be the coal of Yakutia. The Yakut ASSR contains one-third of the USSR's coal resources and almost one-fifth of the world's volume.[33] A relatively small but high-quality share of these resources (1.6 per cent in 1970) is in South Yakutia.

The South Yakutian coalfield can be compared to an elliptical bowl, a geosyncline, with the coal seams outcropping along the periphery.[34] The bowl is filled with overburden that reaches a maximum depth of 315 m (1,025 feet) in the centre of the basin. The surface coal periphery etches a ring that stretches in the west from the upper Aldan river to north of Chul'man to beyond the river Timpton in the east to south of Neryungri. This inscribes an area of 83,000 sq. km (31,250 square miles) (Figure 7.4).[35] An average coal seam is 20 m (65 feet) thick, reaching thicknesses of 70 m (227 feet) at the bottom of the bowl. By 1980 11,000 million tons of coal were proved reserves (A + some B). Simultaneously, Soviet and Japanese geologists prognosticated a total of 40,000 million tons of hypothetical resources (A through D_2), almost two-thirds of which were within 300 m (972 feet) of the surface. At least one-fifth of the coal was coking variety.[36]

The first geological survey of any part of the basin occurred along the

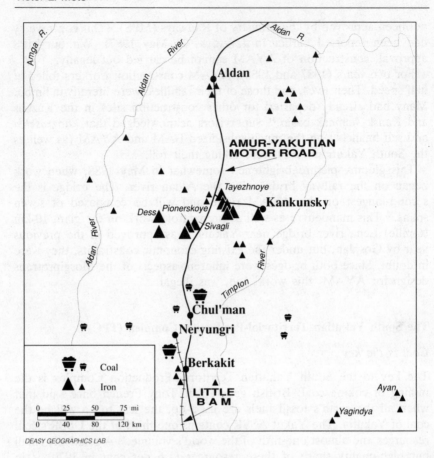

Aldan

AMUR-YAKUTIAN
MOTOR ROAD

Tayezhnoye

Dess Pionerskoye   **Kankunsky**

Sivagli

Chul'man

Neryungri

Coal

Iron

Berkakit

**LITTLE
B AM**

Ayan

Yagindya

0   25   50   75 mi
0   40   80   120 km

DEASY GEOGRAPHICS LAB

7.4 South Yakutian coal and iron deposits

Aldan side in 1851, but the first mine (at Chul'man) did not open until 1934.[37] Pre-war output ranged from 900 tons in 1934 to a peak of 7,000 tons in 1937 to 6,000 tons at the war's outbreak. The mine's production supported gold-mining enterprises along the Aldan river. During the war, output from the Chul'man mine once again fell below 1,000 tons per year, but rose to 35,000 tons in 1955, one year after strip mining began on site and at Neryungri. By 1965, 208,400 tons of bituminous coal were mined in the South Yakutian basin, most coming from shaft mines near Chul'man.

During the 1950s, the industrial demands on the coal mine at Chul'man increased. The mine served the interests of no less than one dozen Yakutian enterprises. The settlement of Chul'man grew into the largest urban area in the South Yakutian basin, and a small state regional power plant (GRES) with a rated capacity of 62.5 mw was built to serve the community. The shaft mines near Chul'man were hard-pressed to meet the new demands and the stripable reserves were limited. Thus, the Neryungri strip mine was forced into service, at first providing fuel for the Chul'man GRES.

Proved reserves at Neryungri approximate 500 million tons, 72 per cent of which are reasonably high-quality heavy (SS- and S-type) coking coal.[38] The mine reached its projected capacity of 13 million tons in 1986.[39] All the coal is strippable and may be taken from twenty seams of 10 to 70 m (30 to 225 feet), averaging 27 m (87 feet) each. The coal is high in ash (9 to 10.2 per cent), which, along with natural mine dust, lately has become the source of a deleterious air pollution problem.[40] (This was already evident during my two-day visit to Neryungri and the Ugol'naya strip mine.[41]) Other negatives are the high mean volatility (18 to 21 per cent) and high moisture content (8 per cent) of the coal. The presence of the ash, volatiles and solubles are the primary reasons for the construction of the Soviet Union's largest coal washery at the site of the strip mine (Ugol'naya). The detritus from the washery operations – itself burnable, as steam coal used at the nearby Neryungri GRES – simply augments local air pollution problems. On the plus side, the sulphur content of the coal is only 0.3 to 0.4 per cent, and phosphorus is a mere 0.004 per cent.

Japanese assistance – mutual compensation or boondoggle?

Soviet officials recognized early that the giant regional development pro-gramme slated for BAM and its environs could be implemented more rapidly with the assistance of foreign investment. During the 1960s, they invited experts from Japan to examine the coalfields of South Yakutia. The results of the fieldwork ultimately culminated in the 1974 joint Soviet-Japanese mutual compensation agreement permitting the exchange of Japanese technology and know-how for Neryungri coking and bitumin-ous coal for a period of twenty years.

According to Mathieson,[42] the Export-Import Bank of Japan loaned the USSR $390 million to buy Japanese plant and equipment, $60 million to purchase winter clothing and consumer goods for Soviet coalfield workers and $135 million for the construction of the Little BAM from Bamovskaya (formerly Bam Station) to Ugol'naya. The total credit and investment came to US $585 million.

A Japanese consortium of fifty-one firms, including eight metallurgical companies, fifteen trading companies, eight machine-building companies, two coke-gas producers, two shipping lines and sixteen banks was formed to oversee Japan's interests in the agreement. By the end of 1982 almost all of the equipment had been purchased including 403 bulldozers and front loaders, 692 cranes and 176 excavators.[43]

Since its signing, the agreement has been hampered by physical factors at Neryungri, the changing world economic conditions and a large dose of Japanese disaffection. Under the terms of the agreement, the Japanese were to import a total of 104 million tons of coking coal for six different steel mills over a twenty-year period beginning in 1979 and persisting through 1998. However, the first coking coal deliveries from the Neryungri mine did not take place until late 1984.[44]

From the beginning, the strip mine was plagued by a variety of physical constraints. By the time that the Little BAM reached the Ugol'naya minehead in 1979, miners had discovered that in order to reach the main coking-coal seams, they had to remove 240 million cu. m of overburden and upper-layer steam coal. The work took its toll on both men and equipment; even the best Japanese and American machinery, which had been tested on the Kola Peninsula, failed in the harsh conditions of South Yakutia.[45] Nevertheless, progress was made, and not all the work was completely in vain. For example, between 1978 and 1983 over 14 million tons of the overlying steam coal was shipped out of Neryungri via the Little BAM for use in regional power plants and boilers. Just as the miners were on the verge of reaching the coking-coal bed, however, they suffered yet another setback. The deposits were covered by a thick rock layer that required the import of Japanese hydraulic excavators in order finally to reach the coal.[46] This in turn proved to be disastrous; the fluid in the hydraulic lines froze up, and the expensive machinery lay deadlined until it could be replaced by electo-mechanical units. (During my tour of the BAM service area in 1985, supervisors were furious over this episode.) The excavators had cost millions and did not last the winter.[47] Because of these delays, the first shipments of any type of Neryungri coal did not reach Japan until 1983, and the first 800,000 tons of coking coal did not arrive until 1984. Even though they had contributed unintentionally to the delays because of the failures of their equipment and intentionally to them by halting investments for almost a year because of the Soviet invasion of Afghanistan, the Japanese were not happy with the slow turn-around times. There was also some question of real need for the Soviet output.

By 1983, the Japanese demand for coking coal had altered considerably. Half of Japan's aluminum refineries had closed because of high energy costs, and its steel firms had taken a back seat to less material-intensive automobile, electronics and ceramics industries.[48] Between 1974 and 1983 the Japanese, like the Americans, had become very energy-conscious, adopting conservation strategies and a reliance on nuclear power. Finally, although Neryungri coal compares in quality to the coal that the Japanese import from the United States, it is inferior to that from Australia. In the end it suffers from competition from not only those two countries but also from Canada, China and South Africa, all of which at any or odd times exceed or meet the Soviet share of Japan's market and all of which have sought to redress their trade balances.

Neryungri coal has at least one comparative advantage over that of many of its competitors: it is extremely cheap at just over $44 per ton versus, for example, the British Columbian price of $94 per ton.[49] During the 1980s, the Soviet share of Japan's coking-coal imports was 3 per cent or less *before* shipments began from Neryungri. This was expected to rise to between 6 and 7 per cent by 1988, outstripping the shares of China and South Africa, but far behind the proportions of Australia, Canada and the US. Bradshaw feels the Soviet price advantage eventually may hurt the Canadian producers.[50]

South Yakutian iron ore

Iron ore was first discovered in South Yakutia at the end of the 1940s.[51] Since that time wildly optimistic predictions about the size and quality of the deposits on the Aldan upland have appeared in Soviet literature. Most of the optimism has been generated by proponents of a South Yakutian-based full-cycle steel mill, prompted, of course, by the propitious location of the ore, close to the Neryungri coking-coal deposit. Regional geological surveys suggest the existence of 2,200 to 3,000 tons of potential resources, including 600 million to 1,500 million tons in the 'identified' category (A + some B).

The problem with the South Yakutian iron ore deposits is that they are widely scattered in no less than forty different locations.[52] The main ore bodies are clustered about 100 km (60 miles) north of Neryungri on both sides of the projected route of the AYAM and the existing Never-Yakutsk highway. The largest of these are the Tayezhnoye, Pionerskoye, Sivagli and Dess magnetite bodies. High in silicates, ranging from 2 to 33 per cent, Aldan ore is unsuitable for direct reduction, the most advanced steelmaking technology in the world today. The four major deposits range from 51 per cent (Sivagli) to 61 per cent (Dess) average iron content, 7 to 10 per cent silicate, and 0.04 to 1.9 per cent sulphur.[53] Currently, the Tayezhnoye deposit is expected to be the first site to be developed, but sometime after 1990.

Original projections called for the open-pit extraction of 13 to 17 million tons of Aldan ore each year, with as much as 5 million tons going for export.[54] Recently, a more conservative figure of 11 million tons total extraction has been mentioned.[55] Feasibility studies of the development of the Aldan iron ore deposits were called for explicitly by the 11th Five-Year Plan (1981–5), and are implicit in the statutes of the current plan (1986–90).[56]

Feasibility studies of a new full-cycle iron and steel mill somewhere in the Soviet Far East are also called for in the 12th Five-Year Plan. This subject has a thirty- to forty-year history. Yakutian politicians have long favoured Chul'man as the site of the mill. Other early-suggested sites were Svobodnyy and Nerchinsk, both of which are on the Trans-Siberian Railway. Lately, Gosplan RSFSR has argued for two large full-cycle plants: one in East Siberia and a second in the Soviet Far East. In each case, Chul'man is still in the running because of its favourable location in the midst of coal and iron resources. Other plans limit Chul'man's role to pig iron production (ingots), simultaneously calling for a sheet metal mill in Tayshet and a diversified rolling mill in the Soviet Far East (perhaps Svobodnyy or Komsomol'sk).[57]

The Chara-Tokko iron ore deposits

Soviet planners have now included the Chara-Tokko iron-ore fields within the projected confines of the South Yakutian TPC. These deposits are located in the Chara-Tokko river basins, some 350 km (220 miles) west of Neryungri on the border between Yakutia and Chita Oblast. The iron-ore district contains 6,000 to 8,000 million tons of large-grained magnetite quartzite that is low in sulphur, phosphorus and metallic impurities. The ore is rich and pure enough to be suitable for direct reduction. The Chara-Tokko fields are no more than about 30 km (18 miles) north of the BAM, and mining costs are allegedly competitive. Geographically, the assignment of the Chara-Tokko complex to South Yakutia is moot inasmuch as the fields are no more than 100 km (60 miles) away from Chara and the Udokan industrial node.[58]

Other resources

Seligdar apatite

The 12th Five-Year Plan also called for initial construction of the Seligdar apatite processing plant west of the projected AYAM route near the city of Aldan, some 300 km (190 miles) north of Neryungri. Seligdar will serve as a local source of fertilizer raw materials, which to date have been imported expensively from the Urals. The costs of shipping a ton of phosphate fertilizer to the Far East from the Urals are two to four times the transport costs of shipment from Western Siberia.

With 3 billion tons of fertilizer-quality apatite in reserve, Seligdar's expected output of raw super-phosphate is estimated at 30 to 60 million tons per year. Originally compared to the ores of Karatau and the Kola Peninsula, recent assays indicate problems with P_2O_5 content (5.0 to 7.0 per cent) and a poor-to-average beneficiation potential.[59] The sulphuric acid needed for the concentrator could conceivably be obtained from the conversion of sulphurous by-products from smelters in the Soviet Far East.

The fate of the Seligdar operation hinges in no small measure on the fortunes of the AYAM railway. Even here, departmentalism has raised its ugly head: the MPS and Ministry of Transport Construction are in charge of building the railway, the construction industry in South Yakutia is under the Coal Ministry and Seligdar is under the Mineral Fertilizer Ministry.[60] With such a farrago of bureaucracy, the Seligdar concentrator may very well be in limbo too. As of 1988, construction on the plant had not begun.

Aldan mica, limestone, building materials and timber

Between 10,000 and 15,000 tons of phlogopite mica are extracted annually in the Aldan region (Kankunskiy), yielding roughly 30 per cent of the total for the entire country.[61] The South Yakutian mines comprise 40 to 60 per cent raw mica. Phlogopite is a magnesium mica, and the Aldan variety has been mined since the Second World War. The raw mica traditionally has been trucked to Irkutsk for processing before shipment to Soviet European markets.

The South Yakutian TPC is relatively well endowed with construction materials, including limestone, dolomite and vermiculite. Just north of Tommot on the left bank of the Aldan river and on site in Chul'man (Chul'man means 'white rock' in Evenk) are limestone deposits. Appropriately processed, these could be used as a fluxing agent in ferrous metallurgy. Otherwise, together with local supplies of dolomite and other building stone in the iron-ore fields north of Neryungri, the limestone may serve as a much-needed building material for the BAM-AYAM service area. Lastly, a large deposit of the versatile insulating material, vermiculite, was discovered in the old gold-mining centre of Aldan. The 700,000 tons of mineable reserves there are a welcome addition to the BAM–AYAM cold-weather construction projects.

Timber in South Yakutia grows both poorly and slowly. Although some 15 per cent of the regional labour force was employed in logging in the TPC in the mid–1980s, it was used chiefly to clear trees from new construction sites. Thus, in 1985, roughly 1.6 million tons (2.3 million cu. m) of roundwood was shipped out of the Neryungri-Chul'man region by means of the Little BAM. This represented about 14 per cent of the total shipments, which were mainly coal.[62] In the final stages of formation of the TPC, the share of loggers in the overall labour force will be 4 per cent or less.[63]

The South Yakutian TPC

As Dienes has indicated, there are more than 100 definitions for what constitutes a territorial-production complex or TPC.[64] Suffice it to say that a TPC is a mechanism for spatial organization that optimizes the use of a given area's human and physical resources for a desired economic effect. Recent work by Soviet planners on the composition of the South Yakutian TPC prognosticates four industrial nodes for the region.[65] These include:

1 The central Aldan industrial node – specializing in non-ferrous metal mining (gold), mica, apatite, logging and wood processing;
2 The Neryungri-Chul'man industrial node – focussing on the mining and agglomeration of coking coal, electric-power production, and, possibly, full- or partial-cycle iron and steelmaking;
3 Tayezhnoye; and
4 Chara-Tokko industrial nodes – both centring on iron-ore mining.

The development of the South Yakutian TPC has been a priority during both the 11th and 12th Five-Year Plans. The current plan exhorts cadres to:[66]

1 continue to develop the South Yakutian TPC;
2 start up the second unit of the Neryungri GRES;
3 begin construction of the Seligdar apatite processing plant;
4 complete feasibility studies on the creation of a ferrous metallurgical base in the Far East that will utilize local coking coal and iron ore;
5 extend the construction of the AYAM railway from Berkakit to Tommot.

The South Yakutian TPC thus represents the first-priority territorial-production complex in the BAM service area, excluding the previously existing one at Komsomol'sk. It is clearly distinguished from other BAM service area development projects because of its diverse and generous endowment of extractable resources, on which a very promising, multifarious industrial base can be established. Indicative of its priority, and enhanced by the magnitude of the Soviet and Japanese investments (well over 1,000 million roubles combined by 1988), the population of South Yakutia tripled in the eight years between 1975 and 1983 as the Neryungri regional population expanded by 70 per cent. With a population of a mere 1,000 persons in 1975, the city of Neryungri topped 56,000 ten years later and became Yakutia's second-largest city (after Yakutsk) in the same year.[67] The composition of the regional labour force is likewise changing radically (Table 7.1). If in 1975 seven out of ten South Yakutian workers were employed in gold and mica mining, then the future labour force should be distributed more uniformly in coal, iron, steel and other industries.

If a TPC is a mechanism for the harmonious organization of space, then South Yakutia possesses excellent potential. As the central activity, coal mining yields a nationally (and internationally) significant product, the steam coal by-products of which feed the furnaces of the nearby

Table 7.1 Composition of the South Yakutian labour force, 1975 and projected, in percentage of total

	1975	Projected
Fuels	2	16
Electric power	4	3
Mining, ferrous metallurgy and machine-building	69	58
Logging and wood-processing	15	4
Building materials	4	10
Light industry	2	3
Food industry	4	6

Source: Krivoborskaya, A. I. (1987) *Prirodno-ekonomicheskiy potentsial zony BAM*, Novosibirsk: Nauka, p. 17.

Neryungri and Chul'man GRESs. The former plant's initial designed capacity will be 630 mw from three 210-mw generating units. The second unit, which is a high priority during the current five-year plan, actually went on stream in late 1984 but suffered an accident in February of the following year. Until then, 340 mw of power were distributed in the interests of the South Yakutian TPC, with the excess going to the Far East power grid. Because of the expected growth of the Neryungri urban zone (to 200,000 by the year 2000), the plant's capacity may be expanded beyond the initial stage of 630 mw.[68]

In any industrial operation, fuels and electric power generation are absolutely essential. The South Yakutian TPC is therefore solidly based on both. The coking-coal output cannot only go for export, providing the Soviet Union with much-needed hard currency, but can also go for use in the projected ferrous metallurgical base, which, like the second unit of the Neryungri power plant, is an integral part of the existing five-year plan (1990).

The new ferrous metallurgical base is also an integral component of Gorbachev's 'Comprehensive Plan for the Far East Major Economic Region, Buryat ASSR, and Chita Oblast to the Year 2000' announced in August of 1987.[69] This plan foresees 'closed-cycle integrated plants', whose output will go exclusively for the needs of the region to reduce its reliance on long-haul finished steel products. Emphasis will be placed on the production of machines required for the extraction of regional raw materials. The first unit of the plant will produce 3 million tons of rolled metal per year and will go into operation sometime before the year 2000.

Given this mandate, should the Chul'man site win out as the location of the full-cycle mill, its raw-material demands will spur the organization of iron-ore and coking-coal production in the TPC. Its product should stimulate metal-fabrication enterprises in the region that will, in turn, supply machinery not only for the entire Far East Major Economic

Region but also for the extractive industries in the TPC (coal, iron, apatite, mica, gold and so forth). The closed-cycle nature of the ferrous-metallurgical process will enable the operators to capture potentially deleterious sulphurous gases and convert them into sulphuric acid for use in the Seligdar super-phosphate operation, the product from which will benefit agricultural activities throughout the Soviet Far East.

All of this should theoretically stimulate population growth in South Yakutia, further spurring the growth of the building materials industries, including limestone quarrying (part of the product of which may go to the steel industry as flux), dolomite extraction, the mining of other building stone and vermiculite enterprises. Already, in 1985, the South Yakutian construction industry was a 250-million-rouble-per-year industry, employing more than 30,000 people.[70] In short, the South Yakutian TPC has the potential to become a model of what Soviet theoreticians had in mind when they first proposed the TPC concept several decades ago.

Conclusion

The economic realities of the modern world, however, may play havoc with South Yakutia's 'potential' development. To begin with, the region is thousands of miles from the principal markets in the European USSR. Despite his comprehensive plan for Far Eastern development, a 232,000-million rouble plan that has received Politburo approval,[71] Gorbachev, in everything else he has proposed, has reflected a bias for the development and renovation of the economy of Soviet Europe, not Siberia. In contradiction of 'third-wave economics' the industries proposed for South Yakutia are essentially extractive and 'smoke-stack' varieties, whose fortunes elsewhere in the world have been in decline. Because of this worldwide decline, the once-promising, hard-currency-earning export potential of the region has also been damaged in the interim. Lastly, and perhaps most important of all, the regional infrastructure, especially in transportation, is still primitive at best. The AYAM is supposed to remedy some of this problem, but, as this chapter has shown, AYAM is being constructed at a snail's pace, to the extent that its very completion is a subject of considerable controversy. Given these problems, the development of the South Yakutian TPC can neither happen quickly nor easily, but it is occurring and will occur inexorably throughout the rest of this century and well into the next.

Epilogue

As this chapter went to press, a major development took place between the USSR and China that may have important implications not only for the regional assimilation of South Yakutia but also for that of the entire Soviet Far East. The most critical problem facing the economic development of the region has been a chronic labour shortage. The territory is both too far away and too harsh for the appetites of most Soviet citizens. Very few people have ever chosen to migrate to the region on a truly voluntary basis. Indeed, the large military presence in the Soviet Far East serves both defensive and civilian purposes; in the absence of Stalinist edict, it is the only large-scale form of 'involuntary labour' left in the Politburo's bag of tricks. Not any more. In late June 1988, the USSR and China agreed 'in principle' on the use of excess Chinese labour in 'the Soviet Far East region'.,[72] This important agreement may very well prove to be the salvation of many Far Eastern (and BAM) development projects.

Notes

1. Mote, Victor L. (1987) 'The Amur–Yakutsk Mainline: A Soviet concept or reality', *The Professional Geographer*, vol. 39, no. 1, pp. 14–15; Dienes, Leslie (1987) *Soviet Asia. Economic Development and National Policy Choices*, Boulder, Colorado: Westview, pp. 104–10; Wood, Alan (ed.) (1987) *Siberia. Problems and Prospects for Regional Development*, London: Croom Helm, pp. 64–5.
2. Mote, Victor L. (1983) 'The Baikal–Amur Mainline and its implications for the Pacific Basin', in Robert G. Jensen, Theodore Shabad and Arthur W. Wright (eds) *Soviet Natural Resources in the World Economy*, Chicago: University of Chicago Press, p. 170.
3. Wood, *Siberia. Problems and Prospects*, op.cit., pp. 88–9.
4. Shabad, Theodore and Mote, Victor L. (1977) *Gateway to Siberian Resources (the BAM)*, New York: Wiley, pp. 85–7.
5. Bond, Andrew (1987) 'Spatial dimensions of Gorbachev's economic strategy', *Soviet Geography*, vol. 28, no. 7, p. 496.
6. Aganbegyan, A. G. and Ibragimova, Z. (1984) *Sibir' na rubezhe vekov*, Moscow: Sovetskaya Rossiya, p. 47.
7. *Narodnoye khozyaystvo RSFSR v 1984 g.* Statistical yearbook, (1985), Moscow, pp. 7, 12, 17.
8. Dienes, *Soviet Asia. Economic Development*, op.cit., pp. 186–7.
9. Wood, *Siberia. Problems and Prospects*, op.cit., p. 149.
10. Dienes, *Soviet Asia. Economic Development*, op.cit., p. 215.
11. *Vodnoy Transport*, 10 June 1977 and V. I. Tonyayev (1984) *Geografiya vnutrennikh putey SSSR*, Moscow: Transport, p. 154.
12. Aganbegyan, A. G. and Mozhin, V. P. (eds) (1984) *BAM stroitel'stvo. Khozyaystvennoye osvoyeniye*, Moscow: Ekonomika, p. 217.
13. In Wood, *Siberia. Problems and Prospects*, op.cit., p. 150.
14. Mote, 'The Amur–Yakutsk Mainline', op.cit., pp. 16–17.
15. *Gudok*, 8, 21 December 1985, 7 January 1986.
16. ibid., 3 February 1988.
17. Swearingen, Rodger (ed.) (1987) *Siberia and the Soviet Far East*, Stanford, California: Hoover, p. 64 and Wood, *Siberia. Problems and Prospects*, op.cit., p. 152.

18. *Gudok*, 26 September 1984.
19. ibid., 18 February 1986.
20. Mote, Victor L. (1985) 'A visit to the Baikal–Amur Mainline and the new Amur–Yakutsk Mainline rail project', *Soviet Geography*, vol. 26, no. 9, p. 697.
21. ibid.
22. *Gudok*, 13 March 1988.
23. ibid., 26 September 1984. Average annual freight traffic expected for AYAM was estimated in 1985 (*Izvestiya*, 20 March 1985). Northbound shipments will fall sharply beyond Berkakit and Neryungri, with 10.4 million tons flowing north over the now-existing Little BAM, 6.3 million tons continuing to Tommot, and 5.9 million tons moving on to Yakutsk (see Figure 7.3 in text). Deadheading (empty cars) of up to 90 per cent will be the chief characteristic of the trains southbound from Yakutsk to Tommot. Only 0.5 million tons of raw materials will be dispatched from the Yakutian capital. These cargoes will be enhanced by as much as 600,000 tons of mica in the Tommot area and will increase gradually as the trains flow through Aldan (apatitie, vermiculite and building stone), the Tayezhnoye, Dess and Pionerskoye iron-ore fields, and the coalfields of South Yakutia. Some 22 million tons, the majority of which will be coal, are slated to pass through Berkakit on the way to the BAM and Trans-Siberian railways. Once completed, the AYAM is expected to absorb two-thirds of the incoming freight (10.4 million tons) and the bulk of the outflowing freight (22 million tons) both to and from Yakutia. The other third of the future shipments will be shared equally by the Lena river and Northern Sea routes. Very little will be carried by truck or aircraft. Mote 'The Amur–Yakutsk Mainline', p. 20.
24. *Pravda*, 5 March 1986.
25. *Gudok*, 1 January 1986.
26. *Izvestiya*, 24 January 1986.
27. ibid., 11 May 1987.
28. *Literaturnaya gazeta*, 12 June 1985, quoted first in Bond, 'Spatial dimensions of Gorbachev's economic strategy', op.cit., p. 493.
29. Having at first seen an abbreviated version of the Gorbachev Vladivostok speech in a *Soviet Life* special supplement, I thought that he had made no mention at all of BAM during his week-long visit (see Mote, 'The Amur–Yakutsk Mainline', op.cit., p. 15). Only later did I find out that the unedited version in *Pravda*, 29 July 1986 contained two modest references to the railway. For this, I am grateful to Roy Kim.
30. *Gudok*, 13 March 1988.
31. *Izvestiya*, 26 December 1986.
32. *Gudok*, 5 May 1988.
33. Chudinov, G. M. and Gotovtsev, I. P. (1969) *Ugol'naya promyshlennost' i toplivnyy balans Yakutskoy ASSR*, Moscow: Nauka, p. 5.
34. Mathieson, Raymond S. (1979) *Japan's Role in Soviet Economic Growth. Transfer of Technology since 1965*, New York: Praeger, p. 72.
35. Serdyuchenko, D. P. (ed.) (1960) *Zheleznyye rudy Yuzhnoy Yakutii*, Moscow: Nauka, p. 49.
36. Lyudogovskiy, G. I. (ed.) (1960) *Razvitiye proizvoditel'nykh sil Vostochnoy Sibiri. Chernaya metallurgiya*, Moscow: Nauka, p. 154.
37. Chudkov and Gotovtsev, *Ugol'naya promyshlennost' i toplivnyy balans*, op.cit., pp. 13, 17.
38. Mathieson, *Japan's Role in Soviet Economic Growth*, op.cit., p. 72.

39. *Gudok*, 8 January 1987.
40. *Izvestiya*, 27 December 1986.
41. Mote, 'A visit to the Baikal–Amur Mainline', op.cit., p. 707.
42. Mathieson, *Japan's Role in Soviet Economic Growth*, op.cit., pp. 71–2.
43. Bradshaw, Michael J. 'Japan–USSR trade relations – The implications for western Canada,' unpublished paper presented at the Canadian Association of Geographers' Meetings, Calgary, Alberta, Canada, June 18–24, 1986 (photocopied), p. 11. I gratefully acknowledge Mr Bradshaw for the receipt of this reference.
44. Bradshaw, Michael J. 'Japan and the economic development of the Soviet Far East. Past, present, and future', unpublished paper presented at the East–West Conference: Gorbachev's 'New Thinking': Social and Economic Reforms, Middlebury, Vermont, 7–9 October, 1987 (photocopied), p. 12 and Mote, 'A visit to the Baikal–Amur Mainline', op.cit., p. 704.
45. Rodgers, Allan, 'Commodity flows, resource potential, and regional economic development: the example of the Soviet Far East', in *Soviet Natural Resources in the World Economy*, op.cit., p. 202.
46. Smith, Gordon B. (1987) 'Recent trends in Japanese–Soviet trade', *Problems of Communism*, vol. 36, no. 1, p. 59.
47. Interviews with Novosti Press correspondent Mikhail T. Morozov in Neryungri and Academician Abel G. Aganbegyan in Novosibirsk, March 1985.
48. Smith, 'Recent trends', op.cit., p. 59.
49. Bradshaw, Michael J. (1987) *Canada and the Changing Economy of the Pacific Basin*, working paper no. 29, Vancouver, B.C.: Institute of Asian Research Publications, pp. 37, 48.
50. ibid., p. 48.
51. Krivoborskaya, A. I. (1987) *Prirodno-ekonomicheskiy potentsial zony BAM*, Novosibirsk: Nauka, p. 10.
52. Serdyuchenko, *Zheleznyye rudy*, op.cit., p. 49.
53. Lyudogovskiy, *Razvitiye proizvoditel'nykh sil*, op.cit., p. 152.
54. D'yakonov, F. V. (1977) 'The BAM as a set of major long-term economic problems', in *Baykalo-Amurskaya magistral'* (in Russian), Moscow: Mysl', p. 30.
55. Mote, 'The Amur–Yakutsk Mainline', op.cit., p. 19.
56. SSSR, KPSS (1986) *Materialy XXVII s"yezda Kommunisticheskoy partii Sovetskogo Soyuza*, Moscow: Politizdat, p. 320.
57. Krivoborskaya, *Prirodno-ekonomicheskiy potentsial*, op.cit., p. 10 and Babun, R. V. (1987) *Razvitiye chernoy metallurgii Sibiri i Dal'nego Vostoka*, Novosibirsk: Nauka, pp. 105–16.
58. Nedeshev, A. A., Bybin, F. F., and Kotel'nikov, A. M. (1979) *BAM: osvoyeniye Zabaykal'ya*, Moscow: Nauka, p. 91.
59. Aganbegyan, A. G. and Kin, A. A. (eds) (1985) *BAM: pervoye desyatiletiye*, Novosibirsk: Nauka, p. 131; Molodenkov, L. V. (ed.) (1987) *Osvoyeniye novykh khozyaystvennykh territoriy v vostochnykh rayonakh RSFSR*, Novosibirsk: Nauka, p. 85; and interview with Richard Levine, US Bureau of Mines, November 1986, New Orleans, Louisiana.
60. *Izvestiya*, 26 December 1986.
61. Minakir, P., Renzin, O. and Chichkanov, V. (1986) *Ekonomika Dal'nego Vostoka*, Khabarovsk: Khabarovskoye knizhnoye izdatel'stvo, p. 99 and Mote, 'The Baikal–Amur Mainline and its implications', op.cit., p. 163.
62. ibid., p. 17.
63. Krivoborskaya, *Prirodno-ekonomicheskiy potentsial*, op.cit., p. 17.

64. Dienes, *Soviet Asia. Economic Development*, op.cit., p. 238.
65. Krivoborskaya, *Prirodno-ekonomicheskiy potentsial*, op.cit., p. 11.
66. SSSR, KPSS, *Materialy XXVII s"yezda*, op.cit., p. 320.
67. Krivoborskaya, *Prirodno-ekonomicheskiy potentsila*, op.cit., p. 16 and Mote, 'A visit to the Baikal-Amur Mainline', op.cit., p. 697.
68. Mote, 'A visit to the Baikal–Amur Mainline', op.cit., pp. 709–10.
69. *Pravda*, 26 August 1987.
70. Aganbegyan and Kin, *BAM: pervoye desyatiletiye*, op.cit., p. 136.
71. Minakir, P. and Syrkin, V. (1987) 'Strategiya kompleksnogo sotsial'no-ekonomicheskogo razvitiya Dal'nego Vostoka', *Ekonomicheskiye nauki*, no. 12, p. 3. Special thanks to Professor Kiichi Mochizuki of the Slavic Research Centre, Sapporo, Japan.
72. *Sankei Shimbun*, 18 June 1988 (in Japanese). Special thanks to Professor Hiroshi Kimura of the Slavic Research Centre, Sapporo, Japan.

Acknowledgement

I wish to acknowledge all the members of the Slavic Research Centre, Hokkaido University, Sapporo, Japan for partial support in producing this chapter.

Chapter eight

The Far Eastern transport system
Robert N. North

Introduction

The Soviet Far East has a very simple transport system. Its backbone is
the Trans-Siberian Railway, which hugs the southern border of the
region, following the Amur and Ussuri valleys to the Pacific. The railway
provides a lifeline for the Far East to the western USSR, a channel for
foreign trade through Pacific ports and a transit route from Europe to
the Pacific. Ever since its completion early in the century it has been
the prime magnet for settlement and economic development, so that now
well over half the region's population and economic activity lie within
its immediate orbit.

Almost all other components of the transport system feed into or
distribute from the Trans-Siberian, unless they handle only local traffic.
Coastal shipping connects the railway ports in the south, from Vladivostok to Vanino, to the islands, coastal settlements and river mouths of
the Soviet Pacific littoral and the Arctic. The railway ports also handle
three-quarters of the region's foreign trade.[1] Yakutia and the north east
have their own transport backbone, the river Lena, but it too relies on
the Trans-Siberian, to which it has been linked since 1951 through the
river port of Osetrovo at the town of Ust'-Kut in Eastern Siberia, and
a railway from there to Tayshet.[2] Where rivers cannot be used and the
railway has not yet penetrated, winter roads and a few long-distance all-
weather roads complete the surface transport system, but they too ultimately feed into the railway.

There are one or two exceptions to this pattern of reliance on the
Trans-Siberian. Direct maritime routes link the Arctic coast to the western USSR, and there are direct exports from minor Pacific ports, but
the scale of traffic is relatively small. Completion of a second east–west
railway north of the Trans-Siberian, the Baykal–Amur Mainline (BAM),
will give the system a double backbone but not radically change its
nature.

As Leslie Dienes points out in Chapter 11, transport plays a more

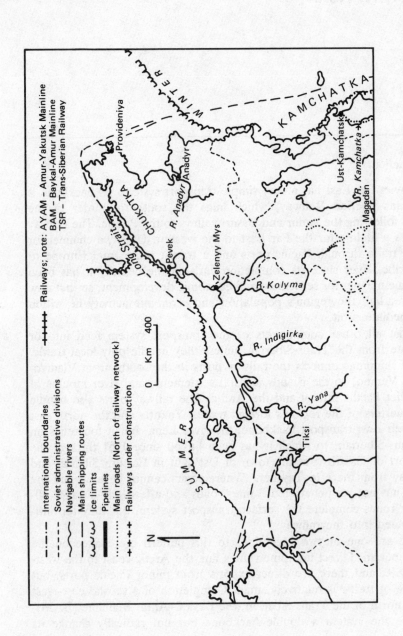

Railways; note: AYAM – Amur-Yakutsk Mainline
BAM – Baykal-Amur Mainline
TSR – Trans-Siberian Railway

International boundaries
Soviet administrative divisions
Navigable rivers
Main shipping routes
Ice limits
Pipelines
Main roads (North of railway network)
Railways under construction

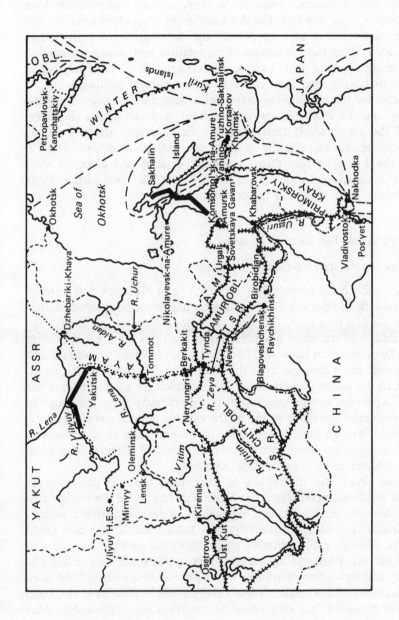

8.1 The Soviet Far East transport network

prominent role in the economy than in any other region of the country. More investment in transport is required to achieve a given economic goal than elsewhere; transport costs are a higher share of total delivered costs for most products; transport workers are an exceptionally high proportion of total workers and the number of trans-shipments per ton of goods delivered is at least double the national average. All these things follow from the sparseness of population and economic activity, the enormously long distances to be covered with relatively small amounts of freight, and the extremely severe natural conditions.

This chapter firstly describes major influences on the evolution of the economy and therefore the transport system, and the spatial characteristics of the economy. It then examines the functions of transport, its spatial characteristics, and the roles of the individual modes. Third it divides the Far East into three major sub-regions and discusses their transport facilities and problems in more detail. The final section looks at the future of transport in the region.

Transport in the Far Eastern economy

Influences on the development of transport

Ever since the early seventeenth century, when Russians first entered what is now the Soviet Far East, the region has attracted interest primarily for its natural resources, and secondarily for its location on the Pacific. The evolution of transport has reflected interactions among a consistent set of influences which have modified the impact of these two attractions. They include great distance from the economic, demographic and political core of the country, west of the Urals, which we shall designate the metropolis; severe natural conditions; transport technology; the range of natural resources attractive to the metropolis for domestic use or export; and national foreign policy. As the last two items suggest, the course of major developments has normally been determined outside the region. Local initiatives have been of little account.

Distance from the metropolis is important because the resource demands which have stimulated Far Eastern development have originated there rather than in the intervening regions. Prime attractive resources have varied over time, but generally they have been those which could repay the costs of exploitation in an inhospitable environment, as well as long-distance transport to markets: furs, gold, some fish, diamonds, antimony and tin. When relations with neighbouring countries have been good, resources with a relatively low value-to-weight ratio have also been attractive, provided that they could be exported over reasonably short distances or by cheap sea transport. Current examples are wood and coal. 'Reasonably short', of course, may mean one thing if profit is the criterion and another if the overriding aim is to obtain hard currency. Over half the forest products exported to Japan in fact come from Eastern Siberia along the Trans-Siberian Railway over distances of 3,000 km and more. Away from the railway and the coast such resources are exploited only to the extent needed for local use. The metropolis can obtain them closer at hand, and the intervening regions have their own.

Foreign relations also affect transit traffic and imports. Transit can be thought of in terms of the region, an example being the East Siberian forest products just mentioned, and the nation. The Trans-Siberian Railway was built partly for the tea and silk trade from China to Europe, while modern international transit traffic is exemplified by the containers moving from Japan to Europe along the same railway. During the Stalin era, by contrast, there was no international transit traffic at all. Imports through Far Eastern ports were also negligible, while in recent decades Japanese manufactures and Canadian wheat have moved through in substantial quantities. Finally, foreign relations affect the government's attitude to settlement and the level of military activity, both of which in turn influence transport. Settlement has always served to demonstrate effective occupance of the land, especially in the south, while the biggest maritime traffic flow of the present day is that between Vladivostok and Petropavlovsk-Kamchatskiy, the big Pacific naval bases.[3]

Spatial characteristics of the economy

The Far Eastern economy can generally be described as 'nodal and linear'. Isolated pockets of activity are located at attractive natural resources and ports. They are also found, because of improved accessibility, at otherwise less-attractive locations along the routes to those places. As might be expected in a region two-thirds larger than western Europe, but with a population only the size of Austria's (or, in North American terms, two-thirds the size of the United States, but with a population only a little larger than that of New Jersey), there are few places where one can speak of areal rather than nodal development. The main examples are the agricultural lands along the railway in Amur Oblast and Khabarovsk Kray, and the more diversified area in the south of Primorskiy Kray.

A second characteristic is the low level of regional integration. The Soviet Far East is typical of pioneering regions in that the individual nodes tend to be linked primarily with the outside world, sometimes over very great distances, rather than with each other. The extreme case is Magadan Oblast, where only 5 per cent of traffic is intra-regional.[4] Less typically of pioneering regions, the general situation is true of manufacturing as well as natural resources. More than ten branches of the machine-building industry have factories in the Far East, but almost everything they produce is exported from the region, while machinery needed locally for logging, mining, the fuel and energy industry and fishing is imported.[5] For example, a factory in Birobidzhan fits caterpillar tracks to rice-harvesting combines from Krasnoyarsk, 4,250 km away, which it then ships back past Krasnoyarsk, 9,000 km to Central Asia and Transcaucasia. Tracked vehicles needed in the Far East, however, come from as far away as the Ukraine and the Baltic republics.[6] Dienes gives other examples in Chapter 11. In 1977 only 18 per cent of inter-regional shipments of all types of freight from the Far East went to the nearby regions of Eastern and Western Siberia, and only 12 per cent of inshipments came from there.[7]

Perhaps the most striking example of poor regional integration is the separateness of Yakutia, together with adjacent areas of the north east, from the rest of the Far East. Yakutia, except for the South Yakutian coalfield, is mainly oriented south-westwards along the river Lena lifeline to the Trans-Siberian Railway and ultimately European USSR. The rest feeds into the railway much closer to the Pacific, focussing on the inland and coastal centres of Khabarovsk and Vladivostok. The separateness is well illustrated by the pattern of current air routes. Yakutsk and Khabarovsk are the two hubs, with ten and nine major routes respectively, but at least up to 1986 there was no direct flight between them.[8]

The functions of transport

Obviously the principal role of transport in the Far East is to move supplies in to natural resource locations and products out, but the implications of that function vary with the nature of the resource and its intended use. Tin and gold, for example, are likely to be processed on site. The resultant products can be moved out, normally to European USSR, by low-capacity surface transport or even by air. The heaviest demands on transport are for moving in supplies – fuel, food, building supplies, machinery and equipment. Building supplies and coal might originate locally or within a few hundred kilometres, but much of the rest must be brought in from outside the region. This situation is typical of Yakutia and the north east, though it is sufficiently prevalent in the whole region that inward freight movements exceed outward by 2.0 times in value terms and 2.5 times in volume terms.[9] In the south, by contrast, the demand for freight capacity outwards from resource locations to the Pacific ports, for wood and coal, can exceed that for the inward transport of supplies, which may also be coming from the opposite direction. This situation is typical of what has developed along the Trans-Siberian and Little BAM railways.

Other functions of transport contribute less to traffic. Military traffic along the Trans-Siberian may contribute substantially to the predominance of freight shipments into over those out of the region, but civilian transit traffic and regional exports of manufactured goods are minor components by weight.

Spatial characteristics of transport

Network density and form

By any measure relating route length to area, the Far East has the sparsest transport network in the country. The density, however, varies greatly within the region. The network of railways and permanent roads taken together is more than twenty times as dense in the south as in the north (Table 8.1), though even the figure for Sakhalin Island is only half the national average. The Engel coefficient, which relates route length to population as well as area, also shows the Far East to be the worst-served region in the country, even if navigable rivers and pipelines (though admittedly not coastal shipping routes) are included.[10] Adding in gross value of production as well (the Uspenskiy coefficient) carries it above North European USSR, though mainly on the strength of navigable rivers of limited capability compared to those west of the Urals.[11]

Table 8.1 Provision of transport routes in the Soviet Far East (km)

	Rail ways	Srfcd roads	Earth roads	All roads	Rail ways	Srfcd roads	Earth roads	All roads
	Per 1,000 sq km				Per 1,000 inhabitants			
Yakutia	0.04	1.0	0.5	1.5	0.1	3.0	1.6	4.6
Magadan Obl.	–	1.4	1.1	2.5	–	3.1	2.3	5.4
Kamchatka Obl.	–	1.2	1.0	2.2	–	1.3	1.0	2.3
Khabarovsk Kray	3.0	3.0	0.4	3.4	1.4	1.4	0.2	1.6
Amur Obl.	8.3	5.7	2.8	8.5	2.9	2.0	1.0	3.0
Primorskiy Kray	7.8	23.8	1.1	24.9	0.6	1.8	0.1	1.9
Sakhalin Obl.	10.2	13.0	12.1	25.1	1.3	1.6	1.5	3.1
Total	1.3[a]	2.4[b]	1.0	3.4	1.0[a]	1.9[b]	0.8	2.7
USSR[c]	6.5	52.4	18.3	70.7	0.5	4.2	1.5	5.7

Sources: *Narodnoye khozyaystvo SSSR za 70 let* (1987) Moscow: Finansy i statistika, pp. 345–55, 699; *Atlas zheleznykh dorog SSSR* (1986) Moscow: GUGK pri Sovete Ministrov SSSR, pp. 132–6; *Atlas avtomobil'nykh dorog SSSR* (1980) Moscow: GUGK pri Sovete Ministrov SSSR, pp. 90–8.

Notes: [a]Based on 7,752 km. Kazanskiy, *Geografiya*, p. 139 gives 7,300 for a density of 1.2 km/1,000 sq. km and 0.9 km/1,000 people.
[b]Based on 15,011 km. Kazanskiy gives 23,700 for a density of 3.8 km/1,000 sq. km and 3.0 km/1,000 people.
[c]Kazanskiy gives a length of 27,200 km of navigable rivers in the Far East, for a density of 4.4 km/1,000 sq. km and 3.5 km/1,000 people. National figures are 5.5 and 0.4.

Most of the Far East has a spine-and-branch transport network, with the Trans-Siberian Railway as the spine and the river Lena as a principal branch. Networks with circuits are rare. There is one circuit on the

Sakhalin Island railways, and BAM has created the first on the mainland. There are small road networks with circuits in the interior agricultural regions and a larger one in the south of Primorskiy Kray. Sakhalin Island joins the category by dint of having two north-south routes and links between them.[12]

Freight traffic patterns

In view of the nature of the network and the importance of external links, certain aspects of the freight traffic pattern may seem surprising at first glance. The Far East has one of the highest ratios of internal to total freight traffic movements in the country. This is true even of movements by railway, the main link with the rest of the country, and the preponderance of internal movements has barely been declining. In 1960, 87.6 per cent of rail loadings terminated within the region, and in 1975 86.8 per cent, both the highest in the country. The share of terminations which originated within the Far East fell from 80.5 to 68.5 per cent, dropping the region from second to fourth place in the country and suggesting an increased dependence on western regions for supplies, which was perhaps associated with the start of work on the BAM.[13]

The high proportion of internal movements is less surprising when examined more closely. The adjacent region, Eastern Siberia, does not have a strongly complementary economy, except for adding to forest product exports and making up deficiencies in local coal supplies. Distances to the rest of the USSR are so great that local supplies of such materials as coal, wood, sand and gravel are used wherever possible. For the same reason mineral exports are processed to reduce weight before shipment. Also, the Pacific basin is so isolated from western Soviet coastal waters that there is scarcely any traffic between them. Finally and most significantly, long-distance movements on the primitive transport system require many trans-shipments. The Siberian average is eight, compared to a national average of three.[14] A piece of machinery for the interior of Magadan Oblast might travel along the Trans-Siberian Railway from the Ukraine to Vladivostok; by sea from there to Pevek on the Arctic coast; by coastal vessel to the mouth of the river Kolyma; by river vessel up the Kolyma; and by truck to the final destination. All segments except the first would be recorded as internal to the region. In 1972 neither Magadan Oblast nor Kamchatka accounted for any direct regional exports or imports at all.[15]

The distribution of freight traffic within the Far East carries fewer surprises. The three administrative regions along the Trans-Siberian Railway accounted for over 80 per cent of tons originated by all forms of transport in 1972, and nearly 80 per cent of terminations (Table 8.2). Sakhalin accounted for 13 and 11 per cent, leaving 5 and 10 per cent for the whole of the north: Yakutia, Magadan Oblast and Kamchatka.

The predominance of the area served by rail does to some extent reflect the measure used. By value of transport operations the three railway regions accounted for only 57 per cent of the Far Eastern total, and the north for 29 per cent. The different picture reflects the tariff system. Railway tariffs are based on average costs for the whole system, with special discounts for moving freight to the Far East. These have applied since 1922. Costs to the railways, however, are much higher in the Far East than the system average. Furthermore the difference has probably grown since 1972. Tariffs have barely changed since 1967, though costs to the railways have risen. For example, ice-cooled freight cars used to move fish from the Far East have been replaced by refrigerated cars. They are much costlier to buy and operate, and the benefits of reduced spoilage have so far accrued entirely to the customers. In sum, Far Eastern railway traffic is heavily subsidized.[16] But river tariffs are based on costs for individual basins, and in the Lena basin and the north east, where river transport has a virtual monopoly, they are set very high.[17]

Table 8.2 Spatial distribution of freight traffic in the Soviet Far East, 1966, %[a]

	Value of transport operations	Originations	Terminations	Exports	Imports	Intraregional hauls	
						Originations	Terminations
Yakutia	14.7	3.6	4.6	3.2	8.0	3.6	3.7
Magadan Obl.	9.4	0.6	2.6	-	-	0.7	3.3
Kamchatka Obl.	4.7	1.0	2.5	-	-	1.0	3.2
Khabarovsk Kray	15.3	28.6	30.2	28.4	25.9	28.6	31.8
Amur Obl.	10.8	22.8	15.9	14.7	16.1	23.6	15.8
Primorskiy Kray	30.9	30.8	33.3	46.3	48.4	29.4	28.8
Sakhalin Obl.	14.2	12.6	10.9	7.4	1.6	13.1	13.7
Total	100.0	100.0	100.0	100.0	100.0	100.0	100.0

Source: Kolesov, L.I. (1982) *Mezhotraslevyye problemy razvitiya transportnoy sistemy Sibiri i Dal'nego Vostoka*, Novosibirsk: Nauka, p. 125.

Note: [a]Includes rail, sea, river and pipeline transport.

Roles of the various transport modes

The Far Eastern transport system represents an intermediate stage in a development process common in the northern and eastern regions of the country. Early explorers, fur traders and settlers used navigable rivers joined by portage roads and supplemented by overland tracks. River steamers, the first representatives of modern high-capacity transport, reinforced the predominance of the rivers. They were soon supplanted by railways in the south, along the west–east belt of maximum settlement, but the railways did not penetrate very far north at first for want of adequate traffic. (In the Far East the Trans-Siberian took an even more southerly route than it does now until the Russo–Japanese war of 1904–5.) With the arrival of motor transport the overland tracks became winter roads, or in a few cases all-weather roads. Further transport development north of the west-east railway has depended on the growth of demand, though the level at which a system of rivers with supporting roads is superseded by one of railways with supporting rivers and roads has varied with the evolution of transport technology. Narrow-gauge railways, which at one time might be built to handle 200,000 tons a year, have disappeared from consideration. A broad-gauge line is likely to be built now only with the expectation of over a million (instead of 500,000) tons a year, or in some cases only if 5,000,000–5,500,000 tons can be expected. Depending on conditions, all-weather roads are currently considered justifiable for traffic flows ranging from 200,000 to 1,500,000 tons, and winter roads for volumes of 50,000 to 200,000 tons.[18] For oil or natural gas, of course, pipelines are built rather than railways, as in north-western Siberia. Air transport serves long-distance passenger and construction-site traffic. In the final stage, both within and north of the railway belt, networks of local roads evolve to take over short-distance traffic from the railways and rivers.

The whole inland transport system, both within and north of the railway belt, has been supplemented by coastal shipping where appropriate. In this case the evolutionary process has been relatively recent, from general-purpose cargo vessels to such specialized types as bulk carriers and container ships, which reduce trans-shipment costs, and most recently to ferries, roll-on roll-off (RoRo) vessels, and barge carriers, which enable mobile equipment to avoid conventional trans-shipment altogether.

North of the BAM the Far East is still at the river-and-road stage, and even in the railway belt it has not advanced very far. There is no continuous highway network, and coastal shipping is at present moving somewhat painfully away from general-purpose vessels.

The relative importance of the various modes depends on the measures used. Despite being confined to the south, railways account for far more traffic than water transport. In 1970, for which comparable figures are available, sea and river transport each carried a little over 14,000,000 tons. The railways originated 71,030,000 tons and terminated 90,920,000 (Table 8.3). Figures for road transport have not been found. In terms of ton-kilometres the railways may be much less prominent, since a 1984 publication states that sea transport accounts for 40 per cent.[19] In terms of the value of work accomplished, too, the railways are less prominent because of the variations among modes in tariff-setting procedures, mentioned earlier, as well as the greater amount of work required for every ton moved when it cannot be moved by rail or sea.[20]

Table 8.3 Modal distribution of freight traffic in the Soviet Far East, 1970, million tons

	Orig- inations	Term- inations	Exports	Imports	Intra- regional hauls
Railways	71.03	90.92	7.22	27.11	63.81
incl. coal, coke	25.54	29.32	0.05	3.83	25.49
crude oil, petroleum products	3.49	12.81	0.11	9.43	3.38
ore	0.31	0.14	0.27	0.10	0.04
forest products	11.43	11.20	1.71	1.48	9.72
cement	2.17	2.62	-	0.45	2.17
Sea transport	14.05	14.43	-	0.38	14.05
incl. coal, coke	1.81	1.92	-	0.11	1.81
crude oil, petroleum products	3.06	3.17	-	0.11	3.06
ore	0.12	0.02	0.10	-	0.02
forest products	1.63	1.78	-	0.15	1.63
cement	0.49	0.53	-	0.04	0.49
River transport	14.23	15.65	0.36	1.78	13.87
incl. coal, coke	2.38	2.19	0.19	-	2.19
crude oil, petroleum products	0.87	1.65	-	0.78	0.87
ore	-	-	-	-	-
forest products	2.68	2.70	0.07	0.09	2.61
cement	0.07	0.22	-	0.15	0.07

Source: *Transport i svyaz' SSSR. Statisticheskiy sbornik* (1972) Moscow: Statistika, pp. 68–85.

Pipeline transport is little developed in the Far East. A gas pipeline in central Yakutia supplies Yakutsk; there are two oil pipelines from fields in northern Sakhalin Island to Komsomol'sk on the mainland; and a gas pipeline was completed late in 1986 along the same route.[21] A project for a gas pipeline from central Yakutia to the southern Pacific coast, to be built with Japanese and American help in order to export liquefied natural gas, was abandoned when the reserves proved inadequate to satisfy the foreign partners. A proposal for an oil pipeline from western Siberia to the Pacific coast, in the 1970s, was rejected in favour of using the BAM. It seems likely that more extensive pipeline building will have to await the proving of larger reserves of oil or gas, a substantial increase in world prices, or an eastward shift from Western Siberia of production for the European market.

Air transport in the Far East has four major roles: to move people to and from the metropolis in less than a tenth of the time taken by surface transport; to move them among the widely scattered Far Eastern cities; to supplement surface transport by, for example, flying freight to Yakutsk that is stranded at Osetrovo at the end of the navigation season; and to provide general transport to particularly remote areas in the north. The last role ranges from servicing construction sites, to transporting reindeer herders and their supplies between herds and permanent villages, to ice reconnaissance in the Arctic. Two recent developments have had opposite impacts. On the one hand a new generation of heavy-lift aircraft has made possible the movement of more and heavier freight than before. On the other hand it appears that there has been a sharp decline in air freight traffic in some areas near the coast with the introduction of more versatile Arctic cargo vessels and of such devices as air-cushion lighters and amphibious tractors.[22]

Regionalization of transport

Clearly the transport system varies a great deal from region to region within the Far East. The rest of this chapter will examine transport facilities, problems and prospects in more detail within each of three regions: the Lena basin and the north east; the Pacific coast and islands; and the 'railway belt' in the south.

The Lena basin and the north east

This is a very large region. Yakutia and Magadan Oblast together are 10 per cent larger than the Canadian North-west Territories and Yukon, and nearly 20 per cent larger than western and northern Europe. Nevertheless the region has important features in common as far as transport is concerned. One is the absence of railways except for the extreme south of Yakutia, which will be considered in a later section. Another is a preponderance of incoming over outgoing freight in the ratio of about six to one by weight, stemming from the reliance, mentioned earlier, on exporting natural resources which are processed before shipment.

The region can be envisaged as a huge storehouse of natural resources with a limited number of entry points. The main one is Osetrovo, the upper Lena river port which tranships freight from the western BAM. Before the railway arrived in 1951 freight was trucked to Osetrovo from the Trans-Siberian and amounted to 60,000–70,000 tons a year.[23] Quantities multiplied thereafter. According to the 1986 plan, which was approximately fulfilled, Osetrovo was to dispatch 2,200,000 tons of dry cargo into Yakutia, as well as 1,750,000 tons of petroleum products through the nearby Ust'-Kut oil terminal.[24] These figures are to rise to 2,800,000 and 2,000,000 tons by 1990.[25] The second entry point, in fact a series of maritime entry points along the Northern Sea Route (NSR), used to account for about 14,000 to 24,000 tons a year in pre-railway days. Since then its relative share has fallen, though quantities have grown. In 1986 the planned amount was 600,000 tons. About a quarter comes in from the west, mainly Murmansk, and three-quarters from the Far East, partly because the navigation season opens from east to west.

The third entry point, the truck route from Never and Berkakit to Yakutsk, used to be the main one, carrying over 100,000 tons a year.

Now it carries about 250,000 tons. No recent exact figures have been found for the last entry point, the port of Magadan on the Sea of Okhotsk, from which another truck route runs to the upper Kolyma and Yakutsk. The route handled over 100,000 tons in 1950. Quantities may have declined temporarily thereafter with the reduction of forced labour and hence the population to be served, but the throughput of Magadan is known to have grown in recent years. From 1978 to 1986 it grew 21.8 per cent.[26] The two truck routes comprise the only year-round surface links between the north east and the outside world, Magadan's port having operated through the winter since 1962. However, movement along the Magadan–Yakutsk route is sporadic in summer, since it is less of an all-weather road than that from Never.[27]

Physical problems of water transport

Much of what comes in by sea along the NSR is trans-shipped to river vessels and moves inland up the rivers draining to the Arctic. Also, the region supplies its own coal, building stone, sand, gravel and wood, and these commodities also move largely by river. Total freight in the Lena basin alone is about 13,000,000 tons a year. Relying primarily on water transport has considerable advantages, because routes do not have to be built from scratch. In Yakutia and the north east, however, there are also drawbacks. The worst are shallowness, ice and drought.

Shallowness Shallowness affects both maritime and river transport. The Arctic coast for the most part is gently shelving, with inadequate depths for large vessels close to shore. Even the icebreakers, whose captains have good knowledge of local conditions, often go aground.[28] There are few good harbours. Much trans-shipment takes place at open roadsteads, a dangerous practice in an area which is also subject to high winds and where a change of wind can blow offshore ice onto the coast. The situation on the rivers is equally unsatisfactory. On major rivers in European USSR, depths of 3.5 or 4.0 metres can normally be guaranteed. On the river Lena, by contrast, there is supposed to be a guaranteed depth of 2.2 m from Osetrovo to the mouth of the river Vitim, and 2.6–2.9 m downstream from there.[29] In fact, from Osetrovo to Kirensk a depth of 1.8 m 96 per cent of the time is the best that can be guaranteed at the moment.[30] There are also rocky shallows in the middle reaches of the river, at Olekminsk for example, which are difficult to dredge. Below Yakutsk dredging is easier, and a channel of 3.5 m is now maintained through to the Arctic, though vessels drawing that much might not be able to visit riverports *en route*. The main difficulty in the lower reaches in the past was shallowness across the Lena delta, and this has been tackled with new ice-capable dredges imported from Finland. Shallowness across river-mouth bars affects several of the other rivers draining to the

Arctic and is being tackled by similar means. Apart from that, however, there has been little effort to improve navigation along them.[31] Depths are guaranteed on some rivers and range from about a metre on the upper Yana and upper Kolyma to 2.0 m or more on the lower reaches of both rivers and the Indigirka.[32] There have also been few improvements to the major tributaries of the river Lena. That section of the Vilyuy below its hydro-electric station is used for navigation, but depths depend on how much water is released through the dam and are therefore unreliable. Dredging could ensure depths of 1.2–1.3 m irrespective of variations in what is released at the dam, but it has not yet been decided whether this would be economic. The river Vitim is navigable downstream from Romanovka, 150 km north of Chita in Eastern Siberia, and there is a guaranteed depth of 1.9 m below Luzhki. As for the river Aldan, it has already been established that deepening on the river Uchur-Tommot stretch would not be justified. The present guaranteed depth is 1.4 m.[33] The river is narrow and winding and the slope is 30–40 cm per km, giving currents of 14–17 km per hour. The bed is rocky. The lower Aldan, up to Dzhebariki-Khaya, is another matter. The present guaranteed depth is 2.0 m, and plans are being prepared to deepen it further for coal traffic.[34]

Winter ice   Winter ice closes all rivers in the region. The ice-free season ranges from 76–93 days on average at the mouths of the northern rivers to 129 days at Osetrovo.[35] Variations can be up to 20 days on either side of the mean, with disastrous effects on the reliability and predictability of transport. River icebreakers can extend the season a little, but they add greatly to expenses and are hampered by the shallowness of the rivers. In winter rivers freeze to a depth of 1.5–2.0 m, which means that smaller rivers freeze to the bottom. Even the deeper rivers are too shallow to take icebreakers powerful enough to break winter ice. The most powerful river icebreakers currently in use draw over 3.0 m and can break 0.6–0.7 m of ice.[36] Even while trying to extend the navigation season, icebreakers are quite likely to press the ice to the river bed rather than break through. In 1984 this happened to one of the river Lena icebreakers, which then slid over the ice and was stranded.[37]

Ice is no less a problem on the Northern Sea Route, where conditions east of the Taymyr Peninsula are much worse than those to the west. They also seem to have been deteriorating during the past thirty-five years.[38] The navigation season on the eastern section of the NSR is often shorter than that on the rivers, not so much because of ice in coastal waters as because of the late opening and early closing of access to them, through Vilkitskiy strait on the west and Long strait on the east. Indeed, early-season traffic to the Yana, Indigirka and other Arctic rivers is more likely to come down the Lena and along the coast than from Murmansk or Vladivostok. By contrast with the waters west of the Taymyr Peninsula, there are as yet no icebreakers which can maintain navigation through the winter. The most powerful atomic icebreakers are 70,000 hp, while estimates of what would be required in the eastern Arctic range upwards from 130,000 hp. There is a proposal for an icebreaker of 150,000 hp to be built by the end of the century, but there is also opposition, on the grounds of too much reliance on brute force as opposed to brains.[39] Note has been taken of the exploits of the Canadian icebreaking freighter, the M.V. *Arctic*, which relies on sophisticated navigational equipment to pick its way through the ice, and there have been complaints about the failure to install such equipment on Soviet Arctic vessels.

Summer drought On the rivers, summer drought is almost as effective as winter ice in reducing the length of the navigation season. The rivers are shallow at the best of times, as we have seen, but in addition the summer fall in level tends to be greater than in other parts of the Soviet north. Virtually all the region is underlain by permafrost, so melting snow does not sink into the ground to form a reservoir which can then seep into the rivers. Instead it forms ponds and bogs or, in the hilly country which characterizes much of the region, runs off rapidly. The rivers typically have a short period of high water during the spring run-off, usually in June or July, followed by very low water through the rest of the summer. Conditions seem to have deteriorated in recent years: one writer states that ten of the past fifteen seasons have seen exceptionally dry conditions.[40] The years 1985 and 1986 were particularly bad. In 1985 the Lena at Osetrovo varied in depth between 1.6 and 1.8 m.[41] In 1986 low water affected the Lena, Aldan, Vitim and Yana directly,[42] and traffic on the Vilyuy was held up because of a lack of water in the Vilyuy hydro-electric station's reservoir.[43] Conditions in 1985 provided an extreme example of another impact of drought. The smoke from forest fires closed down navigation on the Lena completely, necessitating extra shipments along the NSR.

The combined effects of winter ice and summer drought can be devastating for navigation. Smaller rivers may have an average navigation season of a month or less, even if the limiting depth for navigation is taken as 0.6 m. In extreme years the season on such rivers may be as little as 10–15 days.

River vessels

Freight is moved along the northern rivers in a variety of types of self-propelled vessel and tug-barge combination. Some of the former are 'river-sea' vessels, capable of operating in coastal waters as well as inland, though the need for sufficient draft to operate safely at sea restricts their range on the rivers. A typical design in current production, the *Yakutsk* class, has a capacity of 2,100 tons, and a new design of dry-cargo vessel with enhanced capability for working in ice will have a capacity of 2,800 tons. Non self-propelled barges are of somewhat larger capacity. One class produced since 1980 for use on the middle Lena is 85 m long by 16.5 m beam, with a draft of 2.55 m at 2,500 tons capacity or 2.95 m at 3,000 tons capacity. It can operate in 10 cm of complete-cover ice and can also be used at sea to a limited extent. At sea such barges are towed, usually in pairs. On the main rivers they can be used in consists of 18,000–24,000 tons with pusher-tugs, though usually the consists are smaller.[44] Larger sizes are unpopular because of excessive delays in poorly-equipped ports.[45] Most petroleum products are shipped in self-propelled tankers of similar size to the dry-cargo vessels, and some dry-cargo vessels are used as container carriers, with capacities of up to seventy-two 15-ton containers, or ninety with modifications to the cargo holds.[46]

Routes to the north east

It was mentioned above that freight arriving by the NSR is trans-shipped to river vessels which ply the smaller rivers draining to the Arctic. There are in fact several routes by which freight can reach those rivers. From 1951 until recently the established route has been by river vessel from Osetrovo to Tiksi, a seaport near the Lena delta. Trans-shipment to coastal vessel for the trip to the mouths of the other rivers is followed by another trans-shipment for travel upstream. Alternatively, river-sea vessels can be used for the whole route. They have been operating below capacity because of shallow water at Osetrovo and, until very recently, across the delta, as well as across the river-mouth bars of some of the other rivers. This alternative is likely to become increasingly popular as more effective river-mouth dredging and upstream water control reduce the capacity constraints. A third possibility, used more at present because of a shortage of river-sea vessels and the greater efficiency of pure river vessels on the rivers, is to use the river-sea vessels for parts of the route – from Yakutsk downstream or from Tiksi, for example. Fourth, coasters and river-sea vessels can distribute freight delivered to northern seaports along the NSR from Murmansk, Kandalaksha, and even Leningrad in the west, or from Vladivostok. Ocean-going vessels used to deliver directly to some of the river mouths, but in the extremely severe conditions of 1983 several became trapped in the ice, and one, the *Nina Sagaydak*, was lost. Since then trans-shipment has generally taken place at the better-equipped ports of Tiksi and Pevek.[47] Tiksi in particular has been substantially re-equipped since 1979 to handle deep-sea vessels.[48]

Organizational and technical problems

The Osetrovo bottleneck Osetrovo, the main entry port for the north east, seems to lurch from year to year in a permanent state of crisis. Plagued by physical problems, it also faces severe technical and organizational difficulties. To move nearly 5,000,000 tons in four months requires that freight brought in by rail be ready to load throughout the navigation season; that ships be always available for loading; and that trans-shipment be as fast as possible. Despite considerable expansion of port facilities in the past decade, Osetrovo trans-shipped all goods presented by the end of the navigation season in only one recent year, 1987.[49] All three aspects of the trans-shipment process normally cause problems. First, the availability of railborne freight at the dockside is hampered by congestion both in the port and on the railway leading to it. The port lacks storage space. By 1990 it is supposed to be able to accumulate a million tons of freight during the closed season (i.e. over a third of the dry goods planned for trans-shipment in that year),[50] but at present it relies heavily on a continuous flow of freight cars during the navigation season. Unfortunately the railway lacks a large marshalling yard near the port. At times during the 1987 season there were up to 3,500 freight cars backed up along the railway from Tayshet.[51] Even refrigerator cars with perishable goods had to wait months to reach the port. Although, as noted above, the port managed to clear all the traffic offered, the railway did not in fact manage to move in all it should have.

Second, even if the freight is ready, ships are not always available for loading. In the first place there are insufficient dry-goods vessels on the river. It appears that Gosplan works out the number of vessels required on the assumption of ideal operating conditions. No account is taken of the possibility that forest fires may cause delays, that vessels may have to operate below capacity because depths are inadequate, or that an early freeze-up may catch them away from their maintenance yards. Another problem, common to all the eastern rivers, is that investment in vessels tends to run ahead of investment in port facilities. Vessels are frequently held up at their destinations by slow unloading, or because cargo from the previous arrival is still clogging up the facilities.[52]

The most forcefully implemented improvements in the past few years have been in trans-shipment. The great bulk of general cargo for the north is now containerized or packaged. In Osetrovo, which suffers from a labour shortage, the entire improvement in throughput from 1986 to 1987 was due to the reduction of labour-intensive trans-shipment. In 1987, fourteen more freight cars with containers moved through the port every day,[53] but the total number of cars moving through rose by only thirteen a day. By 1990 it is intended that movements in large containers through Osetrovo will amount to 500,000 tons, or half of all such movements on Soviet rivers.[54] Containerization, however, causes its own problems. Many containers fail to return from the north, where they are in demand for storage. At the end of the 1987 navigation season the river authorities owed the railways 63,400 medium-sized containers.[55] Over 40,000 remained in the north over the winter.[56] Those that do return clog up Osetrovo port, because the railways cannot move them out fast enough. At the end of October 1987 there were 28,000 awaiting removal.[57] For these reasons there has been some shift of preference from containers to non-returnable packaging.

Alternatives to the route through Osetrovo Some freight, as we have seen, goes all the way down the Lena to Tiksi, then along the coast and up the Arctic rivers. Because of congestion at Osetrovo, the river authorities have suggested that the NSR, which has been carrying a growing share of freight for the northern coast and Arctic rivers since 1984, take it over entirely. This proposal seems logical at first sight, particularly since some freight is sent to Yakutsk and even Lensk by the NSR. Moving upstream, it passes freight for the north coast moving downstream, so any spare capacity created on the Lena could be used to correct the cross-shipments. However, there are sharply divergent viewpoints on how best to supply the Arctic coast and indeed the whole north east. River transport authorities themselves state that the Lena route is 20 to 60 per cent cheaper than the NSR. Goods from central European USSR and the Urals destined for the river Yana cost 48–61 per cent less to send by rail to Osetrovo and thence by river-sea vessel, than by rail to Murmansk or Arkhangel'sk and thence by ocean-going ship and river vessel. Goods shipped from Eastern Siberia and the Soviet Far East to the river Kolyma cost respectively 35 to 20 per cent less by the river route than by rail to Vladivostok and thence by sea.[58]

If the river route is so much cheaper, why do goods still travel by sea? If congestion at Osetrovo is the reason, why is heavy investment going into Arctic shipping, as it is, while the Lena river authorities are short of funds for, for example, deepening the river at Osetrovo? One reason is certainly that an updated Arctic fleet can provide logistical support for the armed forces and for Arctic mineral exploration. Another, suggested by the writer in an earlier paper, is the international prestige accruing to the first nation to master the Arctic Ocean effectively. However, A.L. Buritko, a planner with the Soviet railways ministry, has recently asserted that there are also reasons connected with the organization of Soviet transport and industry.[59] The Ministry of the Merchant Marine, he claims, is happy with the long voyages involved – 8,000 km from Leningrad to Tiksi or Zelenyy Mys. The Ministry is paid by the ton-km of work performed, and its performance indices are helped by the fact that ships on such long routes spend only a small proportion of their total working time in port, where delays are common. The shippers and receivers of goods are also content. The high transport costs are anticipated and included in production costs. The bigger the money turnover, the better for their organizations in terms of profits (added on as a percentage of production costs, since there is no competition), size of staff which can be employed, and hence prestige. Some freight is even sent deliberately by roundabout routes. An example is asbestos cement, sent from Spassk to Yakutsk by way of Vladivostok rather than Osetrovo.

As a comment on the losses to the national economy caused by this situation, Baritko also points out that the ships which take cargo from Leningrad or Murmansk to Tiksi or Pevek usually return westwards empty. But the railway freight cars which could have taken their cargoes to Osetrovo still travel eastwards, empty, because their prime function is to move Siberian coal, wood and other raw materials westwards. Baritko claims that the logical procedure would be to upgrade the railway to Osetrovo, provide it with a marshalling yard, and build a low dam on the Lena to raise the water level to a guaranteed 2.5 m from Osetrovo to Kirensk. If the dam had been built twenty years ago, when the need for it was first realized, its cost would have been less than the extra transport costs incurred since then to keep Yakutia supplied. The design was in fact completed in 1987, but its prospects are uncertain. River-transport and hydro-electric interests are opposed by an ecological lobby, which wants the Lena left in its natural state; by engineers, who foresee damage to navigation downstream; and by economists, who claim that the transport needs of Yakutia can be met adequately and more efficiently by the Amur–Yakutsk Mainline, to be discussed below, or by a proposed railway along the bank of the Lena to an expanded port at Kirensk.[60]

The use of the NSR is not the only instance of great expense caused by the bottleneck at Osetrovo. Air transport has often had to compensate for the inadequacies of surface transport at enormous cost. For example, sometimes the Lena has frozen before shipments of petroleum products to Yakutsk have been completed, and fuel has been sent in by air. This takes one ton of fuel for every ton transported to Yakutsk.[61] Another example concerns the Never-Yakutsk highway. Before the most recent expansion of Osetrovo was completed, 9,000 extra drivers had to be assigned to the highway to carry the goods which Osetrovo could not handle.[62] And in some years, including 1987, freight has been sent by river-sea vessel from Krasnoyarsk down the river Yenisey, around the Taymyr Peninsula with icebreaker assistance, and up the river Lena.[63]

Departmentalism Baritko believes that the pursuit of narrow departmental interests leads to huge investments without adequate evaluation from a national viewpoint. One example is atomic-powered icebreaking barge-carriers, a much-heralded forthcoming addition to the Soviet Arctic fleet.

They can service poorly-equipped locations along the northern coast, dropping off loaded barges, picking up empties or return cargo, and wasting no time in laborious trans-shipment. During the period 1988–90 a consortium of maritime, river and railway authorities is to work out the relative merits of running the new ships from Murmansk or Leningrad to the west, or Vladivostok to the east. No comparison, apparently, is to be made with the costs of supply via the Lena, either by river-sea vessels or by barge down the Lena and coastal barge-carrier to other river mouths.

Road transport

The final leg of many journeys in Yakutia and the north east is by road. In addition to the urban roads and the few all-weather long-distance roads (which include roads from Yakutsk westwards and from Lensk, a port on the Lena, to Mirnyy in the diamond-mining district as well as those already described), there are thousands of kilometres of winter roads in the region. These are routes used after freezing, partly across country and partly along frozen rivers. They are in use from 100 to 125 days in the south and from 180 to 220 days in the north, and they account for about half of the traffic volume by road.[64] In Chukotka they comprise 80 per cent of all roads.[65]

Though cheap to build compared to a permanent road or a railway,[66] winter roads are very expensive to operate. It is reckoned that they are uneconomic for traffic of 100,000 tons a year over distances greater than 100 km. That amount of traffic would be required by one of the larger mining operations in the north east.[67] Some writers claim that winter roads are used to excess, partly for organizational reasons: common carrier road transport tariffs do not take the type of road into consideration. This means that winter-road transport is underpriced. For some northern enterprises this is sufficient incentive to avoid the building and maintenance of storage facilities, which would be required if all goods arrived by river during the summer.

The Amur–Yakutsk Mainline

So far the question of access to Yakutia and the North-east has been framed in terms of one main route, the river Lena, and a few minor routes which supplement it and in part compete. One current project could change the picture. That is the Amur–Yakutsk Mainline (AYAM). The route was first surveyed at the end of the 1930s, but construction was approved by the Politburo only in March 1985.[68] Completion to Tommot (380 km) is planned for 1990, and to Yakutsk (830 km) for 1995. Conditions for building are hard. Much of the route passes over deep permafrost. Twenty large bridges will be required, but fortunately no tunnels – the bane of the BAM. Winter temperatures along the route range down to −63°C. Track-laying has begun, but the project has already encountered problems. Designated funds appear to have been diverted by the railways ministry to other projects it considers more important, and the builders apparently even lack blueprints. Unless the AYAM is designated a priority project very soon, it seems unlikely to meet its deadlines. In its proposed form, too, it seems unlikely to do more than relieve the strain on Osetrovo. Traffic through Osetrovo has been growing at 4.2 per cent per annum. Central Yakutia is reckoned to be short of needed goods by 250,000 tons a year at present.[69] Total shipments out of Yakutia are expected to be 1.8 times the 1985 figure in the year 2000, and inshipments about 2.5 times. The latter figure implies growth from 6,400,000 to 15,300,000 tons. (This includes shipments into the extreme south by rail, not included in figures quoted earlier.)[70] The Little BAM, of which the AYAM is a continuation northwards, is already seriously overloaded, five years after completion. Despite all this, the AYAM is to be built as a single-track line with low-grade rails.[71] It seems likely that the almost half-hearted attitude on the part of some authorities, particularly the railways ministry, may reflect more than immediate financial problems. First, the preceding big railway project in the Far East, the BAM, is so far proving to be an enormously expensive white elephant, and the ministry may fear a second BAM. Second, *perestroyka* may dampen enthusiasm not only for such transport projects, but also for north-eastern resource development in general. *Perestroyka* and its potential impacts will be discussed at the end of the chapter.

The Pacific coast and islands

This second major sub-region includes the mainland coast of the Sea of Okhotsk, Sakhalin Island and the Kuril Islands, the outer coast of Kamchatka, and the south coast of Chukotka. Notable features of the Soviet eastern coastline are the number of settlements with no landward transport links, and the lack of coastwise roads and railways. Coastal trade is therefore the main *raison d'être* of the Soviet Pacific merchant fleet, supplemented by river-sea vessels of the Amur River Steamship Company around the Sea of Okhotsk. The maritime fleet also serves the Arctic coast of Chukotka. There is little overlap between the Northern Sea Route service areas of Vladivostok and Murmansk east of Pevek, and river-sea vessels from the Lena do not normally penetrate farther east. Indeed, the Anadyr' river fleet is run by the Ministry of the Merchant Marine, which provides the only access to the river through the ports of Anadyr' and Provideniya.

Maritime transport

Traffic patterns In 1986, the nineteen Far Eastern ports belonging to the Ministry of the Merchant Marine had a turnover of 50,000,000 tons of dry cargo (oil tanker figures have not been found), including 20,000,000 tons of imports and exports. If we assume partial double counting of coastal traffic, i.e. deliveries to other Ministry ports rather than to, for example, ports belonging to the Fisheries Ministry, then the total carried may have been about 40,000,000 tons, about half of it coastal. This compares with 25,000,000 tons (including oil), about 60 per cent coastal, in 1970).[72] Coastal trade includes the distribution of supplies from Vladivostok and the other big ports in the south, the collection of resources (fish, wood and minerals), and some movements not involving the big mainland ports. For example, some 550,000 tons of coal a year move from Pos'yet, near the North Korean border, to Petropavlovsk-Kamchatskiy.[73] Because of the absence of other links to coastal settlements, passenger traffic is unusually important. The Far East accounts for about 18 per cent of all passengers carried by sea in the country,[74] and far fewer of them are tourists than elsewhere. Virtually all the domestic maritime traffic, freight and passenger, is confined to the Far East. Inter-basin traffic accounts for 1 or 2 per cent. A characteristic of intra-basin traffic is that the distribution of supplies to outlying ports far outweighs the return traffic to Vladivostok and the other big southern ports – by five times in the case of Kamchatka and ten times in the case of Magadan.[75]

Foreign trade can be divided into three main components. The first is trade with other socialist countries in east and south-east Asia, though much of the trade with south-east Asia is handled through Soviet Black Sea ports. The second is long-distance trade with non-socialist countries around the Pacific (grain imports from Canada, for example) which is largely handled by Soviet vessels. The third is trade with Japan, including coal and wood exports and machinery imports, in which the Amur River Steamship Company, using its river-sea vessels, takes part alongside the maritime fleet. Of total tons carried in foreign trade, exports account for 55 per cent.[76]

Organization Maritime shipping is run by four steamship companies. The Far Eastern Steamship Company, based in Vladivostok, is much the largest. In 1973 it had over 200 dry-cargo vessels, fifteen passenger ships and three large icebreakers. In 1970 it carried 14,600,000 tons, including 8,000,000 in the coastal trade.[77] The Primorskiy Steamship Company, formed in 1972, is based in Nakhodka and had a fleet of oil tankers, numbering forty-six in the early 1970s. The Sakhalin Company, based in Kholmsk, has some sixty dry-cargo vessels used mainly in coastal shipping, and the smallest of the four companies, the Kamchatka Company, had forty-five dry-cargo vessels in the early 1970s and had fifty-three in the late 1980s. Its current annual turnover is 2,300,000 tons. The total Far Eastern fleet, together with the North-eastern Administration of the Merchant Marine (covering the eastern Arctic) accounts for about 17 per cent of the Soviet merchant tonnage, or 2,900,000 registered tons, and 12 per cent of the traffic handled, measured in ton-kilometres.[78]

Though the names of the companies are territorial, the service areas overlap. For example, the Kamchatka Company accounts for 55 per cent of the maritime traffic of the Kamchatka Peninsula,[79] but the Primorskiy Company delivers petroleum products, the Far Eastern Company handles containers, and the Sakhalin Company delivers some coal (though not that from Pos'yet, mentioned above, which is handled by the Kamchatka Company). Foreign trade is handled mainly by the Far Eastern Company, but not exclusively. Even for the relatively insignificant Kamchatka Company, foreign voyages account for 12 per cent of goods carried, though this is partly because its ships enter the Primorskiy Kray–Japan trade in winter after the close of navigation around Kamchatka. They also operate to China and Singapore, while the Sakhalin Company operates to Canada and Australia as well.[80]

Ships and shore facilities   The make-up of the Far Eastern fleet presents sharp contrasts. Lesser companies like the Sakhalin and Kamchatka Companies complain about the high proportion of old vessels, discards from the Black and Baltic Sea fleets in many cases. The Kamchatka Company has the oldest fleet in the country, with an average age close to twenty years.[81] At the same time the region has more new, specialized vessels than the national average.[82] They include several of the *SA–15* or *Noril'sk* class of multi-purpose icebreaking freighters built in Finland. Operating at 15,000 tons deadweight in Arctic ice or 20,000 tons otherwise, they have engines of 21,000 hp and can break ice one metre thick without backing and ramming. They have stern ramps for RoRo cargo, their own cranes, tanks, and in some cases a refrigerated hold. Though technically very successful, they have proved expensive to operate.[83] Other specialized vessels include container carriers and the *Aleksey Kosygin*, a prototype icebreaking barge carrier. However, there are problems associated with the new vessels. One, not a new problem but one brought into focus by the high costs and operating expenses of these new, complex vessels, is related to the fact that the Far Eastern fleet both helps to service the Arctic coast and operates in ice for part of the year in its own region. Ice-capable vessels have to be maintained at 35–40 per cent of the fleet, but their capabilities are put to use only at the beginning and end of the navigation season. The rest of the time they are much more expensive to operate than other vessels doing the same job. The tariff structure offers no compensation.

A second and particularly embarrassing problem with new equipment relates to log exports to Japan. Saw logs are shipped from eighteen ports, most of them in Primorskiy Kray, and pulp logs from fourteen or fifteen of them, mainly from Vanino and Nakhodka. Until 1966 logs were rafted to Hokkaido, but the rafts sometimes broke up, so self-propelled vessels came into use instead. Then it was decided to introduce tugs with self-dumping log barges, of the type used on the Canadian west coast. However the Japanese, who had not been consulted, refused to let the logs be dumped into the water at their destinations, for ecological reasons, and with conventional unloading procedures the new technology is more expensive than the old.[84]

A third problem is that, in a situation analogous to that on the river Lena, the updating of shore facilities has tended to lag several years behind the introduction of new vessels. Thus the *SA–15*s are equipped with ramps for RoRo traffic, but outside the Vladivostok region there is only one port, Magadan, able to handle such traffic.[85] The barge carrier *Aleksey Kosygin* has been operating in the Far East for three years, partly on a run from Vladivostok to Kamchatka and Chukotka. But facilities for handling barges have not yet been provided at most of the ports it is supposed to visit, so the ship has been used much of the time not for its intended purpose, but for carrying imported grain.[86]

Even apart from the special requirements of new vessels, shore facilities are often deficient. Even Vladivostok cannot handle grain in bad weather, because the terminal is not under cover. Also, the port may receive 450 to 500 freight cars a day, but it can only process 100 to 120.[87] Facilities away from the big ports tend to be much more primitive. Northern Sakhalin Island, Kamchatka and the Kuril Islands have many locations where all cargo is trans-shipped at roadsteads.[88] This multiplies the time required and raises the incidence of damage. In the region as a whole about 12 per cent of traffic is handled at roadsteads.[89] One reason for the primitive facilities is that many small ports are operated not by the Ministry of the Merchant Marine but by organizations which have no reason to invest in new equipment for the convenience of visiting ships. Out of twenty-two ports serviced by the Kamchatka Company, for example, eighteen belong to the Ministry of Fisheries and are especially primitive.[90]

Where new shore equipment has been installed, co-ordination among the various elements in the production and transport process is sometimes lacking. The facilities at Vostochnyy for handling wood chips and coal are under-used, and so are those at Nakhodka for handling sawn lumber. The problem with the coal facility is that the railways cannot move freight cars through fast enough for it to work at full capacity. The problem with the lumber facility is lack of lumber.[91]

In another situation analogous to that on the river Lena, containers pose special problems. There has been a determined move towards complete containerization of general freight in coastal shipping, and container vessels have been purchased to that end. But although specialized container facilities have been built in Vladivostok, Petropavlovsk-Kamchatskiy (where the terminal should be able to handle 100,000 containers per annum by 1989), and Magadan, they do not yet exist in Chukotka or along the Arctic coast – areas which account for 15–20 per cent of all container traffic – or in the Kuril Islands. This means that containers have to be unloaded with ordinary cranes, a waste of time and money.[92] Another problem is that most general cargo in the Far East is one-way. Using containers means that provision must be made for picking up empties. There is of course the final problem that they are often not there to be picked up but are being used for storage, just as they are in Yakutia.

Major ports None of the Far Eastern ports handles very large tonnages by world standards. Vladivostok is the oldest, founded in 1860. In the mid–1970s it handled somewhat more than 9,000,000 tons per annum, though in 1986 only 'over 5,000,000 tons'.[93] It requires icebreakers for four to five months in winter and is afflicted by fog in spring and summer, so expansion of capacity at the end of the Trans-Siberian has taken the form of opening new ports rather than enlarging Vladivostok. Nakhodka, east of Vladivostok, received urban status in 1950 and in the mid–1970s handled the largest tonnage in the Far East, well over 10,000,000 tons per annum with an emphasis on coal, wood and general cargo. It handled about two-thirds of all the foreign trade of the Soviet Far East at that time, but it too has come up against natural limitations, related less to ice (which forms in winter, but rarely to such an extent as to hamper shipping) than to a lack of land for industry and storage. Further development has therefore focused on the adjacent port of Vostochnyy. Its first wharves were opened in the early 1970s, and it is intended to handle ultimately over 30,000,000 tons a year, mainly forest products, coal and containers. In 1986 its turnover was 7,800,000 tons, all of it imports or exports and including 5,400,000 tons of coal exports. Nakhodka's turnover was 6,000,000 tons.[94]

One other major port is linked to the railway system directly. Vanino, with the adjacent Sovetskaya Gavan' (Soviet Harbour), is located at the end of a line from Komsomol'sk, which was isolated by the lack of a bridge across the river Amur until 1975.[95] Now it has become prominent firstly as the port at the end of the BAM, and secondly as the mainland terminus for a railway ferry to Sakhalin Island. North of Vanino the

Tatar strait between Sakhalin Island and the mainland is both very shallow (as little as 5.0 m in places) and icebound in winter, but the ferry and the port operate year-round. Vanino is expected to grow when the BAM is in full operation.

Magadan, or rather its outport of Nagayevo, also remains open year-round, though icebreakers or icebreaking freighters are needed for about five months. Korsakov, the main port on Sakhalin Island, Kholmsk, the second port and island terminus for the ferry, and Petropavlovsk-Kamchatskiy, the principal port on the Kamchatka Peninsula, also remain or can be kept free of ice. Petropavlovsk has an exceptionally fine harbour, though one badly hampered by fog, and freight turnover in the early 1970s was 2,500,000 tons. Ninety per cent of that consisted of arrivals, probably reflecting the importance of military supplies.

One other regional port should be mentioned. Provideniya is the most easterly port in the Soviet Union and the main repair and maintenance base for the eastern Arctic. It was founded in 1937 and has a well-sheltered harbour.[96]

River transport So far this section has concentrated on sea transport, since over much of the sub-region there is very little else except for some river traffic on the Kamchatka and Anadyr' rivers and local road traffic. The Anadyr' takes very small amounts of traffic, mostly supplies for local communities, but the Kamchatka is more important. It is navigable for 570 km, though only with depths of 1.0–1.5 m and sometimes less over a river-mouth bar. General supplies move upstream, in the amount of 100,000 tons in 1988 if the year's plan was realized, while fish, furs and, mainly, forest products move downstream.[97]

Sakhalin Island Sakhalin Island is an exception to the dominance of sea transport in the sub-region. Indeed it is almost an extension of the relatively well-serviced southern extremity of Primorskiy Kray. This is to be expected, since although it comprises only 1.4 per cent of the land surface of the Far East, it accounts for 30 per cent of the industrial production and 20 per cent of the construction activity of the region. The island is served by railway, road, sea, river, air and pipeline transport, but its problem is lack of integration, partly the after-effect of shifts between Soviet and Japanese control of the southern half. There are three railway gauges in public use: 1520, 1067, and 750 mm (5', 3'6", and 2'5½"), and even the broad-gauge railways cannot take heavy rolling stock. Rail and road transport are hampered by rough terrain, heavy snowfall, ice and fog. Some 30 per cent of the roads are winter roads, and most of the bridges on the permanent roads are wooden and unable to take heavy vehicles. Furthermore there are few common-carrier vehicles, and since the various organizations owning their own vehicles do not co-operate, there are many empty return hauls. The fundamental climatic problems also affect air and sea transport, and the latter has to cope with the poor port facilities already mentioned. In consequence of all this, ton-kilometre costs are well above the national average: almost 8 times on the railways, 2.6 times on the roads, 2.4 times by sea, and 2.2 times by air.

The picture is not entirely unfavourable, however. The maritime fleet has been undergoing rapid renewal, and the ports of Kholmsk and Korsakov were rebuilt during the last Five-Year Plan. About 300 km of new railways were built during the 10th and 11th Plans, including a major north–south route. Finally, the Vanino–Kholmsk railway ferry has greatly improved links with the mainland, though there remains the delay of switching gauges from 1520 to 1067 mm.[98]

An additional problem until recently was the need for improved transport services to cope with the oil and gas exploration off the north-east coast of the island. The main local land transport is a railway run by the oil and gas industry (on yet another gauge – 760 mm or 2'6") which achieves speeds of 13 km/hr.[99] The recent suspension of activities, however, has removed the problem.

The southern railway belt

This sub-region is examined last because the Trans-Siberian Railway does not only serve its local area and facilitate foreign and transit trade. It also serves the two sub-regions already examined. Indeed most of the developments and problems of the whole Far East are eventually expressed in some form of impact on the Trans-Siberian. The particular interest of the railway belt in recent decades, however, has been as the location of two of the country's major investment projects, both linked to the resurgence of foreign trade. They are the BAM and the development of the new southern ports: Nakhodka, Vostochnyy, and Vanino.

From the standpoint of transport facilities the sub-region can be further divided into three: Primorskiy Kray; the southern interior along the Trans-Siberian in Khabarovsk Kray and Amur Oblast; and the BAM zone. Primorskiy Kray has much the best developed railway, road and port facilities in the Far East. This one might expect, since it has been the terminus of the Trans-Siberian Railway for nearly a century, contains the best agricultural land and forest as well as attractive mineral resources, has the best locations for commercial ports, almost ice-free and close to Japan, and includes in Vladivostok one of the country's principal naval bases. The interior along the Trans-Siberian has similar but poorer natural resources and a much poorer, fragmented road network to supplement the railway. It was favoured during Stalin's time, relative to the Vladivostok area, by the internalization of the economy.

Khabarovsk grew as the main interior centre, the new steel mill of the era was located at Komsomol'sk, and a branch line from the Trans-Siberian joined the two in 1940. But the emphasis in investment switched to the southern coast from the 1950s to the early 1970s, when it again switched inland to the BAM. Finally, the BAM zone can be considered a region in transition from the viewpoint of transport facilities. The railway is unfinished and ancillary facilities have not yet felt its impact: winter roads still account for 40 per cent of the length of roads and half their traffic, measured in ton-km.[100]

The Trans-Siberian Railway

During the Stalin era the Trans-Siberian Railway was the lifeline of the Far East, almost the region's sole link not only with the rest of the Soviet Union but with the outside world in general. Since then its monopoly has gradually declined, but it continues to fill the same basic role. On the Zabaykal'skaya Railway, serving Amur Oblast and its neighbour in Eastern Siberia, Chita Oblast, transit freight still predominates. Petroleum products and grain move eastwards, together with a broad range of machinery and equipment. Westward movements, much smaller, include fish and fish products of the more valuable type, like canned crab and salmon, as well as tin, other non-ferrous concentrates and containers.[101] The bulk of the containers are in transit between the Far East and the COMECON countries of Eastern Europe, Scandinavia, or Iran. The Trans-Siberian Container Landbridge was set up in the early 1970s to compete for traffic previously carried by sea. By 1979 it had succeeded in capturing about 10 per cent of the total traffic and a quarter of that between Japan and north-western Europe. Thereafter the competing maritime lines succeeded in containing its growth to some extent. Traffic in 1988 comprised 98,000 containers (20-foot equivalent units), excluding that with Iran and Afghanistan. Two-thirds were westbound, and Japan accounted for 75,000. Japanese and Soviet ships feed containers into a Japanese-built terminal at Nakhodka from places as far afield as Rangoon and Singapore. Though an important earner of foreign currency the Landbridge contributes only a few per cent to traffic volumes on the Trans-Siberian.[102]

On the Far Eastern Railway, which serves Primorskiy Kray, most of Khabarovsk Kray, and Sakhalin Oblast, local movements make up half the traffic. Coal, wood, construction materials and petroleum products (from Khabarovsk and Komsomol'sk) predominate. This is despite the fact that the railway is the conduit for wood from Eastern Siberia and coal from South Yakutia moving to the Pacific for export, as well as for the container traffic described above. It reflects the need for a remote region in the Soviet Union to be relatively self-sufficient and, by Western standards, the lack of roads for intra-regional traffic.[103]

The Baykal–Amur Mainline

Construction of the BAM began in the 1930s but was abandoned during the Second World War. After the war two lines were built which would be components of the finished project but had independent rationales: the Amur-Sovetskaya Gavan' and Tayshet-Ust'-Kut (Osetrovo) lines, both described already. There were also branches from the Trans-Siberian to places which would eventually be on the BAM: Khabarovsk-Komsomol'sk in 1940 and Izvestkovyy-Urgal, to reach coal deposits, in 1948. When construction re-started on the BAM as a whole in 1974, therefore, it was able to do so at several points along the route simultaneously. A further branch was started at the same time, from BAM station on the Trans-Siberian, the nearest point to the BAM route, to Tynda. This branch was extended to the South Yakutian coalfields at Neryungri and became known as the Little BAM. It was the first part of the BAM project to be completed and, operating as a branch of the Trans-Siberian, accounts for the great bulk of traffic so far generated – about 13,000,000 tons. Its station Berkakit is the starting point for the AYAM, discussed above.

Starting work at several points simultaneously and declaring the BAM a national priority project led the Soviet government to believe that it could complete the line by 1983. Two tunnels remain to be finished, however, and hasty construction in a region which is both seismically unstable and affected by permafrost has already forced reconstruction of parts of the line and of many trackside facilities. The BAM is not now expected to be fully open until 1992, with the opening of the North Muya tunnel east of Lake Baykal.

Tardy completion of the BAM does not only reflect physical problems. Its role is also in question, both as a component of the transport system and in the economy in general. Early plans stated that it would move 35,000,000 tons a year initially, including 25,000,000 tons of oil from Western Siberia for export through Pacific ports. Later increments to traffic would come from exportable minerals along the route, including copper, molybdenum and asbestos. One source states that from 1981 to 1984 the line carried 50,000,000 tons and over 5,000,000 passengers.[104] A recent source, however, claims that the line is now carrying less than 1,000,000 tons a year,[105] which suggests that the earlier traffic may have comprised mainly construction supplies and workers for the railway itself.

World commodity prices declined drastically while the railway was being built. Markets for potential exports from the BAM zone became glutted and remain so, despite some recovery of commodity prices. Even on completion it is questionable how much traffic the line will be required to carry, at least until some of the mineral deposits along the route are exploited. At present levels of traffic it is a severe burden to the railways ministry. In 1985 its ton-kilometre costs were the highest of any railway in the country, 1.8 times those on the next most costly railway to run and 7.2 times those on the cheapest, which was the West Siberian Railway.[106] It must be remembered that the BAM was unique in Soviet railway-building history in receiving enormous investments well ahead of demonstrated demand – before the Trans-Siberian had even been fully electrified, for example. Certainly there may have been strategic reasons for wanting a second east–west line farther away from the Chinese border than the Trans-Siberian, but the BAM was also very much a personal prestige project for L.I. Brezhnev. It is noticeable that the only mention of BAM in a major speech by M.S. Gorbachev in Vladivostok in July 1986 related to the uses to which regional authorities could put the construction crews no longer needed on the route.[107]

The Amur–Ussuri river system

One transport mode not yet discussed, which also runs east–west through the railway belt, is the Amur–Ussuri river system. Its role is in sharp contrast to that of the river Lena. As is typical for rivers in the railway belt throughout the country, the Amur and its tributaries fill a subordinate role. The principal freight is sand and gravel, much of it dredged from the river bed and carried over short distances. Only the chance juxtaposition of the Raychikhinsk coal deposits and their markets (mainly the Komsomol'sk and nearby Amursk thermal power stations) gives the river some longer-distance traffic. The distance to Komsomol'sk is of the order of 1,000 km. In 1975, when coal accounted for 2,000,000 tons out of 16,900,000 tons in all transported by river, the average haul for all freight was 308 km. This compares with 1,067 km on the Lena, Yana, Indigirka and Kolyma together in the same year, when they carried 9,000,000 tons. At present, traffic on the Amur amounts to about 30,000,000 tons a year.[108]

Though less hampered than the north-eastern rivers, the river Amur does have difficult conditions for navigation. Out of 8,700 km navigable, 5,200 km have guaranteed depths. However, these are as little as 0.85 to 1.5 m on tributary rivers, a little more along the Amur itself down to Blagoveshchensk, and substantially more only thereafter.[109] The ice-free period is longer than on the Lena, averaging 186 days on the stretch from Komsomol'sk to Khabarovsk and 178 days from Khabarovsk to Blagoveshchensk. Two things particularly hamper commercial transport. First, because of the monsoon climate the Amur system is subject to large, destructive summer floods, alternating unpredictably with periods of extremely low water. Second, because the river turns north at Khabarovsk, it is far less useful than if it drained south to Vladivostok. Deep-draught vessels cannot enter the river at Nikolayevsk, ice conditions can be bad near the mouth, and the most direct route south by sea is that through the shallow Tatar strait, mentioned earlier.

The problem of flooding has been reduced somewhat by building a hydro-electric dam on the river Zeya. In the 1950s there were plans to control the whole system in co-operation with China, with benefits in the form of power generation, navigation and flood control. Very recently, with relations between China and the USSR improving again, negotiations have resumed.[110] However, it is questionable whether a controlled river would carry substantially more traffic. The ice-free period would probably be shorter on a slower-moving river; locks would slow traffic down; and completion of the BAM will probably free up capacity on the Trans-Siberian Railway. There is a possibility of expanding the international traffic, using navigable Chinese tributaries to deliver Soviet forest products, which has already resumed.

Projects

Two other projects have been under study in the railway belt. The first is a more direct water route from the Amur to the Tatar strait through the Kizi lakes.[111] The second is a railway to run southwards from Selikhin, 63 km east of Komsomol'sk on the line to Vanino, to Nakhodka. It would have a number of branch lines to the Pacific coast, but since its main purpose would be to relieve the Trans-Siberian of extra traffic generated by the BAM, it is unlikely to be built before the year 2000.[112]

Conclusions

Past development

The evolution of the Far Eastern transport system presents an interesting case study of the role of transport in a frontier region under Soviet socialism. By contrast with western practice, transport is seen not as a development leader but as a sector which should receive the minimum investment consistent with meeting a demonstrated demand. As in other parts of the country, this attitude has tended to cause underestimation of the stimulating effects of transport investment and therefore the construction of facilities is soon swamped by growing traffic. The Little BAM is a recent example.

At the same time the region provides examples of deviation from the standard attitude to transport. The BAM is a rare case of heavy investment ahead of demand, and the outcome suggests that such an experiment is unlikely to be repeated. The case of Arctic shipping is similar in that there may have been considerations of military strategy behind both investment decisions, but in other respects it seems different. It is possible that success in keeping the western sector of the NSR open all year led to unjustified enthusiasm for investment farther east. It is also possible that the need to respond in the short term to the bottleneck at Osetrovo has stimulated a level of investment which will be hard to justify in the long term.

Despite the traditionally cautious Soviet attitude towards investment in transport, the scale of investment north of the railway belt seems very large compared to that in northern Canada or Alaska. It must be remembered, however, first that northern development has been stimulated by a national policy of exploiting domestic raw materials rather than relying on world markets, and second that Soviet northern development has been relatively labour- and materials-intensive. Light portable buildings, for example, are a relatively recent addition to the Soviet inventory. In other words, investment in transport is not lavish relative to the demand for transport, with the two exceptions noted above.

The future

At the beginning of the chapter it was suggested that the evolution of transport in the Soviet Far East has reflected the interaction of a set of consistently important influences. Future developments are likely to reflect the same set of influences, but with one or two additions.

Demand for natural resources

As in the past, much will depend on the demand for natural resources. As far as domestic demand is concerned, the apparent determination to become more efficient in the use of hydrocarbons and to pay more attention to reforestation in northern European USSR may slow down the eastward hunt for resources which demand high-capacity transport. As for high-value-to-weight-ratio resources, there is likely to be closer scrutiny of the relative economics of exploiting them in the North-east, exploiting them in more accessible parts of the country if possible (gold in Central Asia, for example) or trading for them in world markets. As far as the exportability of Far Eastern resources is concerned, one can only speculate on trends in world, and especially Pacific-Rim, markets, and on the course and influence of relations with Pacific-Rim neighbours. The BAM when completed, however, should provide spare capacity for some time to come.

Attitudes to regional development

Up to the present, the remoteness of the Far East from the metropolis, coupled with the lack of scope for local initiative, has restricted regional development to resource exploitation, with enough additional settlement to demonstrate effective occupance. Mr Gorbachev's Vladivostok speech in the summer of 1986 and the follow-up plans published a year later imply that the government intends to achieve much more diversified investment and rounded development.[113] If that occurs, transport needs are likely to change: better intra-regional links may have to be forged. However, one views the current proposals with some scepticism. They are, after all, not very different from those published in 1967 in a special decree on the Far East and Chita Oblast.[114] Furthermore, there are clearly influential planners in the Soviet Union who believe that the Far East is a bottomless pit for investment and that the country is in greater need of investment where it can achieve the greatest national returns than it is of remote-area development for its own sake.

In relation to any programme for rounded, integrated Far Eastern development the further question arises of the definition of the region. At present it can scarcely be said that Yakutia and the north east on the one hand, and the rest of the Far East on the other, function as one region. Both are tied to European USSR, but separately. Completion of the AYAM, and its eventual extension to Magadan, which has been proposed for decades,[115] might conceivably give Yakutia more of a Pacific orientation, but the continuing separateness of the two major subdivisions of the Far East seems more likely.

International prestige and foreign policy

Non-economic aspects of foreign policy can affect transport development in two ways: through a growing need for logistical support if the Soviet military presence in the Far East continues to build up; and through a desire for prestige as the dominant nation in Arctic development. The latter seems likely to be mainly responsible if there appear, towards the end of the century, super-powerful icebreakers able to keep open the eastern sector of the NSR.[116]

New facilities and technology

A most noticeable feature of Soviet transport in the past few years has been a drive to improve the technology employed, in part by radical innovation but mainly by the more thorough adoption of known techniques. The latter is exemplified by an increase in the proportion of railway freight cars with roller bearings and by the rapid containerization of freight for the north. Innovation is found mainly in Arctic transport. There has been reaction against the cost of some of this, to be discussed in the final section, but some new ideas are more modest, having to do with the use of ice for building harbour structures, for example, and the preservation of these and of winter roads through the summer.[117] However, really radical technology such as vertostats (a combination of helicopter and dirigible), Arctic monorails and freight-carrying submarines or semi-submersibles, though much written about in the Soviet press,[118] seems to be a long way from implementation.

The impact of *perestroyka*

We come finally to the influence on transport which is of the greatest current interest. *Perestroyka* requires, among other things, a new and vital concern with cost accounting and self-financing at levels ranging from ministries to individual enterprises. It implies a new and different incentive system in the economy and new ways of evaluating options on the part of economic units.

Geography of economic activity We can distinguish two channels through which *perestroyka* can influence transport. The first is through its impact on the geography of economic activity. If the exploitation of natural resources in the North-east were re-examined against an alternative of trading in world markets, some current activities – tin mining for example – might not show up favourably. Even farther south, along the BAM, some of the resources highly lauded when the line was being built are rarely mentioned now. Furthermore some of the Far East's factories, producing specialized machinery for the national market, would probably be candidates for closure or re-orientation of their production to regional needs if evaluated against competitors farther west.

Profits, not volume Second, *perestroyka* affects transport directly. The Soviet railway system is divided into regional railways and divisions. The waterway system is divided into river steamship companies and the maritime merchant fleet, as we have seen, into maritime steamship companies. All of these have started to re-evaluate their roles under *perestroyka*, and the technical journals bear ample evidence of their shifting priorities. First, they are focussing more on what brings in income, rather than on what affects the performance indices which used to determine their status and bonuses, such as ton-kilometres covered during the year. For the river steamship companies this could well lead to increased concentration on dredging and delivering sand and gravel, an unglamorous but highly profitable activity, and a decline of interest in operating river-sea vessels. The latter cover long distances non-stop, which favourably affects the old performance indices, but over-enthusiasm has brought them some quite uneconomic hauls. Under *perestroyka* they are more likely to be restricted to hauls where their true advantage of bypassing river-mouth trans-shipment does indeed outweigh their high line-haul costs relative to pure river or maritime vessels. Similarly the maritime companies using the NSR may well re-evaluate their operations if, eventually, there is oversupply of freight capacity and they have to compete with the Lena route instead of relying on the long distances between ports to produce excellent performance indices.

Attitudes to new equipment Second, a changed attitude to new equipment is already evident. The river and maritime steamship companies are in a difficult position if they have taken delivery of new equipment recently. It seems clear that much equipment has been approved for adoption more because of its technical characteristics than for its economic performance in the existing environment of tariffs and depreciation allowances. Recent journal articles complain about the cost of new, and especially imported, river-sea vessels and the unrealistically long service periods expected of old vessels when they have to compete with newer ones in overseas markets;[119] about the excessively high purchase and operating costs of the Ka–32c ice-reconnaissance helicopter relative to the less-sophisticated Mi–2 it is to replace (which has resulted in an almost unprecedented cancellation of orders); the high operating costs of a new air-cushion lighter with amphibious tractor for unloading ships at roadsteads in the Arctic; and the burden of having to operate ships like the SA–15, which benefit from their capability in ice for only a small portion of the year and are expensive to operate the rest of the time.[120] This enhanced awareness of the economic characteristics of equipment will undoubtedly bring benefits nationally, but it presents difficulties in the Far East, where so many operations require special equipment (including road and rail operations as well as those cited above) but the companies involved may be governed by national tariff schemes.

Tariffs and subsidies It follows that, thirdly, regional and sectoral organizations have started to lobby strongly for changes in national transport tariff systems and in regional and sectoral subsidies. Railway authorities, for example, would like a full-scale national tariff review and, even before the review, the elimination of concessionary tariffs in the Far East, higher tariffs on the BAM and any other new lines in the Far East and the north, and the right to set their own tariffs for specific hauls in special-purpose freight cars.[121] The maritime companies would like to be compensated for having to run icebreaking freighters and to receive initial subsidies for upgrading outdated ports so that they can start their new careers on a competitive footing.[122]

Departmentalism Fourth and finally, *perestroyka* has thrown into relief some of the long-standing problems of departmentalism in Soviet transport. Departmentalism has been claimed to be a major cause of poor, unbalanced transport development in pioneering regions.[123] The government has been trying hard to improve co-operation, both within modes, for example through the creation of a State Production Association linking all the steamship companies in the Far East, and intermodally, especially at seaports and in the sharing of traffic between railways and waterways on overloaded routes. Despite successes in specific cases, an initial impact of *perestroyka* has been to heighten the attitude of 'every department for itself'. Co-operation is seen as an expensive nuisance. However, the old desire to grab all available traffic, to increase volume indices, is being replaced in some cases by a desire to discard unprofitable traffic. This has strengthened the call for tariff reform, to preserve operations which are economic from a national viewpoint but do not bring good rewards to any participant.[124]

Apart from discarding unprofitable routes, transport organizations may wish to discard whole divisions which they perceive as hampering their search for greater profits. One activity which has already suffered, apparently, is Arctic aviation. It has shifted, over several decades, from being a major operation of the NSR administration to being a minor nuisance for the Ministry of Civil Aviation and then, in part, a function of the Ministry of the Merchant Marine. As a result there is no good replacement in sight for the venerable I1–14 aircraft, used for visual ice-reconnaissance and the supplying of drifting Polar ice stations, and the radar equipment used for ice-reconnaissance is in dire need of updating. Prospects look worse under *perestroyka*, unless the service is shifted to some organization for which it will once again have high priority.[125]

In sum, the impact of *perestroyka* on transport in the Far East is quite likely to be negative, in the sense of slowing down development. National sectoral organizations, both in transport and in the economy in general, may try to minimize their investments, concentrating on parts of the country where returns come more readily. Local transport organizations may try to shrink their operations, focussing on profits and cutting out those activities which contribute only extra traffic volume. The future of Far Eastern transport will depend on the extent to which national schemes for regional development, the restructuring of transport tariffs, and perhaps a realization by the transport companies of the potential profits in intermodal operations, can counteract such trends.

The impact of *perestroyka* invites comparisons with the west. Will Far Eastern transport now develop in conditions more akin to those of northern North America? Certainly it appears that the similarity will be greater than in the past. But convergence of transport development may be slow. In the first place Far Eastern transport companies, especially the river and maritime companies, have recently acquired a great deal of equipment which they might have chosen not to buy if *perestroyka* had taken place earlier, but which will have to be part of their operations for another fifteen or twenty years unless they can afford to write it off. In the second place, *perestroyka* up to the present seems heavily concerned with improving the operations of existing transport companies. To reach any degree of similarity with the west, there would have to be provision for the creation of intermodal companies – a different order of change from trying to improve co-operation among companies with a strongly sectoral outlook. And in the third place, greater similarity with northern North America is likely to come only with an enlarged vision of the potential of road transport, indeed, an elevation from mode of last to mode of first resort.

Notes

1. Savin, N.I. (1984) 'Morskoy transport v narodnokhozyaystvennom komplekse Dal'nego Vostoka' in *Territorial'nyye aspekty razvitiya transportnoy infrastruktury*, Vladivostok: AN SSSR, Dal'nevostochnyy nauchnyy tsentr, Tikhookeanskiy institut geografii, p. 64.
2. Kolesov, L.I. (1982) *Mezhotraslevyye problemy razvitiya transportnoy sistemy Sibiri i Dal'nego Vostoka*, Novosibirsk: Nauka, Sibirskoye otdeleniye, p. 159. The line was not fully open until 1957, but by then it was already carrying 1,000,000 tons a year to Osetrovo.
3. Nikol'skiy, I.V., Tonyayev, V.I. and Krasheninnikov, V.G. (1975) *Geografiya vodnogo transportas SSSR*, Moscow: Transport, p. 211.
4. Krasnopol'skiy, B.Kh. (1984) 'Transport v sisteme regional'nogo khozyaystvennogo kompleksa Severo-Vostoka SSSR' in *Territorial'nyye aspekty*, op.cit., p. 56.
5. Chichkanov, V. (1985) 'Problemy i perspektivy razvitiya proizvoditel'nykh sil Dal'nego Vostoka', *Kommunist*, no. 16 (November), p. 99.
6. *Zheleznodorozhnyy Transport* (hereafter *ZT*), 1988, no. 1 (January), pp. 46–50.
7. Kolesov, *Mezhotraslevyye problemy*, op.cit., pp. 108–9.
8. *Geograficheskiy atlas* (1985) 4th edn, Moscow: GUGK pri Sovete Ministrov SSSR, p. 168.
9. Kolesov, *Mezhotraslevyye problemy*, op.cit., p. 100. Figures probably for 1977.
10. ibid., p. 175. Ke $= \frac{L}{\sqrt{5P}}$, where L = route length in km, S = area of region in thousand sq. km, and P = population in thousands.
11. ibid. $K_u = \frac{L}{\sqrt{5PQ}}$, where Q = the total value of gross regional product in million roubles.
12. *Atlas avtomobil'nykh dorog SSSR* (1980) Moscow: GUGK pri Sovete Ministrov SSSR, pp. 90–8; *Atlas zheleznykh dorog SSSR* (1986) Moscow: GUGK pri Sovete Ministrov SSSR, pp. 132–6.
13. Kolesov, *Mezhotraslevyye problemy*, op.cit., pp. 82–3.
14. *Sibir' v yedinom narodnokhozyaystvennom komplekse* (1980) Novosibirsk: Nauka, p. 217.
15. Kolesov, *Mezhotraslevyye problemy*, op.cit., p. 125.
16. *ZT*, 1987, no. 6 (June), pp. 54–9; *ZT*, 1987, no. 8 (August), p. 54.
17. Kolesov, *Mezhotraslevyye problemy*, op.cit., p. 45.
18. Polyakov, Ye.A. (1963) *Sravnitel'naya effektivnost' razlichnykh vidov transporta v maloosvoyennykh rayonakh SSSR*, Moscow: AN SSSR, p. 79; Mitaishvili, A.A. (ed.) (1982) *Ekonomicheskiye problemy razvitiya transporta*, Moscow: Transport, pp. 205–28.
19. Savin, 'Morskoy transport', op.cit., p. 63. The figure is difficult to reconcile with other transport data on the region but may simply reflect the inclusion of foreign, including trans-Pacific, trade.

20. Kolesov, *Mezhotraslevyye problemy*, op.cit., p. 122.
21. Shabad, Theodore (1986) 'News Notes', *Soviet Geography*, vol. 27, no. 10 (December), p. 756.
22. *Soviet Shipping* (1987) no. 4, p. 22.
23. Shabad, Theodore (1981) 'News Notes', *Soviet Geography*, vol. 22, no. 7 (September), p. 454.
24. *Rechnoy transport* (hereafter *RT*) (1986) no. 3 (March), p. 17. The dry-cargo figure rose to 2,300,000 tons in 1987 (*RT* (1988) no. 4 (April), p. 27), but the figure for petroleum products may have fallen, to judge from *Vodnyy transport*, 20 December 1988.
25. Bagrov, L.V. (1986) *Rechnoy transport Rossii na puti intensifikatsii*, Moscow: Transport, p. 71.
26. *Morskoy flot* (hereafter *MF*), 1987, no. 1 (January), p. 15.
27. Gromov, N.N., Burkhanov, V.F. and Chudnovskiy, A.D. (1982) *Transportnoye obsluzhivaniye severnykh rayonov SSSR*, Moscow: Transport, p. 24.
28. *MF*, 1987, no. 3 (March), pp. 42–5.
29. *RT*, 1986, no. 6 (June), p. 38; *RT*, 1986, no. 8 (August), p. 42.
30. *RT*, 1987, no. 9 (September), pp. 2–3.
31. ibid.
32. Tonyayev, V.I. (1977) *Geografiya vnutrennikh vodnykh putey SSSR*, Moscow: Transport, pp. 137–8.
33. *RT*, 1987, no. 2 (February), pp. 32–3.
34. Tonyayev, *Geografiya*, op.cit., p. 136; *RT*, 1987, no. 9 (September), pp. 2–3.
35. Tonyayev, *Geografiya*, op.cit., p. 131.
36. Arikaynen, A.I. (1984) *Transportnaya arteriya sovetskoy Arktiki*, Moscow: Nauka, p. 174.
37. *RT*, 1985, no. 7 (July), p. 40.
38. *MF*, 1985, no. 6 (June), pp. 36–7.
39. *Soviet Shipping*, 1987, no. 4, p. 22; *MF*, 1988, no. 1 (January), pp. 34–7; Gromov et al., *Transportnoye*, op.cit., p. 27.
40. *ZT*, 1987, no. 2 (February), pp. 46–52.
41. *RT*, 1986, no. 8 (August), p. 42.
42. *RT*, 1987, no. 4 (April), pp. 28–9.
43. *RT*, 1987, no. 9 (September), pp. 2–3.
44. *Sudostroyeniye*, 1987, no. 9 (September), pp. 4, 6.
45. *RT*, 1988, no. 1 (January), p. 16.
46. *RT*, 1987, no. 10 (October), p. 21.
47. *RT*, 1987, no. 3 (March), p. 33.
48. Shabad, Theodore (1981) 'News Notes', *Soviet Geography*, vol. 22, no. 7 (September), p. 454.
49. *RT*, 1988, no. 4 (April), p. 27.
50. Bagrov, *Rechnoy transport*, op.cit., p. 71.
51. Baritko, A.L. (1988) 'Mozhno li razvyazat' Lenskiy "uzel"?', *ZT*, no. 4 (April), p. 60.
52. ibid., pp. 60–1.
53. ibid., p. 60.
54. *RT*, 1987, no. 5 (May), p. 5.
55. *RT*, 1988, no. 4 (April), p. 27.
56. Baritko, 'Mozhno', op.cit., p. 60.
57. *RT*, 1988, no. 4 (April), p. 27.
58. *RT*, 1987, no. 3 (March), p. 33.

59. Baritko, 'Mozhno', op.cit., pp. 58–62.
60. *RT*, 1987, no. 9 (September), p. 2; Bagrov, *Rechnoy transport*, op.cit., p. 72; *Vodnyy transport*, 20 December 1988; *Izvestiya*, 10 October 1988, p. 2; *RT*, 1988, no. 12, p. 17; *ZT*, 1988, no. 11, p. 43.
61. Baritko, 'Mozhno', op.cit., p. 58.
62. *Izvestiya*, 12 April 1985, p. 6.
63. Baritko, 'Mozhno', op.cit., p. 60.
64. Gromov *et al.*, *Transportnoye*, op.cit., p. 24.
65. Kobylyanskiy, V.V. and Savin, N.I. (1984) 'Perspektivy i problemy razvitiya seti avtomobil'nykh dorog na Severo-Vostoke SSSR', in *Territorial'nyye aspekty*, op.cit., p. 88.
66. Kolesov, *Mezhotraslevyye problemy*, op.cit., p. 171.
67. Kobylyanskiy and Savin, 'Perspektivy', op.cit., pp. 89, 91.
68. *ZT*, 1987, no. 2 (February), pp. 46–52.
69. ibid.
70. Mote, Victor L. (1987) 'The Amur–Yakutsk Mainline: A Soviet Concept or Reality?', *Professional Geographer*, vol. 39, no. 1, p. 20. A more recent source suggests more rapid growth of inshipments. If the railway is finished by 1995, it is planned that 17,000,000 tons will be shipped into Yakutia: 8,300,000 tons by rail, 2,500,000 by sea, and 5,700,000 by river through Osetrovo. *Vodnyy transport*, 20 December 1988.
71. *ZT*, 1987, no. 4 (April), pp. 46–52.
72. Briliant, L.A. (1975) *Geografiya morskogo sudokhodstva*, Moscow: Transport, p. 113; *Soviet Shipping*, 1988, no. 2, p. 9.
73. *MF*, 1988, no. 4 (April), pp. 13–14.
74. Savin, 'Morskoy transport', op.cit., p. 64.
75. Nikol'skiy, Tonyayev and Krasheninnikov, *Geografiya*, op.cit., p. 211; Briliant, *Geografiya*, op.cit., p. 116; Tonyayev, *Geografiya*, op.cit., p. 145.
76. Savin, 'Morskoy transport', op.cit., p. 64.
77. Briliant, *Geografiya*, op.cit., p. 114.
78. Savin, 'Morskoy transport', op.cit., pp. 63–4.
79. *MF*, 1988, no. 4 (April), p. 14.
80. ibid.; *MF*, 1988, no. 3 (March), p. 17.
81. *MF*, 1988, no. 4 (April), p. 14.
82. *MF*, 1988, no. 4 (April), pp. 39–41.
83. *MF*, 1987, no. 12 (December), pp. 18–22.
84. Fenton, R.T. and Maplesden, F.M. (1986) *The Eastern USSR: Forest Resources and Forest Product Exports to Japan*. FRI Bulletin No. 123, Rotorua, New Zealand: Forest Research Institute, New Zealand Forest Service, p. 30.
85. *MF*, 1988, no. 4 (April), p. 39.
86. ibid.; *MF*, 1987, no. 12 (December), p. 48.
87. *MF*, 1987, no. 4 (April), pp. 12–13.
88. *MF*, 1988, no. 4 (April), p. 13; *ZT*, 1988, no. 3 (March), pp. 44–9.
89. *MF*, 1988, no. 4 (April), p. 39.
90. *MF*, 1988, no. 4 (April), pp. 13–14.
91. *MF*, 1987, no. 4 (April), pp. 12–13; *MF*, 1988, no. 4 (April), pp. 39–41.
92. *MF*, 1987, no. 5 (May), p. 16; *MF*, 1988, no. 4 (April), p. 39.
93. Nadtochiy, G.L. (1979) *Geografiya morskogo sudokhodstva*, Moscow: Transport, p. 114; *Soviet Shipping*, 1988, no. 2, p. 9.
94. Nadtochiy, *Geografiya*, op.cit., p. 114; *Soviet Shipping*, 1988, no. 2, p. 11.
95. Kolesov, *Mezhotraslevyye problemy*, op.cit., p. 159.

96. *MF*, 1987, no. 2 (February), pp. 21–3.
97. Nikol'skiy, Tonyayev and Krasheninnikov, *Geografiya*, op.cit., pp. 208–9; *Vodnyy transport*, 9 June 1988.
98. *ZT*, 1988, no. 3 (March), pp. 44–9.
99. ibid.
100. Prokof'yeva, T.A. and Rozdobud'ko, N.K. (1986) *Effektivnost' razvitiya transporta v rayonakh novogo osvoyeniya*, Moscow: Transport, pp. 163–4.
101. Kazanskiy, N.N. (1987) *Geografiya putey soobshcheniya*, 4th edn, Moscow: Transport, p. 137.
102. Lydolph, Paul E. and Mayer, Harold M. (1983) 'Effect of Soviet Shipping on World Maritime Trade', *Soviet Geography*, vol. 24, no. 2 (February), p. 107; *DVZ*, 1 December 1988, pp. 27–9; *Lloyds List*, 13 October 1988.
103. Kazanskiy, *Geografiya*, op.cit., p. 141.
104. Kazanskiy, *Geografiya*, op.cit., p. 145.
104. *Izvestiya*, 19 August 1987, p. 2.
106. *ZT*, 1987, no. 6 (June), pp. 54–9.
107. *Pravda*, 29 July 1986, pp. 1–3.
108. Tonyayev, *Geografiya*, op.cit., p. 48; *Vodnyy transport*, 14 July 1984.
109. Nikol'skiy, Tonyayev, Krasheninnikov, *Geografiya*, op.cit., p. 204.
110. *Soviet News and Views* (Soviet embassy, Ottawa, Canada), 1987, no. 17 (September), p. 4.
111. Tonyayev, *Geografiya*, op.cit., p. 141.
112. Shabad, Theodore (1984) 'News Notes', *Soviet Geography*, vol. 25, no. 10 (December), p. 773.
113. *Pravda*, 29 July 1986, pp. 1–3; *Pravda*, 24 July 1987, p. 1 and 26 August 1987, p. 2.
114. Rybakovskiy, L.L. (ed.) (1969) *Vosproizvodstvo trudovykh resursov Dal'nego Vostoka*, Moscow, pp. 120–22; Orlov, B.P. (ed.) (1974) *Ekonomicheskiye problemy razvitiya Sibiri*, Novosibirsk, p. 15.
115. The line was marked on a map, by no means new, used in a lecture given by a representative of the railways ministry in the Geography Faculty, Moscow University, in late 1963 and attended by the writer.
116. Savin, 'Morskoy transport', op.cit., p. 65; *MF*, 1988, no. 1 (January), pp. 34–7.
117. *Soviet Shipping*, 1987, no. 4, p. 33; *Soviet News and Views*, 1986, no. 6 (April), p. 4.
118. Arikaynen, *Transportnaya arteriya*, op.cit., pp. 176, 185.
119. *RT*, 1987, no. 4 (April), pp. 16, 17, 29; *RT*, 1987, no. 10 (October), pp. 23–4.
120. *MF*, 1987, no. 8 (August), pp. 5–7; *MF*, 1987, no. 5 (May), pp. 13–15; *MF*, 1988, no. 1 (January), pp. 34–7.
121. *ZT*, 1987, no. 8 (August), p. 54; *ZT*, 1987, no. 6 (June), pp. 54–9.
122. *MF*, 1988, no. 4 (April), pp. 39–41.
123. *ZT*, 1987, no. 5 (May), pp. 53–7.
124. *ZT*, 1988, no. 2 (February), pp. 39–41; *Vodnyy transport*, 8 May 1988.
125. *MF*, 1988, no. 1 (January), pp. 34–7.

Chapter nine

Commodity movements and regional economic development

Allan Rodgers

These materials are designed to complement North's discussion of transportation. They are principally concerned with interregional commodity movements for the Far East as these linkages have mirrored changing regional development patterns since the 1960s. There follows speculation as to possible economic change by the turn of the century and its reflection on future commodity flows.

The evolution of interregional linkages

Both the Dienes and North chapters have treated the question of commodity movements in the Far East. However, their treatment was necessarily cursory. I propose now to probe these patterns in somewhat greater depth: as the economy of the Far East matures, economic growth will be reflected in changing commodity linkages.

In an earlier paper (Rodgers 1983, pp. 133–213), I analysed interregional freight flows from 1940 to 1970. I propose now to summarize those findings as an historical base level, for it is the 1970s and 1980s which are the principal focus of this chapter.

Commodity flows in the 1950s and 1960s

As is common knowledge in academic circles, the 1950s and 1960s were extraordinarily fruitful periods for analysis of Soviet regional economic development. In some instances in the past, more data were available for the USSR than for the United States and Western Europe.

Table 9.1 demonstrates the sectoral pattern of interregional commodity flows for 1966. However, the statistics for some sectors include foreign trade, while those for other branches do not. The 1966 percentages have been supplemented with absolute tonnage data for 1965 and 1970.

In Table 9.2, however, the sectoral detail is quite restricted. Between 1950 and 1970 the net balance in favour of inbound movements increased nearly five fold. The growth indirectly reflects the expanding Soviet trade

Table 9.1 Volume and value of interregional commodity movements for the Soviet Far East, by sector, in 1966*

| | Tonnage | | Value | |
| | In-bound % | Out-bound % | In-bound % | Out-bound % |
Sector*				
Ferrous metallurgy	11.8	19.1	4.4	4.1
Non-ferrous metallurgy	0.1	0.7	?	?
Fuels	47.2	38.4	8.2	7.4
Machinery and metal fabrication	3.8	1.4	9.5	12.0
Chemicals	1.4	0.0	5.3	0.6
Forest products	2.7	23.4	0.5	8.4
Construction materials	3.0	1.1	1.4	0.9
Ceramics and glass	0.1	0.0	0.1	0.1
Light industry	1.0	0.3	27.6	4.1
Food industry	7.5	6.6	40.4	61.6
Agriculture	4.1	2.0	2.6	0.8
Other	17.3	7.0	-	-
Total	100.0	100.0	100.0	100.0

Source: Transport i Svyaz [Transportation and Communication] (1972) Moscow, pp. 68–85.

Note: *For machinery, chemicals, ceramics and glass, light industry and food industry, foreign trade was included; for ferrous and non-ferrous metallurgy, forest industries and agriculture, foreign trade was excluded. The authors also indicate that foreign trade for the fuel sector was excluded, but the evidence contradicts that statement and I have proceeded accordingly in my analysis.

with nations in the Pacific basin. For example, Shniper and Denisova (1974) argue that the net inbound flow of 11 to 12 million tons in 1966 would decrease to 1 to 2 millions once export and import movements were included. In contrast to the foreign trade pattern, the outbound movements to other regions of the USSR were not low-value bulk goods; rather these flows were dominated by foods, particularly fish and comparatively high value, added by manufacture, goods such as fabricated metals and machinery as well as costly wood products. Such commodities command prices high enough to absorb the costs of rail transport over vast distances. The only major bulk goods moving westwards from the Far East were probably grain imported from the United States, Canada and Australia, non-ferrous and rare metals mined in the northern segments of the Far East and lumber derived from the vast forest resources of the region.

Inbound flows, not destined for export, included petroleum (to supplement the output of the Sakhalin oil fields), coal, chemical fertilizers, scrap and pig iron (to supply the needs of the Amurstal' steel works at Komsomol'sk), steel products, machinery, foods and consumer goods.

Turning from a commodity-sector approach to an examination of temporal changes in interregional flows, Table 9.3 is a summary, by macroregions, of the inbound and outbound movements to and from the Far East from 1950 to 1970. Aside from the absolute growth in tonnage of

Table 9.2 Volume of interregional commodity flows for the Soviet Far East, by product, 1965 and 1970

	1965				1970			
	Inbound		Outbound		Inbound		Outbound	
	tons ('000s)	%	tons ('000s)	%	tons ('000s)	%	tons ('000s)	%
Coal	2,530	14.8	740	10.2	3,940	13.5	240	3.2
Oil and oil products	5,890	34.4	470	6.5	10,320	35.3	110	1.5
Ores	10	0.1	170	2.4	100	0.3	370	4.9
Forest products	470	2.7	1,150	15.9	1,720	5.9	1,780	23.5
Cement	370	2.2	60	0.8	640	2.2	-	-
Other	7,839	45.8	4,640	64.2	12,550	42.8	5,080	66.9
Total	17,109	100.0	7,230	100.0	29,270	100.0	7,580	100.0

Source: *Transport i Svyaz* [Transportation and Communication] (1972), Moscow, pp. 68–86. These data exclude foreign trade.

both the inward and outward streams, there was a rapidly increasing negative balance that has already been attributed mainly to goods ultimately destined for export. However, it is relative and absolute changes, by region, over time that are of concern here.

Table 9.3 Volume of interregional commodity movements for the Far East, by region, 1950–1970*

	Inbound				Outbound			
	1950		1970		1950		1970	
Region	tons ('000s)	%	tons ('000s)	%	tons ('000s)	%	tons ('000s)	%
Eastern Siberia	1,742	26.8	11,854	40.5	1,120	50.0	2,509	33.1
Western Siberia	812	12.5	8,488	29.0	304	13.6	955	12.6
Urals	1,398	21.5	2,722	9.3	204	9.1	758	10.0
Kazakhstan and Central Asia	689	10.6	1,581	5.4	103	4.6	1,213	16.0
European Russia	1,859	28.6	4,625	15.8	508	22.7	2,145	28.3
USSR	6,500	100.0	29,270	100.0	2,239	100.0	7,580	100.0

Sources: Khachaturov, T.S. (ed.) (1966) *Povysheniye Effektivnosti Transporta v SSSR* [The Increase in the effectiveness of transportation in the USSR] Moscow, p. 32; Shniper, R. and Denisova, L. (1974) *Mezhotrasleveye Svyazi i Narodnokhozyaistvennye Proportsii Vostochnoi Sibiri i Dalnego Vostoka*, Novosibirsk, p. 102.

Note: *Foreign trade is not included. Tonnage data are my estimates based on index numbers, percentages and absolute values published in the Soviet press.

Note the growth, in percentage terms, of the inward flows from Eastern and Western Siberia (from 39 per cent to 70 per cent) complemented by a decline in the shares of inbound movements originating in the western regions of the nation. In contrast, while there was a major percentage decline in the outbound flows from the Far East destined for Eastern Siberia, it was offset by marked increases in the outward flows to Kazakhstan, Central Asia and European Russia. It must be emphasized, however, that the decreases in percentage values only reflect relative changes; for in absolute terms tonnage increases were registered to and from all of these macro-regions during this two-decade interval. Despite these geographical shifts, the length of interregional hauls for the Far East remained, by any standard, enormous. The average distance of the outbound shipments in 1970 was 5,800 kilometres and the value for the inbound flows was 5,200 kilometres. Comparable distances for Eastern Siberia for the same year, were 3,300 and 3,400 kilometres, while the average lengths of outbound and inbound flows for Western Siberia were 2,400 and 2,100 kilometres respectively. The values for the interregional flows for the Far East are solely those for rail hauls, while the data for Siberia probably include transport by pipeline as well as minimal interregional movements by road and river.

Table 9.4 demonstrates the interregional flows for the Far East in 1966. Its utility lies not only in the level of regional detail but in the juxtaposition of tonnage and value data for these economic regions.

Table 9.4 The volume and value of interregional commodity movements for the Soviet Far East, by region, 1966

| | Tonnage | | Value | |
| | In-bound | Out-bound | In-bound | Out-bound |
Region				
	%	%	%	%
North-west	1.2	2.7	3.6	4.8
Central	3.2	7.0	20.9	15.5
Volga-Vyatka	1.7	2.3	3.7	2.7
Central Chernozem	1.3	1.2	3.2	3.3
Volga	3.8	3.6	5.3	2.2
North Caucasus	1.8	1.3	5.0	3.1
Urals	15.3	9.1	8.3	5.4
Western Siberia	24.7	15.3	13.1	13.0
Eastern Siberia	39.5	32.5	23.2	21.4
Ukraine	3.0	4.5	4.4	9.6
Baltic Republics	0.2	1.1	0.3	1.2
Transcaucasia	0.6	0.6	1.5	3.2
Central Asia	1.3	4.4	3.4	5.7
Kazakhstan	1.8	13.8	0.9	6.6
Belorussia	0.4	0.5	–	–
Moldavia	0.2	0.1	2.2	0.2
USSR (%)	100.0	100.0	99.0	97.9
tonnage ('000s)	17,100	7,230	–	–

Source: Shniper, R. and Denisova, L. (1974) *Mezhotrasleveye svyazi i narodnokhozyaistvennye proportsii Vostochnoi Sibiri i Dalnego Vostoka*, Novosibirsk, p. 102.

While the contrasts between the tonnage and value data are self-evident, an explanation of the key differences is in order. Note the variance in the tonnage and value percentages for the Central Region and the north west. These mature industrialized regions, dominated by

Moscow and Leningrad, produce a wide range of producer and consumer-oriented goods. Essentially, all of their products are high in value per unit of weight. Similarly, the Far East ships them foods like fish products and relatively costly manufactured goods. Only goods like these can absorb the transportation costs incurred in such long-haul movements. In contrast, notice the reduction in the percentages from the Urals and Siberia if you compare their tonnage and value shares.

An analysis of interregional flows by region for 1966 appears in Table 9.4. Unfortunately, such detailed data are only available for a limited number of commodities: coal, crude oil, iron and steel and forest products, as shown in Table 9.5.

Table 9.5 Volume of interregional movements, by commodity, 1966

Region	Coal		Petroleum		Forest products		Iron and steel products	
	In-bound	Out-bound	In-bound	Out-bound	In-bound	Out-bound	In-bound	Out-bound
North-west	–	–	–	–	–	1.6	1.7	2.2
Central	–	0.6	–	–	–	4.4	1.0	3.3
Volga-Vyatka	–	–	–	–	–	1.0	0.4	2.9
Central Chernozem	–	–	–	–	–	–	0.9	0.2
Volga	–	–	–	–	–	2.0	0.2	2.8
North Caucasus	–	–	–	–	–	0.2	0.3	3.1
Urals	–	–	–	–	1.2	3.3	28.8	4.8
Western Siberia	56.0	–	–	–	–	7.7	43.4	34.9
Eastern Siberia	44.0	99.4	100.0	–	98.5	24.5	12.5	36.2
Ukraine	–	–	–	–	0.3	–	9.3	3.7
Baltic Republics	–	–	–	–	–	1.6	–	–
Transcaucasia	–	–	–	–	–	–	1.3	1.8
Central Asia	–	–	–	–	–	5.8	·	1.1
Kazakhstan	–	–	–	–	–	47.9	0.2	2.3
Belorussia	–	–	–	–	–	–	–	0.7
Moldavia	–	–	–	–	–	–	–	–
USSR (%)	100.0	100.0	100.0	100.0	100.0	100.0	100.0	100.0
tonnage ('000s)*	2,530	740	5,890**	470**	470	1,150	1,000	500

Source: Shniper, R. and Denisova, L. (1974) *Mezhotrasleveye svyazi i narodnokhozyaistvennye proportsii Vostochnoi Sibiri i Dalnego Vostoka,* Novosibirsk, pp. 103–6.

Notes: *1965 tonnages while percentages are for 1966; tonnages were from *Transport i Svyaz* (1972) Moscow, pp. 71–83.
**Tonnage includes oil products, while the percentages are for crude oil alone.

In the cases of coal and petroleum the patterns are relatively simple. The flows of coal to the Far East totalled roughly 2.5 million tons in 1965 (4 million tons by 1970). Its sources were Western and Eastern Siberia, and it is my supposition that this was a movement of high-calorific low-impurity coking coal ultimately destined for export to Japan. Eastern Siberia was the sole source of inbound shipments of crude petroleum totalling nearly 6 million tons in 1965 (10 million by 1970). This oil was extracted in Western Siberia; it then moved by pipeline to the terminal at Angarsk, near Irkutsk in Eastern Siberia, where it was transferred to tank cars for shipment by rail eastwards. Again the immediate terminus was the Far East, but a portion was destined for Japan. Oil accounted for 40 per cent of the exports of the Far East (by weight) in 1966.

With respect to forest products, more were dispatched than received. Of the roughly 1 million tons that left the Far East in 1966 (almost 2 million in 1970) over half went to the largely unforested regions of Kazakhstan and Central Asia, and one-third was destined for Western and Eastern Siberia. Shniper notes that one-fifth of the exports of our study region (by weight) were wood products, but the main source was the forest area of the Far East itself. The outbound movements of forest products tended to be pulpwood and fabricated wood products rather than low-value roundwood or lumber.

Finally, there are detailed regional data for the flows of iron and steel products. Dibb has estimated the inward flows at 1 million tons; this value would presumably include pig iron destined for the Amurstal' mill and a wide range of steel products (pipe from the Ukraine, beams, bars and rails from the Urals and rails from Western Siberia) designed for producer and consumer markets in the Far East. Yet at the same time about 0.5 million tons of steel products moved westwards. Practically every economic region of the USSR was a destination for a share of the steel output of the Far East. Such movements have been described by some Soviet economists as irrational cross hauls. Yet because of scale economies it must be assumed that there would be an excess of certain steel products over local needs. About 10 to 15 per cent of the exports of the Far East (in value terms) were steel products destined for markets in the Pacific.

Materials exist on inbound and outbound flows of other commodities for the Far East, but these fragmentary data vary as to their time frame, volume (rarely published) and their degree of regional detail. Thus, they can be used only as crude supplements to the more precise data in Table 9.5.

Other inbound movements include grain (1.7 million tons in the late 1960s) from Western Siberia, Eastern Siberia and Kazakhstan, and sugar from the Ukraine and the Central Chernozem region. Deficit food production in this region requires the inward shipment of potatoes, vegetables, meat and dairy products. Since the manufacture of consumer durable and non-durable goods is minimally important in the Far East, such products as textile fabrics, clothing, shoes and appliances must be purchased elsewhere. Seventy-five per cent of these goods are received from European Russia. Local chemical fertilizer production, a relatively new industry in this region, must be supplemented by supplies from the Urals and Western Siberia.

As for the outbound movements not covered in Table 9.5 the highest value shipment is fish which is processed and shipped to every populous centre in the nation. Non-ferrous and rare metals rank high in value terms but less so in tonnage. Their destinations are Western Siberia and the Ukraine. Finally, the most complex inbound and outbound flows are those for machinery. Plants in this sector supply only half of the machinery requirements of the region, yet more than half of the output of the machinery sector of the Far East is shipped westwards thousands of kilometres. In several instances (depending on the product) two-thirds to three-quarters of the products of establishments in this industrial branch are transported to other regions of the nation including European Russia. Such movements were predictably termed irrational in the Soviet economic and transport literature. Gladyshev argues that the cost of rail transport westwards adds about 10 per cent to the ultimate cost of the product. In contrast, transport costs are estimated by the same author to account for 2.4 per cent of the overall cost of machinery for the nation (based on average length of haul).

Commodity flows in the 1970s and 1980s

It is the 1970s and early 1980s which are the principal focus of this chapter. Unfortunately, it was during this period that the Soviet authorities largely eliminated the publication of detailed regional statistical data. A new handbook on transportation was promised for 1989 (Treml 1988, pp. 65–94). Until then, fragmentary data in Soviet journals and books have to be the basis for studies of Soviet commodity linkages.

As reported by Kolesov in 1982, inbound freight movements to the Far East exceeded out-movements by 1.8 times in value and 2.5 times in tonnage. This is a more balanced pattern than tonnage values for 1970 when inward flows were nearly 4 times outbound shipments. Nevertheless, these data reflect how dependent the region is on extra-regional ties. Dienes reports that every rouble of gross economic output requires imports of 30 kopeks. Then too, as we shall see, there is a very poor correlation between Far Eastern production and local needs. The dominance of inbound traffic is particularly striking when compared with Western and Eastern Siberia. The region can only ship out limited amounts of bulk cargo because of sheer distance and remoteness from markets. It is the high value products that can best sustain long-distance transportation costs. In recent years outbound flows grew faster than receipts because of the maturation of the economy and its increasing role in Soviet development.

Table 9.6 (Kolesov 1982, p. 109) demonstrates the evolution of the linkages of the Far East with other economic regions of the USSR. He draws the conclusion that between 1970 and 1977 the region's ties with Siberia have declined drastically because of their lack of complementarity. In turn, he argues that linkages with European Russia have risen markedly. However, his data and his analysis merit further examination. First, the time frame is very short considering the dramatic shifts that would appear to have taken place. Insufficient time alone is an inadequate argument, but there is also contrary evidence. Danilova (1983, p. 361) reports that four-fifths of the inbound and outbound freight

Table 9.6 Origin and destination of freight shipment to and from the Far East (as percentage of all tonnage shipped to and from Far East by rail, sea and river)

Regions of origin and destination	1966 Origin of in- shipment	1966 Destination of outshipment	1970 Origin of in- shipment	1970 Destination of outshipment	1977 Origin of in- shipment	1977 Destination of outshipment
North-west	1.2	2.7	1.8	2.4	5.59	6.60
Centre	3.2	7.5	3.0	7.4	11.60	6.56
Volga-Vyatka	1.7	2.3	1.1	1.8	3.47	3.16
Central Chernozem	1.3	1.2	0.9	1.4	5.24	4.07
Volga	3.8	3.6	3.6	4.1	7.73	10.90
North Caucasus	1.8	1.3	1.3	2.2	5.75	6.84
Urals	15.3	9.1	9.3	10.0	13.80	11.00
West Siberia	24.7	15.3	29.0	12.6	8.76	9.78
East Siberia	39.5	32.5	40.5	33.1	2.79	7.80
Ukraine	3.0	4.5	2.8	5.5	11.47	11.19
Baltic	0.2	1.1	0.2	1.3	4.96	1.60
Transcaucasia	0.6	0.6	0.5	0.7	3.05	1.79
Central Asia	1.3	4.4	1.4	4.8	3.87	1.64
Kazakhstan	1.8	13.8	4.0	11.2	4.50	8.02
Belorussia	0.4	0.5	0.3	1.2	3.62	2.12
Moldavia	0.2	0.1	0.3	0.3	1.44	5.33
USSR	100.0	100.0	100.0	100.0	100.0	100.0

Source: Kolesov, L.I. (1982) *Mezhotrasleviye problemy razvitiya transportnoy sistemy Sibiri i Dal'nego Vostoka*, Novosibirsk, pp. 108–9 (Cited by Dienes, 1985, p. 160).

Table 9.7 Siberian linkages with the Far East, by rail, sea and river, as a share of total flows*

	Kolesov (82) 1970	Kolesov (82) 1977	Minakir (86) Presumably the 1980s
	%	%	%
Inbound	69.5	11.5	67.0
Outbound	55.7	17.6	45.0

Sources: Kolesov, L. I. (1982) *Mezhotrasleviye problemy razvitiya transportnoy sistemy Sibirii i Dalnego Vostoka*, Novosibirsk, pp. 108–9; Minakir, R., Renzin, O., Chichkanov, V. (1986) *Ekonomika Dalnego Vostoka*, Khabarovsk, p. 94.

Note: *computed from tonnage data.

movements of the Far East were with the eastern areas of the USSR. These regions presumably included Kazakhstan and Central Asia as well as Siberia. Minakir and his colleagues (1986, p. 94) also diverge from Kolesov as evidenced by Table 9.7.

The Minakir data, in particular, appear to belie the Kolesov statistics. Another Soviet scholar (Klin 1982, p. 15) reports that, in terms of inbound shipments, the Far East receives from Siberia 35 per cent of its rolled steel, 38 per cent of its cement, 56 per cent of its chemical fertilizers, 48 per cent of its tractors, 39 per cent of its grain, 17 per cent of its meat and 39 per cent of its dairy products and an array of other goods. To these must be added receipts of Siberian petroleum, petroleum products, coal, pig iron, scrap, machinery and roundwood. These data too are in conflict with the Kolesov data.

Although these discrepancies will not be resolved until the release of new transport statistics, what does appear clear is that over time inter-regional linkages with European Russia have increased, but perhaps only marginally. With further growth and the maturation of the Far Eastern economy, I assume that external ties will be reinforced. Then too, there is Dienes's discussion of the ever increasing needs of the military who continue to require material flows from all parts of the USSR, especially European Russia and the Urals.

Product flows

Table 9.8 (Klin, 1982) reports the inbound commodity linkages of the Far East by economic sector in terms of shares of regional consumption.

Table 9.8 Sectoral linkages of the Soviet Far East in value terms

Sector	Inbound movement as a percentage of regional consumption
Ferrous metallurgy	95.1
Non-ferrous metallurgy	73.4
Fuel industry	26.1
Machinery industry	50.8
Chemical industry	89.3
Forest products	18.1
Construction materials	14.2
Light industry	70.6
Food products	22.7
Other industries	52.1
Crops	17.4
Dairy products	4.6

Source: Klin, P., Zabolotnikova, T., Pronina, T., Maslenko, L., and Stolyarova, L. (1982) 'Mezhotrasleviy sviyazi i proportsii v narodno khozyaistve Dalnego Vostoka', in *Retrospektivny analiz ekonomikii Dalnego Vostoka*, Vladivostok, p. 14.

Note: No indication of date, but may be 'recent'.

A discussion of some of these movements follows, their implications will then be analysed. Specific regional origins and destinations, by product, for the 1970s and 1980s were not available.

Fuels

There are only two mineral fuels of consequence extracted in the Far East: coal and petroleum. These have been discussed in the ZumBrunnen chapter on resources. To date, in neither case is the region self-sufficient in these resources.

With regard to coal, there has been a drastic reduction of inbound movements in recent years. Kolesov noted the contrast with Eastern Siberia which shipped out 6 million tons, while the Far East received 4 million tons (Kolesov 1982, p. 101). With the expansion of capacity at Neryungri in the Yakut ASSR (9 million tons of cleaned coal), which produces both energy- and coking-quality coals, there has been a reduction in the need for Siberian coal. Commitments to Japan of coking coal have apparently been met. Coal is now being used in a number of thermo-electric stations. Yet it was recently reported (Singur 1988, p. 94) that 2.5 million tons of energy coal was still being imported from Eastern Siberia.

In contrast, oil production in the Far East, which is confined to Sakhalin Island, is far too low to meet regional demands. It provides only 20 per cent of local needs. It was recently reported that 6.5 million tons of oil and 10 million tons of oil products were shipped into the Far East in 1970. If one includes coal, the cost is 300 million roubles per year (Singur 1988, p. 94). The crude oil moves in by rail from Western Siberia (Tyumen') presumably to the Khabarovsk refinery, while the oil products, from the same source, are refined at Angarsk in Eastern Siberia and railed into the Far East. Foreign trade in oil and oil products is discussed in the chapter by Bradshaw. When the BAM was planned it was predicted that it would carry 10 million tons of oil per year yet hauls of petroleum, so far, are minimal (Mote 1987, p. 379). The original vision is no longer viable because of the slowing down of the rate of growth of oil production in Western Siberia and increased demand in the USSR and its satellites. To date with this rail line still not complete (one tunnel remains to be bored and track to be laid) little, if any, oil moves east to Komsomol'sk, its eastern terminus.

Iron and steel

The Far East still lacks an integrated steel mill although there were earlier discussions of such a mill to be located possibly at Chul'man in southern Yakutia using local coking coals and Chara-Tokko iron ores.

A non-integrated mill is located at Komsomol'sk (Amurstal') which produces about 2.5 million tons of rolled steel a year. Its open hearth furnaces use pig iron imported from the Kuznetsk basin and local scrap iron and steel. Another, reported to have begun operations in the 11th Five-Year Plan (1981–5), is a mini-mill using scrap as its raw material base and should have electric furnaces and possibly continuous casting technology. Its capacity has yet to be divulged, but typically such plants are small compared to those of the nearby plant at Komsomol'sk. It has been reported (Klin 1982) that Amurstal' supplies only 30 per cent of the needs of the Far East. This results in imports of about 7 million tons of rolled steel from the Kuzbas (Western Siberia), the Urals and the Ukraine at a cost of 20 million roubles per year (Chichkanov 1985, pp. 83–4). An earlier source (*Ekonomicheskaya gazyeta*, July 1984, p. 9) cited by Dienes, reports that even the reconstruction of the Komsomolosk steel mill and the building of a new plant will not satisfy more than 50–60 per cent of regional demands.

Machinery industry

This industry is heavily dependent on imports from other regions of the USSR. Apparently 70 to 80 per cent of local consumption is met in this fashion and 50 per cent comes from 'western' regions (Singur 1988, p. 95). To cite examples: *all* tractors, cars and trucks are imported, foundry equipment – 70 per cent, cranes – 56 per cent, lathes – 80 per cent, and diesel and diesel generators – 50 per cent. Eighty per cent of its needs of machine tools come from other regions with half from European Russia alone.

Machinery production in the Far East is 23 per cent of the value of industrial production (half of that is machine repair), yet a major share of local machinery output goes to other regions. Even Soviet authors argue that these outbound movements are irrational. The following extracts illustrate Soviet thoughts on this problem.

> There are ten machine building branches in the Far East, but they play only a negligible role in the economy, because their output does not conform to local needs.
>
> (Chichkanov, 1985, p. 84)

Another writes:

> There is a clear shortage of machinery production in this region for the forest industry, mining and fish processing, etc. This necessitates costly shipments over long distances. Finally, there is a need to reorient the local machinery industry to regional needs, for savings in transportation costs would pay for this investment in 4–5 years.
>
> (Minakir *et al.* 1986, p. 152)

Forest products

The Far East, despite its remoteness, is a supplier for European Russia and particularly for the wood-deficient regions of Kazakhstan and Central Asia. A recent source (Singur 1988) says 10 million cu. m of roundwood and 1 million cu. m of lumber are sent out annually from the Far East. Other sources report that products like mine props, railway ties, paper and particle board are also part of the outbound movements. Conversely, there is a regular inbound movement of roundwood from Eastern Siberia. Nevertheless, most of the Far East's forest product output is destined for local consumption, with a share destined for export, particularly to Japanese markets.

Chemicals

The chemical industry of the Far East is limited. Its products are mainly petrochemicals derived from the oil refinery at Khabarovsk. Klin reports in Table 9.3 that 89 per cent of the Far East's needs are moved in from other regions of the USSR. Given the problem of poor soils in the region, there is an increasing need for chemical fertilizers. These are all imported at a cost of 10 million roubles per year (Chichkanov 1985, p. 85). According to ZumBrunnen, the Far East has an abundance of phosphate resources which could be the raw material base for a mineral fertilizer industry. Natural gas which comes in by pipeline from Sakhalin could, in turn, be the base for a nitrogen fertilizer complex.

Foods

Although there are environmentally favoured areas in the southern provinces of the Far East (including Sakhalin) where most of the population lives, there is still a deficiency of food production. Thus consumption of meat and milk products exceeds local production by almost two fold while vegetable consumption is 1.4 times regional output. This low productivity is surely caused by land quality and insufficient chemical fertilizers, but it has been exacerbated by reductions in farm population as the inevitable draw to the cities progresses, a phenomenon also true of Eastern and Western Siberia.

Unquestionably, a significant share of the region's food requirements comes from other regions of the country. Grain, vegetables, potatoes, meat and dairy products are important inbound movements. The leading source region, as noted earlier, is Siberia, clearly because of its proximity. This is particularly true of products that are perishable. In contrast, foods like grain can come from as far away as Kazahkstan or the Volga region as well as south-western Siberia.

Contrary to the deficit position of the Far East in most food products, its fishing industry provides a major surplus. According to Barr, the Far East accounts for 40 per cent of the nation's fish catch. Processed fish provide significant revenues to the region, and their high value can absorb the costs of shipment to all regions of the USSR.

Other commodities

The Far East has a remarkably limited array of light industries. Most of the consumer products are brought from European Russia. Estimates by various sources report that 70 to 83 per cent of these goods are imported.

With regard to construction materials, most of these products appear to be produced within the region because of their low value per unit weight. One product that does move in, presumably from Siberia, is cement, but here too the region provides the bulk of its needs.

Finally, in the case of non-ferrous ores, rare metals (gold) and diamonds, it is reported that 98 per cent was shipped out to extra-regional markets (Klin *et al*. 1982, p. 14).

Prospect

If we were to take Gorbachev's programme for the Soviet Far East at face value and assume a very rapid rate of growth over the next decade, the repercussive effects on commodity flows would presumably be spectacular. But I agree with Dienes and other scholars that these hopes or dreams are unrealistic. Growth rates may, however, be greater than during the 1980s and these would be reflected in a modest increase in interregional flows. I suspect that there will be a relative increase in movements to and from the industrialized areas of the west (European Russia) and a concomitant decline in interactions with Siberia. There will probably also be changes in the structure or composition of these interactions.

Regional self-sufficiency would clearly be unrealistic, but irrational hauls must be reduced or conceivably eliminated. There may be a reduction of the disparity between inbound and outbound hauls, but the former will still continue to predominate both in tonnage and value terms. More fabricated goods will leave the region; this would particularly be true of forest products as well as processed metals. A rise in outbound movements of machinery appears highly unlikely. In other branches, I would anticipate minimal change. As far as inbound movements are concerned, I forecast a sharp decline in coal and chemical fertilizers receipts. Given the failure to increase Sakhalin oil production, heavy flows will continue to come from Western and Eastern Siberia. As the economy matures, the region will no longer be, in Dienes's

Allan Rodgers

terminology, a 'resource frontier'. Thus, in my view, there will be change in the Far East, but Moscow can scarcely afford the luxury of massive capital investments. Other priorities are more pressing.

Bibliography

Bond, A. (1987) 'Spatial Dimensions of Gorbachev's Economic Strategy', *Soviet Geography*, September, pp. 490–523.
Chichkanov, V. (1986) *Problems of Economics*, November, pp. 77–85.
Danilova, A. (1983) *Ekonomicheskaya geografiya SSSR*, [*Economic Geography of the USSR*], Moscow.
Dibb, P. (1972) *Siberia and the Pacific*, New York.
Dienes, L. (1985) 'Economic and Strategic Position of the Soviet Far East', *Soviet Economy*, pp. 146–76.
—— (1985) 'Soviet–Japanese Economic Relations: Have They Begun to Fade?', *Soviet Geography*, September, pp. 1–30.
Ekonomicheskaya gazeta, July 1984, p. 9.
Ekonomicheskaya gazeta, August 1986, p. 3.
Gladyshev, A., Kulikov, A. and Shapalin, B. (1974) *Problemy razvitiya i razmeshcheniya proizvoditelnikh sil Dalnego Vostoka* [*Problems of Development and Location of Productive Forces in the Far East*], Moscow, p. 88.
Klin, P., Zabolotnikova, T., Pronina, T., Maslenko, L., and Stolyarova, L. (1982) 'Mezhotrasleviy sviyazi i proportsii v narodno khozyaistve Dalnego Vostoka' in *Retrospektivny analiz ekonomikii Dalnego Vostoka* [*Retrospective Analysis of the Economy of the Far East*], Vladivostok, pp. 13–32.
Kolesov, L. (1982) *Mezhotrasleviye problemy razvitiya transportnoy sistemy Sibirii i Dalnego Vostoka* [*Intersectoral Problems of the Development of the Transport System of Siberia and the Far East*] Novosibirsk.
Mikheyeva, N. (1983) 'Analiz mezhotraslevykh i mezhregionalnykh balansov Dalnego Vostoka', *Izvestiya S.O. A.N. SSSR, Ser. Obshch Nauk*, pp. 76–93.
Minakir, R., Renzin, O., Chichkanov, V. (1986) *Ekonomika Dalnego Vostoka* [*Economics of the Far East*], Khabarovsk.
Mote, V. (1987) 'The Amur–Yakutsk Mainline', *The Professional Geographer*, February, pp. 13–23.
Pravda, 24 July 1987, p. 1.
Pravda, 26 August 1987, p. 2.
Rodgers, A. (1983) 'Commodity Flows, Resource Potential and Regional Economic Development: The Example of the Far East', in Jensen, R., Shabad, T. and Wright A. (eds) *Soviet Natural Resources in the World Economy*, Chicago, pp. 188–210.
Shniper, R. and Denisova, L. (1974) *Mezhotrasleveye svyazi i narodnokhozyaistvennye proportsii Vostochnoi Sibiri i Dalnego Vostoka* [*Intersectoral Relations and Economic Proportions in East Siberia and the Far East*], Novosibirsk, p. 77.
Treml, V. G. (1988) 'Perestroyka and Soviet Statistics', *Soviet Economy*, pp. 65–94.

Chapter ten

Soviet Far Eastern trade
Michael J. Bradshaw

The regional development of the Soviet Far East, perhaps more than that of any other region in the Soviet Union, has been influenced as much by changes in the international economy, as by domestic economic policy. The region's remoteness from the European heartland of the Soviet Union has promoted a development strategy based on the exploitation of high-value minerals, such as gold and diamonds, and the export of natural resources to the Pacific basin. This chapter aims to answer two questions: first, what is the role of the Soviet Far East in Soviet foreign trade relations; and second, what is the role of foreign trade in promoting the regional development of the Soviet Far East. The first part provides an introduction to Soviet foreign trade, its organization, structure and dynamics. Because Japan has been the most important trading partner for the Soviet Far East, the bulk of the chapter focuses on Japanese participation in Far Eastern development. The chapter concludes by assessing the relationship between trade and development and discusses the likely impact of Gorbachev's economic reforms on Far Eastern trade.

Foreign trade under central planning

In a command economy, where economic activity is directed by central planners, it is necessary to control foreign economic relations; imports and exports to and from Soviet enterprises have to be accounted for in national economic plans. Prior to the recent reforms, the Ministry of Foreign Trade and the Committee on Foreign Economic Relations exercised a virtual monopoly over Soviet foreign trade (Gardner 1982). The Ministry of Foreign Trade has been responsible for integrating the foreign trade policy dictated by the Party and Gosplan with the operations of the industrial ministries and their associated enterprises. The most important agents in this process have been the Foreign Trade Organizations (FTOs) which act as intermediaries between the Ministry of Foreign Trade and the industrial ministries. Most FTOs are organized by sector and deal with the import and/or export of products from a particular industry. For example, 'Exportles' is responsible for the export of forest products. There are two FTOs responsible for border trade in particular regions: 'Lenintorg' and 'Dal'intorg'. Dal'intorg operates out of the port city of Nakhodka and is responsible for the management of border trade between the Soviet Far East and neighbouring countries. The most important feature of the FTOs is that they have served to insulate the Soviet industrial enterprise from the international market. Prior to recent reforms an industrial enterprise producing goods for export had little or no contact with the foreign purchaser, because the FTO handled the transaction. Likewise the FTO handled the purchase of foreign plant and equipment. Because of the complex bureaucratic system regulating trade, problems of poor integration and departmentalism plague foreign economic relations as much as they did the domestic economy. It is these types of problem that the 1986 reforms of the foreign trade system are intended to address (Hanson 1987).

The recent reforms have relaxed central control over foreign trade by allowing seventy-six enterprises and twenty-two ministries to enter directly into transactions with foreign markets (McIntyre 1987, p. 498). The Ministry of Foreign Trade has retained about forty FTOs and is still responsible for commodities of national significance such as fuels, raw materials and foodstuffs (Ivanov 1987, p. 119). A new central agency, the 'State Commission on Foreign Economic Relations', has been created to promote foreign economic relations and to enhance their contribution to economic restructuring. At enterprise level, the individual enterprises are permitted to retain a portion of the proceeds from trade, thereby providing an incentive to participate in foreign trade. These new measures came into effect on 1 January 1987, and so, as yet, have had little impact upon the trade and development process.

The centralized management of Soviet foreign trade has important implications for regional development. First, because the majority of foreign trade is organized by sector there is little opportunity to integrate foreign trade into regional development strategies, or for regional bodies to promote trade. Hence regional participation in foreign trade is a consequence of the correlation between regional industrial structure and the commodity structure of Soviet foreign trade. For example, since oil and gas are the most important export commodities, regions such as West Siberia must play a key role in foreign trade. The second implication for regional development prior to the recent reforms was that the proceeds from exports (with the exception of border trade) were retained by the central agencies and used to purchase imports. With no regional control or accounting of export participation, the regions had no claim on funds. (Even now it is only particular enterprises within a region that can retain foreign trade proceeds for their own use.) This meant that export revenues generated in one region could be used to purchase imports to promote economic development in another region. For example, oil and gas exports from Siberia have generated revenue to purchase grain and machinery and equipment to sustain the economy of the European regions. A certain amount of Western equipment has been imported to develop oil and gas production in Siberia, but overall the region generates a sizeable foreign trade surplus (Bradshaw 1987). For the central planners, one of the major advantages of the foreign trade monopoly is that it enables foreign trade to be used to address specific national economic problems, and indeed it may be far more cost effective and politically expedient to export oil and import grain rather than to try to increase domestic grain production further (Vanous 1982). From the viewpoint of the exporting region, in this case Siberia, it represents an outflow of resources for which (because of the pricing system) it is not fully compensated (Granberg 1980). Clearly the organization of the foreign trade system, and reforms to that system, can have important implications for a region such as the Soviet Far East where the expansion of the export base has been a key component of economic development strategy.

Because the system of prices used in the Soviet economy serves a different purpose from prices in market economies, it is necessary for Soviet planners to have separate external foreign trade prices and internal domestic prices. For any good traded in foreign markets or imported into the Soviet Union there are at least three prices: first, the price paid or received for the good in the world market; second, the same price converted in foreign trade roubles on the basis of an official exchange rate; and third, the domestic price paid to the producer of an export item or paid by the recipient of an imported good. In some cases domestic prices reflect or may even be higher than world prices, in others domestic prices are considerably lower. Dienes (1982) has examined the export profile of the Soviet Far East and concluded that there has been little undervaluing of Far Eastern exports because the domestic prices of commodities exported were at or higher than world prices. In sum, the nature of the price system makes it very difficult to assess the economic effectiveness of Soviet foreign trade.

The non-convertible status of the rouble and the desire to balance trade and avoid large foreign currency deficits has promoted the use of barter-type agreements to finance imports from the industrialized west. In the context of the current study the most important type of barter agreement is the 'compensation agreement'. The UN Economic Commission for Europe (1982, pp. 172–3) has defined a compensation agreement as:

A specific kind of long-term economic contract between two or more partners from different countries providing deliveries, as a rule on credit, from one partner of complete equipment, licences and know-how for the construction of industrial installations and, from the other partner, for counter-deliveries over several years of products resulting from these installations (or of related products resulting from other installations) in total or part payment for the previously imported equipment and technical documentation.

Soviet compensation agreements with the west are commonly large-scale and capital intensive; they operate predominantly in the resource and primary-processing sectors. As we shall see, such agreements have played a very important role in Japanese trade with the Soviet Far East, providing imports of Japanese equipment to develop the forest and energy sectors in return for exports of forest products and energy resources.

The supply of western plant and equipment under compensation agreements provides a number of advantages; such agreements provide imports on a credit basis which is repaid with the export of products, thereby conserving foreign currency. In addition the western technology is usually supplied in the form of a 'turnkey plant', which means that the western partner is responsible for the construction of the production facilities and all the Soviet partner need do is 'turn the key' to start operation. By the early 1980s, sixty-five projects were being built in the Soviet Union on the basis of compensation agreements (Nirsha 1981, p. 38). One Soviet source estimates that since 1970 more than 10,000 million roubles' worth of equipment, technical documentation, pipe and materials has been delivered under compensation agreements (Zhebrovoskiy and Ponomarev, 1986, p. 18). During the early 1980s compensation agreements accounted for 15 per cent of east–west trade (Ognev 1986, p. 88). However, in recent years compensation agreements have fallen from favour. Western companies are no longer so keen to commit to long-term import of Soviet natural resources. At the same time the Soviet Union wishes to increase the level of 'value added' in its exports and to reduce the share of resources in exports because of their vulnerability to price fluctuations.

Perhaps as a response to the declining popularity of compensation agreements and the desire to improve the structure of Soviet exports, the 1986 reforms of the foreign trade systems have enabled the establishment of joint ventures between western firms and Soviet enterprises. The Soviet partner is to have majority ownership and thus control of such ventures, which are designed to export goods from the Soviet Union rather than provide western firms with access to the domestic market. At the time of writing, no joint ventures have been established in the Soviet Far East, although a Soviet–Japanese venture has been set up in the city of Irkutsk in Eastern Siberia. Japanese companies have also shown an interest in setting up joint ventures in the fishing and service sectors. Given the recent reforms it is clear that the foreign trade system is in a period of transition. This chapter must evaluate the relationship between foreign trade and regional development in the Soviet Far East under the traditional system of management (the foreign trade monopoly) and speculate as to the impact of the recent reforms on the trade and development process.

The dynamics and structure of Soviet foreign trade

Since the Second World War the Soviet Union has become a more active participant in the international trade arena, and the total value of Soviet foreign trade has increased from 2,900 million roubles in 1950 to 130,900 million roubles in 1986 (USSR, Ministry of Foreign Trade 1987, p. 6). However, the Soviet Union cannot be considered a major world trading nation. According to western calculations, in 1986 the share of exports as a percentage of Gross National Product was 4.1 per cent, the equivalent measures were 12.7 per cent for Japan and 11.5 per cent for China (CIA 1987, pp. 24–5). In the immediate post-war years the growth of Soviet foreign trade was associated with the expansion and consolidation of the socialist bloc, while the potential for east–west trade was severely reduced by the Cold-War blockade. During the 1960s there was a change in attitude on the part of the Soviet leadership towards the role of foreign trade in the Soviet economy. The so-called 'international socialist division of labour' became a banner for increased economic co-operation within the Council for Mutual Economic Assistance (CMEA). The signing of a large technology trade agreement with the Italian automobile firm Fiat in 1965 marked a new era in east–west trade. The Brezhnev-Kosygin leadership perceived increased trade with the west as a means of revitalizing a stagnating domestic economy; western leaders, especially in the United States, saw increased trade as a means of fostering an economic interdependence that would enable the west to influence Soviet behaviour. The combination of these attitudes contributed to the first Brezhnev-Nixon summit in May 1972 and the era of *dètente*.

From the early 1970s until 1985 there was a steady increase in the value of Soviet foreign trade (Figure 10.1). While part of this increase can be ascribed to inflation and the manipulation of exchange rates, Soviet foreign trade accounting practices distort the value of CMEA trade and the manipulation of the rouble/dollar exchange rate may undervalue trade with the west, there can be no doubt that foreign trade is more important to the Soviet economy in the 1980s than it was in the 1960s. During the 1970s trade with the west became more important while the share of trade with the developing nations remained relatively stable (Table 10.1). In all cases there was a large increase in the total value of trade.

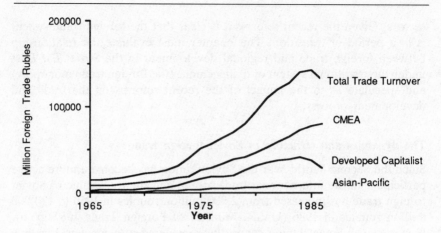

10.1 The growth of Soviet foreign trade, 1965–85

Table 10.1 The geographical distribution of Soviet foreign trade, 1965–86

Trade groups	1965	1970	1975	1980	1985	1986
Total trade turnover, million roubles						
CMEA members	8,473	12,284	26,248	52,185	78,108	79,989
Exports	4,212	6,261	13,363	28,566	40,224	42,193
Imports	4,261	6,023	12,885	23,619	37,884	37,796
Other socialist nations	1,577	2,125	2,304	5,749	8,851	7,511
Exports	789	1,269	1,221	2,626	4,243	3,498
Imports	788	856	1,083	3,123	4,608	4,043
Industrially developed capitalist nations	2,816	4,694	15,844	35,359	37,875	28,989
Exports	1,346	2,154	6,140	17,247	18,581	13,136
Imports	1,470	2,540	9,704	18,112	19,294	15,853
Developing nations	1,744	2,982	6,309	16,446	17,258	14,444
Exports	1,010	1,836	3,310	8,669	9,615	9,550
Imports	734	1,146	2,999	7,777	7,643	4,894
Total trade[a]	14,610	22,085	50,705	109,739	142,092	130,936
Exports	7,357	11,520	24,034	57,108	72,663	68,347
Imports	7,253	10,565	26,671	52,631	69,429	62,589
As a percentage of total trade turnover						
CMEA members	58.0	55.6	51.8	47.6	55.0	61.1
Other socialist nations	10.8	9.6	4.5	5.2	6.2	5.8
Industrially developed capitalist nations	19.3	21.3	31.3	32.2	26.7	22.1
Developing nations	11.9	13.5	12.4	15.0	12.1	11.0
Total trade	100.0	100.0	100.0	100.0	100.0	100.0

Source: Soviet Foreign Trade Yearbooks.

Note [a]the figures for each trade group have been rounded up.

Since the decline in world oil prices in late 1985 there has been a change in the dynamics of Soviet foreign trade. Oil and gas sales to the industrialized west play a very important role in Soviet exports. During 1985 Soviet oil exports earned US$13,000 million; in 1986 earnings fell to US$7,900 million (Gorst 1987). After the first nine months of 1987 the volume of Soviet trade with the west had fallen by 4.6 per cent when compared with the same period in 1985.[1] Despite increasing the volume of oil exports and the level of gold sales, the Soviet Union has not been able to sustain the value of its exports to the industrialized west and between 1985 and 1986 the total value of Soviet trade actually declined. While heavy reliance on oil exports brought substantial windfall profits during the 1970s it is now proving to be a major weakness (Smith 1987). To illustrate the degree of Soviet dependence upon energy and provide a context for Far Eastern exports, the next section examines the structure of Soviet trade.

The general commodity structure of Soviet foreign trade is shown in Table 10.2. These data include all trading groups: CMEA members, other socialist nations, the industrialized capitalist nations and the developing nations. The export of energy resources and raw materials has come to dominate trade since 1970, and is particularly important to Soviet trade with Eastern Europe and the industrialized west. Machinery exports are more important in trade with the developing nations, reflecting the inability of Soviet manufactured goods to compete successfully in western markets. The most important items imported into the Soviet Union are machinery and equipment and food products. The majority of Soviet machinery imports comes from other CMEA nations, with the industrialized west (Western Europe and Japan) accounting for 25–30 per cent of such imports. Food products are imported from the major western grain producers (United States, Canada, Argentina and the European Community) as well as the developing nations. The commodity composition of Soviet hard currency trade (Figure 10.2) shows a pattern similar to that of total trade, with energy and raw materials dominating exports and machinery and agricultural products making up the majority of imports. The very rapid increase in the value of energy exports is illustrated by the fact that in 1970 they earned US$388 million and accounted for 16.1 per cent of hard currency exports, while by 1984 they had increased in value to US$18,900 million and in share of exports to 53.2 per cent (CIA 1987, p. 72). Other important foreign currency exports include: coal, wood and wood products, diamonds, non-fuel minerals, gold and arms sales. In sum, the commodity structure of Soviet foreign trade reflects a desire to use exports of natural resources to finance purchases of machinery and equipment and agricultural products. Thus, one would expect the resource-producing regions of the Soviet Union (such as Siberia and the Soviet Far East) to play the greatest role in export production, and the European regions (where the majority of the population resides and the bulk of industry is located) to consume the majority of Soviet imports.

Soviet Foreign Currency Exports, 1985

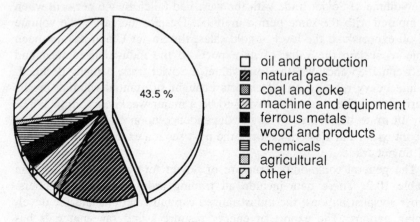

43.5 %

☐ oil and production
▨ natural gas
▦ coal and coke
▨ machine and equipment
■ ferrous metals
■ wood and products
☰ chemicals
▦ agricultural
▨ other

Soviet Foreign Currency Imports, 1985

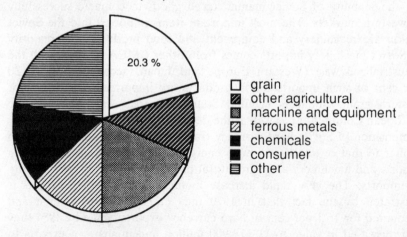

20.3 %

☐ grain
▨ other agricultural
▦ machine and equipment
▨ ferrous metals
■ chemicals
■ consumer
☰ other

10.2 The commodity structure of Soviet foreign currency trade, 1985

To evaluate the role of foreign trade in the economic development of the Soviet Far East it is necessary to identify the region's major trading partners and the commodity structure of such trade. There are a number of different ways in which one can disaggregate Soviet trade into trade blocs. The divisions in the Soviet Foreign Trade Yearbooks reflect a combination of financial and political considerations. The most important distinction made there is between trade with the socialist nations, including the CMEA, and trade with the industrialized west. Apart from the

Table 10.2 Commodity structure of Soviet foreign trade, 1970–86

	1970	1980	1985	1986
Exports (Percentage of total)				
Machinery and equipment, means of transport	21.5	15.8	13.9	15.0
Fuels and electricity	15.6	46.9	52.7	47.3
Ores, concentrates, metals and metal products	19.6	8.8	7.5	8.4
Chemical products, fertilizers, rubber	3.5	3.3	3.9	3.5
Forest products and cellulose, paper products	6.5	4.1	3.0	3.4
Textile raw materials and semi-manufactures[a]	3.8	1.9	1.3	1.4
Food and raw materials for their production	8.4	1.9	1.5	1.6
Consumer manufactures	2.7	2.5	2.0	2.4
Imports (Percentage of total)				
Machinery and equipment, means of transport	35.5	33.9	37.1	40.7
Fuels and electricity	2.0	3.0	5.3	4.6
Ores, concentrates, metals and metal products	9.6	10.8	8.3	8.3
Chemical products, fertilizers, rubber	5.7	5.3	5.0	5.1
Forest products and cellulose, paper products	2.1	2.0	1.3	1.3
Textile raw materials and semi-manufactures	4.8	2.2	1.7	1.3
Food and raw materials for their production	15.8	24.2	21.1	17.1
Consumer manufactures	18.3	12.1	12.6	13.4

Source: Soviet Foreign Trade Yearbooks.

Note: [a]includes furs and skins.

obvious political differences, the other important difference is the different accounting systems used in trade with each bloc. Soviet trade with other socialist nations is usually based on a clearing system (much like barter) and does not involve the use of hard currencies. Soviet trade with the industrialized west, with the notable exception of Finland, is conducted using hard currency. Because of the different accounting systems, a surplus in trade with the CMEA cannot be used to balance a trade deficit with North America: in fact the Soviet Union has tended to use its trade surplus with Western Europe to finance imports from North America and Japan. As we have seen, a further consequence of this situation is that trade with the west is far more susceptible than intra-CMEA trade to price fluctuations in international markets. This means that to a large degree the level of east–west trade is beyond the direct control of Soviet planners and politicians. At the same time, trade with the socialist nations, which is relatively isolated from the world economy, is often used to further the foreign policy goals of the Soviet Union.

An alternative way of disaggregating Soviet trade is to examine the geographical distribution of trade by world region. In the context of the present study, because of its geographical proximity to the Soviet Far East, we are particularly interested in the role of trade with the Asian–Pacific region. In 1970, 63.6 per cent of Soviet trade was with Eastern Europe, 18.5 per cent with Western Europe, 1.5 per cent with North America and 8.5 per cent with the Asian–Pacific region. During the

1970s *détente* promoted increased trade with Western Europe and the United States; however, political manipulation of trade relations by the United States reduced the level of trade between the United States and the Soviet Union. By 1980 Eastern Europe's share of Soviet foreign trade had fallen to 46.6 per cent, while Western Europe accounted for 21.4 per cent of trade, North America 2.7 per cent and the Asian–Pacific region 7.2 per cent. By 1986 declining oil prices had served to reduce the share of Western Europe to 17.6 per cent and increase the share of Eastern Europe to 56.2 per cent, while trade with the Asian–Pacific region reached 8.0 per cent. Throughout the 1970s and 1980s Europe has dominated Soviet foreign trade and the Asian–Pacific region has played a rather modest role. Nevertheless, the value of Asian–Pacific trade has increased from 1,680 million roubles in 1970 to 10,500 million roubles in 1986 (see Figure 10.1). Despite its relatively modest role in Soviet foreign economic relations, the Asian–Pacific region is the most important market for the Soviet Far East.

The structure and dynamics of Soviet Asian–Pacific trade

Soviet trade with the Asian–Pacific region can be divided into three trading groups: the socialist nations of Vietnam, Kampuchea, Laos, Mongolia, North Korea and China; the industrially developed capitalist economies of Japan, Australia and New Zealand; and the member states of the Association of South-east Asian Nations. Since the 1970s the socialist nations and Japan together have accounted, on average, for 85 per cent of Soviet trade with the Asian–Pacific region. The socialist nations account for the majority of Soviet exports to the region (80.4 per cent in 1986), while Japan has been the single most important source of imports (40 per cent in 1986). It is possible to further subdivide the socialist nations into two groups: first, Vietnam, Kampuchea, Laos and Mongolia with whom the Soviet Union has a sizeable trade surplus; second, China and North Korea with whom trade is fairly balanced. Trade with the first group represents a form of foreign economic assistance and is composed of exports of oil and oil products and machinery and equipment; imports are comprised of agricultural products and industrial raw materials. Trade with China and North Korea reflect the often tense relationships between these nations and the Soviet Union. In recent years Sino–Soviet trade has experienced a rejuvenation; the Soviet Union, due to improved relations between Moscow and Beijing, is again providing technical assistance and machinery and equipment, while China is providing agricultural products. The level border trade between China and the Soviet Union has also increased with the Soviet Far East supplying natural resources in return for agricultural products. In recent years Soviet relations with North Korea have also improved and this has prompted increased trade and technical co-operation. The commodity structure of Soviet–Japanese trade reflects the general structure of hard currency trade; according to Japanese foreign trade statistics heavy industrial products (metals and metal products and machinery and equipment) accounted for 81.6 per cent of imports from Japan in 1985, while raw materials, mineral fuels and non-ferrous metals (including gold) accounted for 83.8 per cent of Soviet exports to Japan. The prevailing political and economic environment in the Asian–Pacific region has promoted an increase in trade with the socialist nations of the region, while Soviet–Japanese trade is experiencing relative decline (Bradshaw 1988). Given that the regional economy of the Soviet Far East specializes in the extraction and primary processing of natural resources one would expect the region to play a greater role in Soviet trade with Japan than with the socialist nations of the region. The limited anecdotal evidence available suggests that this is in fact the case. Shlyk (1981, p. 35) has suggested that the Soviet Far East accounts for only 20 per cent of Soviet exports to the Pacific basin, but 40 per cent of Soviet exports to Japan. The reason for this is that lumber and fish account for 83.5 per cent of Far Eastern exports, commodities that play an important part in Soviet–Japanese trade (Chichkanov 1986). According to Japanese trade statistics fish and shellfish and wood (including woodchips) accounted for 30.5 per cent of Soviet exports to Japan in 1985. Stolyarov and Pevzner (1984, p. 180) suggest that the Soviet Far East supplies 44 per cent of the roundwood, 4 per cent of the sawn timber, 8 per cent of the pulp and 77 per cent of the fish exported to Japan. Whilst the Soviet Far East is not the only region providing resource exports to Japan, Japan is by far the most important trading partner for the Soviet Far East.

Japanese participation in Far Eastern development

There are at least three different types of trade between Japan and the Soviet Union. First there is 'big trade' which is controlled by the Soviet Ministry of Foreign Trade through its various foreign trade organizations. Despite the recent reforms, the bulk of Soviet natural resource exports remain under the central control of the Ministry of Foreign Trade (Franklin 1988, p. 214). The second type of trade, which is really a component of big trade, involves the use of compensation agreements to finance Far Eastern development projects. Under these agreements consortia of Japanese companies have supplied machinery and equipment for exploration and development of natural resources in the Soviet Far East. These deliveries are then paid off with exports from the project once they are operational. Between 1969 and 1979, 12 per cent of Soviet imports from Japan and 11 per cent of Soviet exports were under compensation agreements. By 1981 the figures had risen to 32 per cent of

Michael J. Bradshaw

Soviet imports and 25 per cent of Soviet exports (Dienes 1985, p. 512). It is likely that the share of Soviet–Japanese trade under compensation agreements has fallen during the 1980s, because no new agreements have been reached since 1981; however, the share of such agreements in exports has probably increased as projects initiated during the 1970s have begun to reach the pay-off stage. The final type of trade is 'border trade' which is permitted between the Pacific regions of the Soviet Far East and Japan, China, North Korea and Australia. The following discussion of the role of Japanese trade in Far Eastern development gives greatest attention to compensation agreements, because these have significantly influenced the pattern of regional development, but also considers resource exports and border trade.

Soviet–Japanese compensation agreements

Since 1967 over 2,500 million dollars of credit have been provided by Japanese compensation agreements to aid in the development of Far Eastern resources. These agreements have concentrated on the development of forest, coal and oil and gas resources, and also the improvement of port facilities at Nakhodka/Vostochniy (Figure 10.3 and Table 10.3). In addition to the projects that have been pursued a number of other projects have been under discussion at one time or another including: copper production at Udokan in East Siberia, asbestos mining and the construction of a steel mill (Nobuhara and Akao 1983, pp. 202–3; Ogawa 1987, pp. 167–71).

Forestry projects

The possibility of Japanese co-operation in the development of the forest resource of the Soviet Far East was first mentioned in 1962, but it was not until July 1968 that the first large-scale agreement was signed (Judd 1971, p. 691). Since 1968 there have been a number of general agreements on the development of Far Eastern forest resources. Under the first KS Soviet Far East Forest Development Project (KS stands for the initials of the two negotiators, Kawai and Semichastov) a credit of $163 million was provided for the Soviet Union to purchase Japanese machinery and equipment to develop the forest resource of the Sikhote-Alin' mountain range in Primorskiy Kray. In return the Soviet Union exported 8.0 million cu. m of timber between 1969 and 1975. The price of timber was set at the 1968 value, with a 1 per cent increase each year. The credit provided under the agreement was used to purchase equipment to enable increased harvesting and included: 2,300 trailer trucks for hauling logs worth $43 million and supplied by Komatsu, 1,000 bulldozers for building logging roads worth $24 million and purchased from Caterpillar-

250

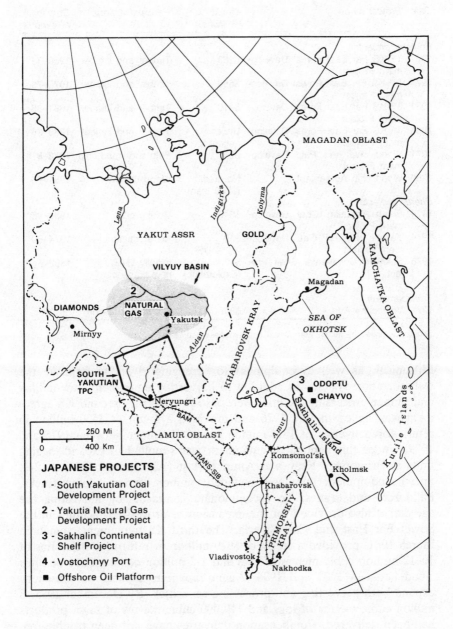

10.3 Japanese involvement in Soviet Far Eastern development

Table 10.3 Soviet–Japanese development projects in the Soviet Far East

Date	Project name	Credit (Millions)	Principal exports	Duration of exports
Forestry projects				
1967	1st KS Far East Forest Development Project	$163	Timber and lumber	1969–73
1974	2nd KS Far East Forest Development Project	$550	Timber and lumber	1975–79
1981	3rd KS Far East Forest Development Project	$910	Timber and lumber	1981–86
1988	4th KS Far East Forest Development Project	Under negotiation	Timber and lumber	1988–96
1971	Wood chip and Pulp Development Project	$50	Pulp and wood chip	1972–81
1985	Wood chip Agreement	No credit ($200 loan)	Wood chips	1981–95
Energy projects				
1974	South Yakutian Coal Development Project	$540	Coking coal	1983–98
1974	Yakutia Natural Gas Development Project	$50 (50% US)	No production	1974–80
1975	Sakhalin Continental Shelf Project	$170 (Exploration)	Initially LNG, now crude oil	1992-?
Transportation				
1971	Vostochny Port	$80	Not compensation	

Source: Compiled by author from Soviet and Japanese sources.

Mitsubishi, as well as equipment for dredging rivers to improve the transportation system. In addition to harvesting equipment, saw-milling and timber-processing facilities were also provided. A second KS agreement was signed in July 1976 and provided a credit of $550 million in return for deliveries of 18.4 million cu. m of timber between 1975 and 1979. Under this agreement equipment was supplied to expand logging units in Primorskiy Kray and Amur Oblast (Alexandrov 1982, p. 24). Specific equipment purchases included: timber carriers, dump trucks, bulldozers and cranes. To accommodate increased log harvesting the agreement also provided for the expansion of processing facilities in the Soviet Far East and East Siberia. The third KS agreement, signed in March 1981, provided a credit of $910 million in return for deliveries of 10–12 million cubic metres of logs and 1.2 million cubic metres of sawn wood between 1981 and 1986. Again Japanese imports were used to expand both harvesting and processing capacities. By December 1985, 8 million cubic metres of logs and 110,000 cubic metres of sawn products had been delivered. Compensation deliveries have not been trouble-free – Soviet producers have experienced problems due to forest fires and poor co-ordination between logging units, sawmills and shippers; the Japanese have consistently complained about the inability of Soviet saw mills to provide lumber in standard dimensions and about the high amount of larch in deliveries. During 1987 negotiations were held on a fourth KS agreement. In January 1988 the Japanese press reported that the Soviet Union had made certain concessions over the level of timber exports and that an agreement was possible.[2] The Soviet Union has proposed that a total of 12 million cubic metres of lumber would be delivered between 1988 and 1996. The Japanese would buy 6 million cubic metres during the first five years, following which the price of the remaining 6 million cubic metres would be renegotiated. The Soviet Union has agreed to purchase bulldozers and other Japanese construction equipment for use in the forest industry. In early February 1988 at a meeting of the Japanese and Soviet Committees for Economic Co-operation it was announced that 'the two nations have agreed on outlines of basic contracts for the fourth forest resource development project'.[3] At the time of writing no further details have been announced concerning the nature of the project. The general forest development agreements have focused on the extensive harvesting of the Far Eastern forest, there have also been attempts to increase the level of resource processing in the region.

In December 1971 agreement was reached on the First Wood Chip and Pulp Development Project. Under this agreement $50 million of credit was provided, $45 million of which was used to purchase chip and pulp facilities, machinery and special ships to transport chips. The remaining $5 million was used to purchase consumer goods for the workforce. In return the Soviet Union was to supply 4.7 million cubic metres of pulp and 8 million cubic metres of wood chip between 1971 and 1981. A combination of production problems and slack demand in Japan meant that by 1981 only 71 per cent of the pulp and 59 per cent of the chips had been delivered. Perhaps as a result of these problems and also as a consequence of the sanctions following the Soviet invasion of Afghanistan, agreement on a second wood chip project was delayed until December 1985. Under the second agreement the Japan Chipwood Trade Company, which consists of nineteen Japanese paper manufacturers, will take delivery of 8.2 million cubic metres of wood chips and 3 million cubic metres of hardwood pulp worth $500 million between 1985 and 1995. The species composition of the wood chips will be 60 per cent larch, 15 per cent white and silver fir and 25 per cent hardwood. At the same time as this agreement was signed a $100 million credit was made available for the Soviet Union to purchase wood chip production equipment.

For the Japanese these forestry projects have provided a source of unprocessed wood, predominantly in the form of logs, which can be used as a raw material in the Japanese forest products industry. For the Soviet Union these projects have provided machinery and equipment for the development of an industry and a region that is low in domestic investment priorities. There is therefore a basic complementarity that promotes Soviet–Japanese forest products trade.

Energy projects

Since 1973 there have been three Soviet–Japanese projects to explore and develop the energy resources of the Soviet Far East. The South Yakutian Coal Development Project represents the first large-scale long-term agreement reached following the visit of the Prime Minister of Japan to Moscow in July 1973. An agreement was signed in June 1974 between Japan and the Soviet Union on the joint development of the South Yakutian coal basin. The development of this coal basin has been the centrepiece of the South Yakutian Territorial Production complex which was discussed in Chapter 7. Under the terms of the agreement a Japanese credit of $450 million was provided: $390 million for the purchase of machinery and equipment for the development of the coal mine and $60 million for consumer goods and Soviet expenses in financing the project (Spandaryan 1975, p. 27). In return the Japanese partners were to receive 104 million tons of coking coal between 1979 and 1998, including 84.4 million tons from South Yakutia and 20 million tons from the Kuznetsk basin in West Siberia. In Japan a special company, the South Yakutia Coal Development Corporation, was created to manage the project. This consortium consists of fifty-one Japanese companies including: eight metallurgical companies, fifteen trading companies, two coke and gas companies, two shipping companies and sixteen banks. The credit supplied by the Japanese has been used to purchase mining machinery for the removal of overburden and the excavation and transportation of coal, machinery for the construction of the necessary infrastructure and equipment for the construction of a coal treatment plant. A separate credit has been provided for the construction of a coal handling facility at the port of Vostochnyy capable of handling 6.2 million tons of coal a year. Most of the machinery and equipment, such as dump trucks, cranes and bulldozers, have been provided by Japanese companies including Komatsu, Mitsubishi, Kato and Sumitomo Heavy Industry. There were, however, some specialized items that the Japanese produced under licence from US companies. The equipment for the coal preparation plant which has a capacity of 9 millions tons of coal a year was purchased from the McNally Corporation of Pittsburgh, Kansas (Mote 1985, p. 709). Ten US-designed Marion Power Shovels were built under licence by Sumitomo. There were a limited number of orders placed directly with US companies including an order of thirty 180-ton dump trucks and twenty-nine 120-ton dump trucks worth $29 million which was awarded to Unit Rig; Unit Rig in turn instructed its Canadian subsidiary to fulfil the order. Coal production was delayed, in part, by the one-year-late delivery of the Marion Power Shovels, but once they arrived they experienced problems because the hydraulic-lift system froze and shattered in the extreme cold (Mote 1985, p. 707). The cold also caused the tyres of the heavy duty trucks to split. In 1981 an additional $90 million credit was supplied to compensate for the inflationary rise in the price of equipment. While $490 million of credit represents a substantial investment, Soviet authors are quick to point out that the total cost of developing the South Yakutia complex is much higher. Agafonov (1985) notes that the book value of the Neryungri truck fleet alone is 500 million roubles (nearly $600 million at 1985 rates). In other words, a large amount of Soviet-built equipment works side-by-side with imported machinery. The Soviet machine-building industry has not been able to produce equipment that will operate efficiently under the harsh conditions in Siberia and the Soviet Far East. In addition much of the foreign equipment was not designed to operate under such extreme conditions. Equipment problems at Neryungri have led the Ministry of Heavy Transport and Machine Building to create a special servicing unit (Dyker 1982, p. 2).

Since the completion of the Little BAM in 1978, steam coal has been produced at Neryungri. But it was not until 1984, a year behind schedule, that miners reached coking coal seams and started to deliver coal to Japan. In August 1985 a ceremony was held in Neryungri to mark the delivery of the first million tons of coking coal (Agafonov 1985). The coal travels some 2,600 km by train to the coal port at Vostochnyy, a journey that takes four days; it is then loaded aboard ships at a special coal terminal for the two-day voyage to Japan. Because of the lack of regional economic data, it is very difficult to estimate the level of South Yakutian coal exports to Japan. According to Soviet sources 0.8 million tons of coal were delivered from South Yakutia during 1984 (Mote 1985, p. 704). The Japan External Trade Organization (JETRO) reported that Japan received 1.6 million tons of coking coal from the Soviet Union in 1984; deliveries from the Kuznetsk basin and possibly Sakhalin Island contributed the balance. In 1985 the compensation agreement called for 3.2 million tons of coking coal to be delivered to Japan. However, according to JETRO figures, only 2.9 million tons of coal were exported to Japan by the Soviet Union. Western sources have reported that during 1986 4.2 million tons of coal were exported to Japan and that an agreement was signed for the 1987–8 period to deliver 4.9 million tons.[4] The Soviet Union had asked the Japanese steel mills to accept 6.5 million tons (as scheduled under the compensation agreement), the Japanese responded with a cut back to 3.7 million tons, therefore 4.9 million tons represents a compromise. The reason for the reduction from 6.5 million tons is that the Japanese, whether intentionally or by miscalculation, have promoted the overproduction of coking coal in the Pacific basin by encouraging producers in Australia, Japan and North America as well as the Soviet Union. For example, a very similar Japanese-sponsored coal project came on stream at the same time as South Yakutia in North-eastern British Columbia in Western Canada. Although the South Yakutian project has been carried through to the pay-off stage its future is by no means secure. The Soviet Union occupies a rather marginal position in Japanese coal imports: in 1985 it ranked fifth supplying 3.7 per cent of Japanese imports; Australia ranked first with 43.3 per cent followed by Canada with 22.3 per cent and the US with 17.8 per cent. The restructuring of the Japanese economy is continuing to reduce demand for steel and competition for contracts is increasing, particularly as China and South Africa are also in the market. Therefore, South Yakutian coal will experience extreme competition in Japanese markets and it is unlikely that the compensation deliveries will be completed by 1998.

The second energy project to be initiated was the Yakutia Natural Gas Project, a tripartite scheme involving the Soviet Union with Japanese and American interests to develop the Vilyuy gas deposits in the Yakut ASSR (Egyed 1983, pp. 60–89). Once sufficient reserves had been proved, the intention was to deliver liquefied natural gas (LNG) to Japanese and West Coast US markets. The Yakutia project was initially proposed by the Soviet Union at a meeting of the Japanese and Soviet Committees for Economic Co-operation in 1968. At first the Japanese showed little interest, but during the early 1970s with the prospect of American involvement and the impact of the energy crisis Japanese companies decided to participate. The American company El Paso Natural Gas (75 per cent of the US interest) and Occidental Petroleum (25 per cent); Bechtel Inc. were also involved in a consultative capacity. Japanese participation was organized into a consortium of twenty-one companies known as Siberian Natural Gas which was headed by the Chairperson of Tokyo Gas. The initial exploration phase had to prove reserves of 1 trillion (10^{12}) cubic metres of natural gas. This amount of gas was necessary to guarantee compensation deliveries of 20,000 million (10^9) cubic metres (1.5 million tons of LNG) a year for twenty years, half of which would go to Japan and half to the US. In return Japan and the US would provide $3,000 to $4,000 million for the development phase of the project. The project immediately fell foul of actions by the US Congress to use trade with the Soviet Union for economic leverage. Particularly important was the Stevenson Amendment which placed a limit of $300 million over four years on United States Export–Import Bank credit to the Soviet Union, and more importantly placed a $40 million limit on credits for energy projects. This presented obvious problems for the American partners who had to raise sufficient funds for the exploration phase of the project. The Japanese insisted that the project could only continue with American participation. In March 1976 agreement was reached on a financial package for the exploration phase of the project. In total $100 million was provided for the purchase of western equipment: $50 million from the Soviet Union, $20 million from the Japanese EXIM Bank, $5 million from a consortium of twenty-three Japanese banks and $25 million from the Bank of America.

The Soviet Union quickly used funds from the project to purchase 'high-tech' exploration equipment. The US company Geosource Inc. was contracted to outfit five geological exploration crews with tracked vehicles worth over $13 million. The vehicles were supplied by Canadian Foremost Industries Ltd of Calgary, Alberta. In addition to this contract Geosource provided $11 million worth of seismic instrumentation and data processing systems. The Carter administration delayed the delivery of two self-propelled seismic stations. By late 1979 the exploration phase of the project was nearing completion with 890,000 million cubic metres of natural gas discovered. The Soviet invasion of Afghanistan, in the last week of 1979, dramatically changed the situation and the project has been on hold ever since. The Japanese are now directing their limited enthusiasm for Soviet energy projects towards the Sakhalin Continental Shelf Project. However, it would be wrong to think of the Yakutia project as a total failure. It did provide access to western seismic exploration technology and enabled the delineation of the reserves of the Vilyuy basin. At present the development of the Vilyuy field is limited to local needs, but future plans call for the expansion of production to help meet the energy needs of the Soviet Far East.

A joint Soviet–Japanese venture to develop oil and gas production on Sakhalin Island was first proposed in 1965, but political problems over the Northern Territories (Kuril Islands) dampened Japanese enthusiasm towards the project. In the early 1970s proposals shifted to the development of oil and gas potential off the coast of Sakhalin Island. In October 1974, after protracted negotiations, the 'Sakhalin Oil Development Company' (SODECO) was created. It consists of eighteen shareholders of which the largest is the Japan National Oil Corporation. In the early stages of the project the US company Gulf Oil had a small share, but it withdrew in the late 1970s. On 28 January 1975 a general agreement was signed on Sakhalin Oil and Gas Development. Under the terms of the agreement $100 million of credit was advanced by the Japanese for exploration for oil and gas off the coast of Sakhalin. In return the Japanese would receive half of the production from the project at a discount rate of 8.4 per cent as payment on the loan (Stern 1985, p. 198). Once the development phase was reached it was understood that the Japanese would contribute half the investment required, at that time $600–800 million (Egyed 1983, pp. 48–9). Conditions off Sakhalin are very difficult with severe cold and ice, and each year drilling activities can only take place between 1 July and 15 October. Such conditions require equipment beyond the capabilities of both the Soviet Union and (at the time that the project was initiated) Japan to produce.

During the first exploration season in 1976 geophysical surveys were conducted by two Japanese vessels and a vessel leased from CIE Générale Géophysique of France. During 1975–7 SODECO purchased equipment to enable it to continue exploration. The Mitsubishi Corporation was contracted to construct two geophysical ships worth $2.5 million, while CIE Générale Géophysique provided $14 million worth of geophysical equipment for the two ships. At the same time contracts worth $5 million were signed to create an operational base on the south-west coast of Sakhalin at Kholmsk. In addition the US company Control Data supplied a Cyber–172 computer for use in the project.

The results of the first season of exploration were encouraging and in October 1977 the first major strike was made at the Odoptu field off the north-east coast of Sakhalin. In February 1978 SODECO announced its intention to extend the $100 million credit, since inflation and strengthening of the yen had increased costs, but it was not until July 1979 that agreement was reached on the terms of the extension. A further $70 million was provided, a large share of which ($35–6 million) was used to purchase a drilling rig from Mitsui Ocean Development especially designed for Arctic conditions. The rig was built to US design and Armco Inc. was contracted to provide equipment for it. In October 1979 the Chayvo field off the northern coast was discovered. It marked the fifth successful drilling site out of ten drilled since the project had begun. Because much of the equipment used in the project was of US design, economic sanctions enforced to protest at Soviet actions in Afghanistan affected equipment and service contracts for the project. However, in April 1980 the Sakhalin project was exempted from the sanctions. While the 1980 exploration schedule was unaffected, the construction of equipment for the project was delayed.

Despite equipment problems and severe conditions, the exploration efforts were proving successful. But the fact that more gas than oil was being discovered complicated the development phase. In the summer of 1981 agreement was reached on the development phase; SODECO was to participate in the construction of a natural gas liquefication plant on Sakhalin from which LNG would be sent by pipeline to Japanese markets. In December 1981 the Reagan administration enforced economic sanctions to protest at martial law in Poland. Despite Japanese protests US sanctions against the Sakhalin project were not lifted until January 1983. As a result drilling was suspended in June 1982 but resumed in August using one Soviet-built rig. In response to these events the Soviet Union threatened it would cancel the project and hold the Japanese responsible. By the end of 1983, when the exploration phase was scheduled to be complete, fifteen wells had been drilled and the project had confirmed reserves of 140,500 million cubic metres of gas, 86.3 million tons of crude oil and 19.5 million tons of oil condensate. During 1984 basic technical plans were agreed upon, but LNG would now be produced on the Soviet mainland at De Kastri rather than on Sakhalin and LNG would be moved by tanker rather than pipeline. The Japanese were concerned that the construction of a pipeline would make Hokkaido too dependent on Soviet imports. The cost of the development phase was put at $3,800 million and the Japanese partners were to provide half of the investment, in return for which they would receive 1.25 million tons of crude oil a year and 3 million tons of LNG a year over a twenty-year period starting in 1993.

Having survived US sanctions the Sakhalin project has now fallen victim of declining world prices. By late 1986 both sides were reconsidering the future of the project. The Soviet Union was concerned with the economic viability of the project. The Japanese felt that energy prices in the 1990s would be high enough to make the project viable, but SODECO could not find markets for the LNG. The power utility companies, the major consumers of LNG, having already secured large supplies of LNG elsewhere, were no longer interested in Sakhalin LNG. During the summer of 1987 it became evident that the Sakhalin project was unlikely to continue in its original format and might even be cancelled. In November 1987 it was announced, however, that the project was to resume, but in a revised form.[5] The Soviet Union has agreed to purchase all Sakhalin-produced natural gas, while the Japanese will buy its crude oil. Although no details have been released, this will probably mean the cancellation of the LNG plant and therefore a much cheaper development phase. Additional pipelines are already being constructed from Sakhalin to the Soviet mainland and the natural gas will serve to reduce the Soviet Far East's reliance on imports of West Siberian oil. It will also provide a feed stock for the development of a petrochemicals industry in the region. Shipments of crude oil may start as early as 1992 and initially Japan will import 1 million tons of oil a year. While the revised project may not provide the Soviet Union with access to LNG technology, it will certainly increase energy production in the Soviet Far East. The failure to sell LNG to Japan also means that the project is unlikely to become a major source of export revenue.

Transportation

As trade between Japan and the Soviet Union has expanded so has the demand for port facilities in the Soviet Far East. To facilitate timber exports related to the various forest agreements, coal deliveries from

South Yakutia and the expansion of traffic on the Trans-Siberian Container Landbridge a special port has been built at Vostochnyy. In December 1970 Japan provided $80 million worth of credit to the Soviet Union to expand port facilities in the Nakhodka region. Between 1972 and 1978 the credit was used to complete the first stage of the port which consists of a timber-loading complex capable of handling 400,000 tons a year, a wood-chip production and loading facility with a capacity of 800,000 tons a year, a container port with a capacity of 700,000 tons a year and a coal terminal capable of loading 6 million tons of coal a year. Japanese companies provided equipment for the construction of the port, including bulldozers, excavators, cranes and steel products. Mitsui provided $27.2 million worth of equipment for the coal terminal, conveying and ship-loading equipment for the wood-chip terminal, container-handling equipment and a computer centre for the container terminal (Zhebelev 1982, p. 27). With COCOM permission, OKI Electric Industry Co. provided a ship traffic-control system for the port complex at Nakhodka-Vostochnyy. In their early stages operations at Vostochnyy were problematic; timber did not arrive in standard packages and had to be repacked at the port, there were insufficient logs to operate the chip plant at capacity and storage space was insufficient to handle incoming containers. These problems have since been resolved and thanks to western equipment Vostochnyy now acts as a model of Soviet efficiency. The second stage of development presently underway will add an additional 7 million tons to the capacity of the coal terminal and expand the container terminal and general cargo facilities. Eventually Vostochnyy will have sixty-four berths with a total length of 18 km and will be capable of handling 40 million tons a year, making it the biggest port in the Soviet Union.

For the Soviet Union compensation agreements have provided credits to purchase machinery and equipment to develop the resource base of the Soviet Far East. Some of these credits have been used to purchase high-technology processing technologies, such as the coal preparation plant at Neryungri. At the same time a substantial part of the credits have been used to purchase relatively low-technology products such as bulldozers and cranes. Imports of process technologies may be compensating for problems with the domestic innovation process, providing technologies that are not readily available in the domestic economy. Imports of products compensate for the inability of the machine-building industry in the Soviet Far East, and in the Soviet Union as a whole, to produce equipment better suited to working in harsh environments, although at Neryungri much of the imported equipment broke down. For the Japanese these agreements have served to diversify their sources of raw material supply. The Japanese machine-building industry has also benefitted from sales to the Soviet Union. Thus the Japanese have imported low-value-added natural resources from the Soviet Far East and exported high-value-added machinery and equipment. The logic of such trade remains intact so long as the Soviet Union requires Japanese machinery and Japan desires Soviet natural resources. Recent events would suggest that, while the Soviet Union continues to pursue Japanese imports, Japanese companies are far less interested in importing Soviet resources. The changing structure of the Japanese economy as it shifts away from resource-intensive industries, has led to reduced demand for natural resources and increased competition among resource suppliers. The current status of the South Yakutia Coal Project and the Sakhalin Continental Shelf Project suggest that the complementarity that promoted large-scale compensation agreements in the early 1970s is fast disappearing. While it is likely that a fourth KS Forest Development Agreement will be reached, it is unlikely that Japanese companies will wish to get involved in totally new large-scale resource development projects. Therefore, the Soviet Union will have to promote other forms of economic co-operation, including border trade and joint ventures.

Natural resource exports

Compensation agreements are not the only means by which Far Eastern resources are exported to Japan. In fact the majority of trade in Soviet forest products falls outside the KS compensation agreements. For example, during 1983 27.8 per cent of Soviet forest product exports to Japan were under a KS agreement, 63.4 per cent was under general trade managed by Exportles, and the remainder under local barter trade.[6] In 1985 forest products comprised 25.7 per cent of Soviet exports to Japan. The other natural resources worthy of consideration are diamonds (1.5 per cent of exports to Japan in 1985), non-ferrous metals (16.1 per cent), gold (10.8 per cent) and fish and shellfish (4.8 per cent). Aluminium and platinum group metals are also important but neither is produced in the Soviet Far East.

Diamonds

Until the 1950s the Soviet Union did not produce diamonds but in response to the Cold War strategic embargo Stalin started a search for a domestic source of diamonds. In 1955 the Mirnyy kimberlite pipe was discovered in Yakutia, and by the early 1960s diamonds were produced from Yakutian mines. In 1962 the Soviet Union agreed to sell virtually all of its uncut diamonds to DeBeers, the South African diamond cartel, thus benefitting from the monopoly position of DeBeers. The Soviet Far East remains the only source of natural diamonds in the Soviet Union. At present the Far Eastern diamond industry is based on three production centres: Mirnyy in the upper reaches of the Vilyuy basin; the Aykal-Udachnyy district on the upper reaches of the Markha river; and the Anabar on the Ebelyakh river, a tributary of the Anabar river. All of these mining operations are in very remote locations and the provision of adequate power and socio-economic infrastructure has been a persistent problem. Statistics on Soviet diamond exports are difficult to obtain since the Soviet foreign trade yearbooks do not report information on many non-ferrous metals and minerals. According to the CIA (1987, p. 72), in 1970 diamond exports earned $175 million and accounted for 7.2 per cent of Soviet foreign currency exports. By 1975 exports had risen to $478 million but only accounted for 5.8 per cent of total exports. In 1980, the last year for which data are available, diamonds earned $1,300 million and comprised 4.7 per cent of exports. Japanese trade statistics report that in 1985 diamond imports from the Soviet Union were worth $300 million. Because of the relationship with DeBeers, the bulk of Soviet diamonds are exported to Great Britain and Belgium. While of limited importance in Soviet–Japanese trade, Far Eastern diamond exports are a very important source of foreign currency and generate far more revenue than exports of forest products.

Gold

After South Africa, the Soviet Union is the world's second-largest producer of gold, and has increasingly used gold to obtain foreign currency. The Soviet Far East has traditionally been the centre of the Soviet gold mining industry. Kaser (1983) has estimated that the gold-fields of the Soviet Far East accounted for 57.8 per cent of Soviet production in 1970, but by 1980 their share had fallen to 48.5 per cent. As production in Soviet Central Asia increases so the share of the Soviet Far East will continue to decline. The most important gold-producing region in the Soviet Far East is the Severovostokzoloto district in Magadan Oblast. However, because of poor management and harsh environmental conditions gold production in this region has been difficult to maintain. The other major gold producing region in the Soviet Far East is the Yakutzoloto district in the Yakut ASSR. According to Kaser's estimates, in 1980 the Magadan and Yakut fields together produced 134 tons of gold which represented 38.9 per cent of total Soviet gold production.

Over the period 1960–79 gold sales, on average, accounted for 14 per cent of the total value of foreign currency exports (Hewett 1983, p. 286). The level of Soviet gold sales in any one year is difficult to predict because the Soviet Union uses gold to meet financial requirements as they arise. For example, in response to the decline in oil prices the Soviet Union increased gold sales from $1.8 billion in 1985 to $4.0 billion in 1986 (CIA 1987, p. 74). In 1985 Japan purchased $154 million worth of Soviet gold which represented 10.8 per cent of total Soviet exports to Japan. While gold sales play an important part in Soviet–Japanese trade, they are far more important in the wider context of east–west trade. If one assumes that the Far East's share of gold exports is equal to its share of domestic production, and combines gold sales with diamond sales, then in 1980 Far Eastern exports of these two commodities generated just over $2 billion or 7.4 per cent of total foreign currency exports. By way of comparison, the total value of Soviet exports to Japan in 1980 was $1,800 million. Thus, while in the context of Soviet–Japanese trade lumber, coal and fish are the most important natural resources, in the more general setting of foreign currency exports diamonds and gold are far more important.

Fish

The Far Eastern fishing fleet is operated by Dalryba and operates mainly in the Pacific and seas adjoining it (see Chapter 6). Prior to the Second World War the Far Eastern fishing fleet accounted for 10 per cent of the national catch, but since the war it has come to national prominence and now accounts for 40 per cent of the national catch. During the 1960s the Far Eastern fishing fleet operated in nearby waters, but during the 1970s it moved into the North and South Pacific. The institution of the 200-mile economic zone, acknowledged by the Soviet Union, has complicated the management of waters under Soviet jurisdiction as well as Soviet offshore fishing operations. The Soviet Union is now able to control the operations of Japanese fishermen within the Soviet economic zone, and under various fishing agreements the Soviet Union receives compensation for Japanese landings of salmon and sea trout which originate in Soviet inland waters. Unfortunately, the negotiation of fishing agreements has become another source of friction between Japan and the Soviet Union. Over 60 per cent of the Far Eastern catch is exported to other regions of the Soviet Union and foreign markets, but the most important export markets are the Eastern Bloc and the developing nations. Nevertheless, the fishing industry plays an important role in the coastal regions of the Soviet Far East and fish exports are an important component of border trade between Japan and the Soviet Union.

Border trade

Border trade first started in 1963 as barter trade in the fishing industry between Japan and Primorskiy Kray and Khabarovsk Kray. Border trade is now administered by the Ministry of Foreign Trade through Dal'intorg. The Ministry of Foreign Trade stipulates which commodities can be traded and to whom. All trade is conducted on a barter basis. Exports represent surplus production (above plan), by-products or products for which there is little local demand. The value of Soviet–Japanese border trade has increased from 43.6 million roubles in 1976 to 70.8 million in 1982. However, in 1982 it only represented 1.9 per cent of total Soviet–Japanese trade (Stolyarov and Pevzner 1984, p. 187). Over 300 small- and medium-sized Soviet enterprises in the Soviet Far East are engaged in border trade. In 1979 47.9 per cent of the exports were lumber and 29.2 per cent fish; thus border trade replicates the commodity structure of 'big trade' (Miller 1981). Products imported from Japan include textiles, consumer goods and machinery and equipment. Half of the imports are used to provide equipment for local enterprises and the other half provides consumer goods for the local population. While border trade is of limited national significance it provides access to imports of equipment and consumer goods without having to pass through the administrative channels in Moscow.

In recent years Soviet Far Eastern border trade with China and North Korea has also gained in importance. In 1983 the Soviet Union and China reached agreement on the resumption of border trade. During 1983 the Soviet Union and China reached agreement on the resumption of border trade. During 1983 the value of border trade with China was 6.3 million roubles, by 1985 the value had risen to 24.2 million roubles, but accounted for only 1.5 per cent of total trade between the two nations (Kuznetsov 1986, p. 65). Border trade between the Soviet Far East and the Chinese border province of Heilongjiang has involved Soviet exports of building materials, agricultural inputs and consumer goods; in return Chinese enterprises have provided bean and meat products, fruit, textiles and consumer goods. Similarly, Soviet–North Korean border trade has involved the exchange of Soviet natural resources and equipment for imports of fruit, vegetables and consumer goods. Soviet–North Korean border trade represents only 2 per cent of total trade. The Soviet Union is anxious to expand border trade in the Far East but the structure of the region's economy limits trade.

The growth of Soviet–Japanese border trade has been hampered by the small market in the Soviet Far East and the limited number of products the Japanese wish to purchase from the Soviet Union. There have also been logistical problems of providing transportation for border trade. In contrast, Sino–Soviet and Soviet–North Korean border trade suits the needs of both partners and so there remains the potential for further increases in the level of trade. In all cases border trade plays a very modest role in trade relations, but is an important source of goods for the border regions. In the Soviet Far East border trade provides much needed agricultural products and consumer goods to improve the standard of living.

Conclusions: Far Eastern trade and regional development

In conclusion we return to the two questions posed in the introduction: what is the role of the Soviet Far East in Soviet foreign trade?; what is the role of foreign trade in the regional development of Siberia and the Soviet Far East?

Due to a lack of data it is not possible to reach precise conclusions concerning the place of Far Eastern exports in Soviet foreign trade. Nevertheless there would seem to be two scales of consideration: first the role of the region in Soviet economic relations as a whole, and second, the role of the region in trade with the Pacific basin. In the more general context of Soviet foreign trade, the mining of gold and diamonds are the most important export-oriented activities. While the region's share of Soviet foreign currency exports is overshadowed by the oil and gas exports of West Siberia, revenues generated by the sale of diamonds and gold are important in funding imports of western technology and agricultural products. Because of the region's small population and narrow economic base, it is unlikely that the Soviet Far East is a major beneficiary of imported machinery but, because of the environmental limitations placed on agricultural production, the region may benefit from imports of foodstuffs, particularly through border trade. In the narrower context of Soviet trade with the nations of the Pacific basin, the resource-oriented export production of the Soviet Far East has meant that the region's most important trading partner is Japan. During the 1970s the Soviet Far East benefitted from a basic complementarity with the resource-poor industrial economy of Japan. Now as we approach the 1990s highly competitive resource markets and the changing structure of the Japanese economy have undermined the complementarity that initially promoted trade. At present Soviet trade with Japan is in decline while trade with the socialist nations of the Asian–Pacific region is rapidly expanding. Because of the limited correlation between the commodity structure of Soviet trade with the socialist nations and the structure of the Far Eastern economy, the region's role in Pacific basin trade is unlikely to grow. Border trade with China and North Korea will probably continue to expand, but in the near future there is little chance of large-scale resource exports to the Pacific basin.

The impact of Soviet foreign trade on the economy of the Soviet Far East can be considered in two dimensions: sectoral and regional. Sectorally trade participation has promoted the development of the natural resources of the region. Compensation agreements have supplemented domestic investment and provided access to western technology to exploit the forest and energy resources of the region. Increased resource exports have also promoted the improvement of the transport infrastructure; clearly the potential expansion of the export base was an important consideration in the construction of the BAM. Because Far Eastern trade has concentrated upon the exploitation of a small number of resources it has done very little to diversify the regional economy, in fact it has probably aggravated the situation by focussing attention upon the resource sector. In addition, because most of the resources leave the region in a relatively unprocessed form, there is only a limited multiplier effect within the regional economy. Regionally foreign trade participation has promoted economic development in the remote mining centres of the Far North, such as Mirnyy. Soviet–Japanese trade and border trade have both tended to benefit Primorskiy Kray, Khabarovsk Kray and the border regions of Amur Oblast. The exception to this pattern is the development of the South Yakutia TPC which represents the first BAM-related industrial complex (see Chapter 7). Japanese participation has been important in the development of this complex, but the current economic environment does not favour additional foreign involvement in BAM-related resource development projects. In sum, foreign trade participation has served to perpetuate the existing narrow sectoral and regional specializations of the Soviet Far East.

Recent plans and policy statements have tended to down play the role of foreign trade in the regional development of the Soviet Far East. Gorbachev's Vladivostok speech spoke of the need to promote new forms of economic co-operation such as border trade and joint ventures. While border trade is growing it still plays a very modest role in foreign trade as a whole. At present no joint ventures have been established in the Soviet Far East, and while the Japanese are interested in setting up ventures in the region they are still unhappy about the terms of joint venture operations. More generally, the reforms of the Soviet foreign trade system are aimed at improving the structure of Soviet exports by limiting the share of natural resources and promoting higher-value-added products. When combined with economic reforms which seek to improve the technological level of Soviet industry, Gorbachev's policies do not favour remote area resource development projects. The 'Long-term State Programme for the Complex Development of the Far Eastern Economic Region, the Buryat ASSR and Chita Oblast to the Year 2000', announced in August 1987, seems to represent a domestic solution to the regional development problems of the Soviet Far East. There may be a realization that a substantial expansion of the region's exports base is neither desirable nor practical given the current domestic and international economic environments. In the future foreign trade will play a more modest role in the economy of the Soviet Far East, relying upon the expansion of border trade and joint ventures rather than resource-based mega projects.

Notes

1. *New York Times*, 1 February 1987.
2. *Japan Economic Journal*, 22 January 1988.
3. *Japan Economic Journal*, 13 February 1988.
4. *Mining Journal*, 27 March 1987.
5. *Japan Economic Journal*, 21 November 1987.
6. *Japan Lumber Journal*, 20 April 1984.

Bibliography

Alexandrov, V. (1982) 'Siberia and the Soviet Far East in Soviet Japanese Economic Relations', *Far Eastern Affairs*, no. 2, pp. 21–32.

Agafonov, S. (1985) 'The Neryungri Project', *Current Digest of the Soviet Press*, vol. 37, no. 36, pp. 16–17.

Bradshaw, M.J. (1987) 'East–West Trade and the Regional Development of Siberia and the Soviet Far East', unpublished PhD thesis, The University of British Columbia.

_____ (1988) 'Soviet Asian–Pacific Trade and the Regional Development of the Soviet Far East', *Soviet Geography: Review and Translation*, vol. 29 no. 4, pp. 367–93.

CIA (1987) *Handbook of Economic Statistics, 1987*, Washington, D.C.: US Government Printing Office.

Chichkanov, V. (1986) 'Problems and Prospects of the Development of Productive Forces in the Far East', *Problems of Economics*, vol. 29, no. 7, pp. 74–91.

Dienes, L. (1982) 'The Development of Siberian Regions: Economic Profiles, Income Flows and the Strategies for Growth', *Soviet Geography: Review and Translation*, vol. 33 no. 4, pp. 205–44.

_____ (1985) 'Soviet–Japanese Economic Relations: Are They Beginning to Fade?', *Soviet Geography: Review and Translation*, vol. 26, no. 7, pp. 509–25.

Dyker, D. (1982) 'Technological Progress and the Development of Siberia and the Soviet Far East', *Radio Liberty Research Bulletin*, RL 443/82.

Egyed, P. (1983) *Western Participation in the Development of Siberian Energy Resources: Case Studies*, East–West Commercial Relations Series, no. 22, Ottawa: Institute of Soviet and East European Studies, Carleton University.

Franklin, D. (1988) 'Soviet Trade with the Industrialized West', in R. Weichhardt (ed.), *The Soviet Economy: A New Course?*, Brussels: NATO Economics Directorate, pp. 275–88.

Gardner, H.S. (1982) *Soviet Foreign Trade: The Decision Process*, The Hague: Kluwer-Nijhoff.

Gorst, I. (1987) 'Oil exports rise in 1986', *Petroleum Economist*, vol. LIV, no. 3, pp. 93–4.

Granberg, A.G. (1980) 'Siberia in the National Economic Complex', *Soviet Review*, vol. 22, no. 2, pp. 44–68.

Hanson, P. (1987) 'Reforming the Foreign Trade System', *Radio Liberty Research Bulletin*, RL 104/87.

Hewett, E.A. (1983) 'Foreign Economic Relations', in A. Bergson and H.S. Levine (eds), *The Soviet Union: Towards the Year 2000*, London: George & Unwin, pp. 269–310.

Ivanov, I.D. (1987) 'Restructuring the Mechanism of Foreign Economic Relations in the USSR', *Soviet Economy*, vol. 3, no. 3, pp. 192–218.

Judd, D. (1971) 'Japanese Economic Interests in Siberia, 1962–70', *Polar Record*, vol. 15, no. 98, pp. 691–8.

Kaser, M. (1983) 'The Soviet Gold Mining Industry', in R.G. Jensen, T. Shabad and A.W. Wright (eds), *Soviet Natural Resources in the World Economy*, Chicago: University of Chicago Press, pp. 556–96.

Kuznetsov, A. (1986) 'USSR-PRC: Trade and Economic Relations', *Far Eastern Affairs*, no. 3, pp. 63–9.

McIntyre, J. (1987) 'Soviet Efforts to Revamp the Foreign Trade Sector', in *Gorbachev's Economic Plans, vol. 2*, Washington, D.C.: Joint Economic Committee, US Government Printing Office, pp. 489–503.

Miller, E.B. (1981) 'Soviet Participation in the Emerging Pacific Basin Economy: The Role of Border Trade', *Asian Survey*, vol. 21, no. 5, pp. 565–78.

Mote, V.L. (1985) 'A Visit to the Baikal–Amur Mainline and the New Amur–Yakutsk Rail Project', *Soviet Geography: Review and Translation*, vol. 26, no. 9, pp. 691–716.

Nirsha, T. (1981) 'Compensation-based Cooperation – A Mutually Advantageous Form of Economic Relations Between East and West', *Foreign Trade*, no. 8, pp. 36–40.

Nobuhara, N. and Akao, N. (1983) 'The Politics of Siberian Development', in N. Akao (ed.), *Japan's Economic Security: Resources as a Factor in Foreign Policy*, Aldershot, Hants: Gower, pp. 197–215.

Ogawa, K. (1987) 'Economic Relations with Japan', in R. Swearingen (ed.), *Siberia and the Soviet Far East: Strategic Dimensions in Multinational Perspective*, Stanford: Hoover Institution Press, pp. 158–78.

Ognev, A.P. (1986) *Ekonomicheskiye otnosheniya Vostok-Zapad v 80-e gody*, Moscow: Nauka.

Shlyk, N.I. (1981) 'Eksportnaya spetsializatsiya Dal'nego Vostoka: osnovnyye napravleniya razvitiya', *Izvestiya Sibirskogo otdeleniya Akadamii Nauk. seriya obshch. nauk*, no. 6, pp. 34–7.

Smith, A.H. (1987) 'Foreign Trade', in M. McCauley (ed.), *The Soviet Union Under Gorbachev*, New York: St. Martin's Press, pp. 135–55.

Spandaryan, V. (1975) 'The Development of Soviet–Japanese Economic Relations', *Foreign Trade*, no. 4, pp. 23–9.

Stern, J.P. (1985) *Natural Gas Trade in North America and Asia*, Aldershot, Hants.: Gower.

Stolyarov, Yu. S. and Pevzner, Ya. A. (eds) (1984) *SSSR-Yaponiya: problemy torgovo-ekonomicheskikh otnosheniy*, Moscow: Mezhdunarodnyye otnosheniya.

UN Economic Commission for Europe (1982) 'Large-scale and Long-term Compensation Agreements in East–West Trade', *Economic Bulletin for Europe*, vol. 34, no. 2, pp. 171–95.

USSR, Ministry of Foreign Trade (1987) *Vneshnyaya torgovlya v SSSR v 1986 g.*, Moscow: Finansy i Statistika.

Vanous, J. (1982) 'Comparative Advantage in Soviet Grain and Energy Trade', *Centrally Planned Economies, Current Analysis*, 10 September, Wharton Econometric Forecasting Associates, Washington, D.C.

Zhebelev, G. (1982) 'Port Vostochnyy', *Soviet Shipping*, no. 2, pp. 26–8.

Zhebrovoskiy, B. and Ponomarev, A. (1986) 'Sotrudnichestvo s zapadnymi firmami na kompensatsionnoy osnove v proshedshey pyatiletke', *Vneshnyaya torgovlya*, no. 10, pp. 18–20.

Chapter eleven

Economic and strategic position of the Soviet Far East: development and prospect*

Leslie Dienes

Soviet policy towards the development of the Far East has always been influenced by the remote location of this huge region from the country's economic heartland and by its strategic position. The Soviet Far East abuts on the Pacific basin, where in recent years the most vigorous growth in world trade has taken place and where some of the most dynamic economies are found. At the same time, the extreme distances from the country's main population centres, the harsh physical environ- ment and lack of infrastructure present greater barriers to effective econ- omic integration into the country's mainsteam than for the regions of Siberia further west. Equally important, nowhere in the USSR do issues of development intertwine with strategic and security concerns over so extensive a territory than east of Lake Baykal. Such links indeed go back to the mid-nineteenth century, when the Amur lands were annexed. In Soviet times, however, critical developments in political and military relationships with China and Japan have heightened these strategic con- cerns. Still more recently, the increasingly global ocean policy of the USSR sharply raised the importance of the Pacific coast, its land and sea approaches, for the Soviet leadership.

These issues and their geographic manifestations need further elabor- ation. The Soviet Far East is by far the largest major economic region of the USSR, covering almost 28 per cent of the country's land area. If Transbaykalia (i.e., the Buryat ASSR and Chita Oblast, which represent that awkward transitional area wedged between Lake Baykal and the northern bulge of Manchuria) is added, that share rises to over 31 per cent (*Nar. khoz. SSSR v 1985 g.*, pp. 12–14). Roughly nine-tenths of this vast region lies in the Northland, the bulk of it in the Far North (Slavin 1982, pp. 22 and 174). On the one hand, this means that huge internal distances and a particularly harsh environment severely limit economic interaction and linkages even among provinces of the Far East. Sub-regional barriers and isolation, therefore, would restrict the possibility of self-reinforcing growth for many decades, even without the well-known rigidities of centrally controlled, Moscow-based planning. On the other hand, inaccessibility and remoteness from the country's econ-omic heartland and, except for the coastal provinces, from foreign indus-trial centres as well, burden the vast territory with inordinate transport costs in its attempt to develop its natural riches. Intervening opportunities from Soviet regions nearer to the European USSR frustrate the Far East on the domestic front; rival resource frontiers, with more accessible, often higher quality and cheaper primary products frustrate it on the world market.

The size and geographic location of the Far East compels not only the USSR but also its Asian neighbours to confront the strategic issues from contrasting positions and in several dimensions. First, with more than 31 per cent of its land surface east of Lake Baykal, the Soviet Union is a permanent East Asian presence and could not disengage from this world region even if it were intent on doing so. Indeed, Gorbachev in his July 1986 Vladivostok address strongly stressed that point (*Ekon. gazeta*, no. 32, August 1986, p. 3). Second, this impressive areal presence is coupled with a distressing asymmetry of population and distance from the coun-try's economic heartland, in sharp contrast with other East Asian powers. Within a radius of 2,500 km from Vladivostok (which just includes the Irkutsk basin and Lake Baykal), 12 million Soviet citizens face 1,000 million Asians – two-thirds of China's and Taiwan's population and all those in the two Koreas and Japan. The Amur–Ussuri outposts of the USSR lie in greater geographic proximity to Manchuria, the North China Plain and Japan than to the Soviet industrial centres of the Kuzbas, let alone to those of the Urals or European Russia. Third, the geographic configuration of the Sino–Soviet-Mongolian boundary has its reverse stra-tegic side. Manchuria is also a salient, wedged between the Transbayka-lian and Maritime Provinces of the Soviet Union. Similarly, the five Soviet divisions in Mongolia, with about 75,000 men, account for roughly 15 per cent of all Soviet ground forces east of the Yenisey and bring Moscow's military might within 800 km of Beijing (Tan Eng Bok 1986, p. 49 and IISS, *The Military Balance* 1986, p. 29).

Lastly, the strategic dimension is not merely regional but also hemi-spherical and global. Until the acquisition of nuclear capability, the status of Russia/USSR 'as a Great Power has rested for the last hundred years on two factors: the control of the largest landmass on our planet and the sheer size of her military might' (Hauner 1985, p. 14). Its evolution from a mere great power into one of the two superpowers was only partly based on the attainment of nuclear parity with the US. The other pillar of this transformation was an expanding global reach, an assured, sustainable presence on all three of the world's greatest oceans. For such a presence, a comprehensive array of fully co-ordinated military facilities on the Soviet Pacific coast (with well-defended approaches, civilian sup-port and strengthened transport links) was the essential requirement and tool. Writes one expert, 'the build-up of Soviet military power on the Far East has been neither in pursuit of a single objective nor on a regular scale. It has been across the board and its evolution has proceeded *pari passu* with the improvement of the USSR's global posture over the past several decades' (Tan Eng Bok 1986, p. 40).

Moscow has no direct foothold on the third great (i.e. the Indian) ocean, though from inside its Transcaspian and Central Asian regions Soviet strategic proximity, in comparison with that of other great powers, is enviable. However, from their Pacific ports, Soviet ships can reach the Indian Ocean more securely than from the Mediterranian via the Suez Canal. Moreover, since the *de facto* acquisition of Vietnam's Cam Ranh naval base, where today thirty Soviet warships are permanently stationed with squadrons of aircraft, a base which is now equipped with the second-largest Soviet intelligence communication network outside the USSR,[1] Moscow has also planted a firm strategic post at the very gateway to the Indian Ocean (Tan Eng Bok 1986, p. 52).

Whatever normalization and arms reduction may be achieved under Gorbachev, I find it inconceivable that the USSR would retreat from being a full-fledged Pacific power. At best it will lower its profile in a *proportional* retrenchment. However, a disproportionate reduction of its Pacific naval/military presence would mean a contraction of its global reach, hence a shrinking of its superpower status. Because of this geostrategic reality, continued high priority will be accorded by the Soviet leadership to the Far Eastern *coastal* provinces in the years ahead. In my view, such a priority remains quite unaffected by any increase or decrease in commitment to the economic development of the Far East. It will also remain independent of the general development strategy employed and of Gorbachev's overall policy of restructuring. However, the rate of growth of the region's civilian economy, its sectoral and geographic proportions, its linkages with the rest of the country and the outside world do not have such autonomy. They have long been influenced by shifts in industrial and regional policy and decisions on foreign trade. Such policy decisions, more prone to change than the strategic importance of the region to the Soviet leadership, will continue to be crucial for the prospects of the Far East.

Economic development: the recent past

The remote and far-flung positions of the Far Eastern provinces suggest consideration of a development strategy rather different from that applied to more accessible regions. The theoretical options include, though perhaps are not limited to, the following: a) a more or less autonomous line of development, with a large degree of local control over investment, other budgetary allocations and trade; b) a foreign-trade-oriented strategy, whereby the region's rich natural resources are developed by the centre overwhelmingly for export (preferably with the help of foreign capital), and the revenues earned are disposed of by that same centre in accordance with its overall sectoral and regional policy; c) continued massive subsidies to the region for reasons of its crucial strategic and potential economic value, but acceptance of a very immature, narrow economic profile and below-average growth for the near future.

The export-oriented strategy: modest results

It is clear that in the early and mid–1970s the Soviet leadership anticipated and appeared very sanguine about the success of the second option. An energy crisis has just hit a totally unprepared advanced industrial world. The spectre of OPEC-like cartels over other natural resources has resulted in panic-buying and a radical increase in the price of energy and raw materials on the world market. The vast resource-base of the Soviet Far East in relative proximity to resource-hungry Japan, whose nervousness about supplies was only too palpable, seemed like a combination highly favourable to the USSR. The policy of east–west *détente*, the easier availability of western credit and an apparent willingness on the Soviet side to accept a much greater degree of integration with the world economy, all seemed to mark out the eastern half of Siberia for a largely foreign-trade driven economic boom. To hasten that future, Soviet leaders themselves were prepared to make huge investments in that vast area to improve accessibility. The decision to build the BAM (Baykal–Amur Mainline) several hundred miles north of the Trans-Siberian Railway was the most obvious manifestation of that commitment. Besides its strategic value, BAM was also regarded as a gateway to the interior of the Far East. The vast and, with the railway, more accessible natural riches would would vest the USSR with a strong bargaining power *vis-à-vis* resource-poor capitalist countries and especially Japan.

These expectations appeared reasonable in the anxious days of the first oil crisis. However, from the perspective of a dozen years, it is now clear that this bargaining power has not materialized and is most unlikely to do so in this century. The second oil shock already failed to drive up the prices of non-energy minerals and other raw materials: indeed the deep recession it helped to create forced these prices downwards. And, just as OPEC, the USSR has failed to foresee the long-term conservation measures put into effect and the deep structural changes experienced throughout the industrial world. The shift away from heavy industries in the advanced economies is exemplified by a 34 per cent drop in steel production between 1974 and 1986 throughout the OECD, one-sixth of that absolute decline taking place in Japan (IISI, October 1987). As a result of this shift and new technologies applied to save materials as well, world demand for a host of metals (aluminium, copper, zinc, lead, nickel, etc.) have been stagnant and even declining).

These momentous changes in the non-communist industrial countries, not the least in Japan, have reduced the attractiveness of investment in resource projects of the Soviet Far East and Eastern Siberia. From the late 1960s through to the first half of the 1970s Japan and the USSR have undertaken eight joint projects, though the expansion of the port of Vostochnyy did not involve natural resources. Co-operation in these projects, however, has proved rockier than had been anticipated. Unpredictable changes of plans and inadequate technical information on the Soviet side, serious differences over financing have frustrated the Japanese, while the harsh physical environment and the lack of infrastructure in the region caused serious delays in some of the ventures. These economic difficulties have been exacerbated by the deteriorating political climate, caused by Soviet refusal even to discuss the issue of the occupied, and later fortified, Japanese northern islands, by the huge military build-up in the Pacific provinces of the USSR and the invasion of Afghanistan (Dienes September 1985, pp. 511–16).

In the early 1980s, about a tenth of the gross regional product generated in the Far East was exported abroad. The value of this shipment amounted only to a small fraction of Taiwanese or South Korean exports at that time[2] and much less today. In addition, socialist countries took 62 per cent of Soviet shipments to the Asian–Pacific region in 1980 and more than 80 per cent in 1986. With the exception of exports to China today, much of that trade represents a form of economic assistance and military supplies and is economically more a burden than a benefit to the USSR (Bradshaw 1988). More than four-fifths of all Far Eastern exports originate from the coastal provinces of Maritime and Khabarovsk Krays (Minakir *et al.* 1986, p. 216). Most of the balance consists of coking coal, shipped from the South Yakutian basin opened up by the Little BAM. The rest of the Far East and the two provinces of Transbaykalia are so distant or inaccessible to the Pacific that their contribution to exports is negligible.[3] More recently, Japan is responding with caution to the new Soviet investment law and to Soviet plans that have been floated about renewed efforts to speed up the development of the Far East. The inhibitions generated by the unfavourable experiences of the last two decades are today compounded by fall-out from the Toshiba affairs. Leading Japanese toolmakers breeched COCOM regulations which resulted in threats of retaliation in the United States.

Elsewhere I have analysed at length Japanese reluctance to further involve themselves with massive investment projects in the Soviet Far East (Dienes September 1985, pp. 516–19). Since the promulgation of the new Soviet law on joint ownership and management participation by foreign companies, Japan has agreed to half a dozen such large revenue projects, some singly with the USSR, others in consortia with western companies. They are huge petrochemical projects and one forest conversion plant but none of them is in the Far East. The forest conversion mill will be near Irkutsk; the petrochemical projects in West Siberia, the Caspian Lowland, and the European USSR (*Interflo*, July 1987, p. 25; *Financial Times*, 20 May 1987, p. 28; *Platt's Oilgram News*, 14 November 1988, p. 5; 15 November 1988, p. 5). After lengthy and difficult negotiations, Japan and the USSR agreed to suspend the offshore Sakhalin LNG project, the largest of all ventures considered by the two states but geographically closest to Japan (*POGN*, 30 June 1987, pp. 2–3).

As I wrote a few years ago, 'the oft-repeated argument of geographic proximity between the two countries is misleading and only partly correct. The vaunted idea of economic complementarity has proved to be a simplistic and meaningless concept, of little value detached from the dynamics of international demand and supply' (Dienes September 1985, p. 521). The hopes placed on a foreign-trade-oriented strategy for the development of the Far East have remained unrealized, for they were largely misconceived. The Soviet Far East is only one of several resource frontiers that industrial nations on the Pacific may turn to and, clearly, it is not the most attractive one. In addition, after a temporary appreciation, primary products are again 'devalued' by the world market against more advanced manufactures. A significant improvement in Soviet–Japanese political relations and tangible manifestation of Soviet goodwill (e.g. the return of some of the occupied islands which the Japanese call their Northern Territory and/or a substantial lowering of the military profile in Russia's Pacific provinces) would almost certainly induce greater Japanese co-operation in developing the Far East. Even in such a 'best case' scenario, however, the objective conditions of the existence of competing and often more attractive raw material supplies available to Japan would have to be faced by the USSR. As in the past, the Soviet Far East must continue to rely overwhelmingly on Soviet efforts for its development even for capital and technology. As before, its prospects will depend on domestic priorities and policy, and on the relative importance Moscow assigns to this far-away region against the capital and manpower needs of regions more accessible to the country's heartland.

Economic structure, performance, and dependency on the European heartland

Notwithstanding the enormous area of the Far East, the region's overall economic output is very small. Its seven provinces generated a net material product (NMP) of only 2.8 per cent of the USSR (Mozhin 1980, p. 78). The addition of Transbaykalia (Chita Oblast and the Buryat

ASSR, officially in East Siberia) barely lifts this share above 3 per cent. In industrial production alone, its contribution is marginally greater, 3 per cent without Transbaykalia, still the smallest among all twenty Major Economic Regions of the country. Over 60 per cent of NMP produced in the Far East originates in the extreme south of the Mainland – the plain along the Amur–Ussuri rivers with their chief tributaries, by Lake Khanka and the southern littoral. Sakhalin and Kamchatka, the Yakut ASSR, Magadan Oblast and the whole northern half of Amur Oblast as well (covering more than nine-tenths of the Far East's territory) contribute less than 40 per cent of the region's NMP (Mozhin 1980, pp. 76–7). Their share in the region's population is very similar (*Nar. khoz. SSSR v 1985 g.*, pp. 12–14).

The economic structure of the Far East is that of a classic resource frontier in its early stages of development. Three primary branches: fishing and fish processing, non-ferrous metals and forest products comprise two-fifths of total industrial output, a share four times as large as in the country as a whole. *Per capita*, wood production in the Far East exceeds the Soviet average four times, the output of non-ferrous metals five times and that of the fishing industry fifteen times. In addition, much of the rest of the economy is tied, directly or indirectly, to these three basic branches. During the Soviet period, the number of different minerals extracted in the region increased from seven to almost eighty (Minakir *et al.* 1986, p. 33).

Another structural feature of the economy of the Soviet Far East is found in the vital importance of the transport sector and in its enormous burden. That sector contributed 11 per cent of the region's gross social product but accounted for 30 per cent of its fixed capital, 19 per cent of its capital investment and 19 per cent of its labour force at the beginning of the 1980s (Proskuryakova 1982). At that time the respective figures for the USSR as a whole were 4.4, 13.6, 12 and 9 per cent (*Nar. khoz. SSSR v 1985 g.*, pp. 45, 48, 366–7 and 389). Of all major Soviet regions, the Far East has the sparsest, most skeletal transport network, which adds further poignancy to the overweening share of transport in the total value of fixed capital in that huge area.

A crucial characteristic of Far Eastern transport linkages is the result of the maritime-continental dichotomy, which appears almost as sharp as the division between the more settled zone along the Chinese border and the empty north. Less than half of the population of the Soviet Far East lives within 80 km (50 miles) of the coast (Treyvish 1982, p. 112), and the immediate littoral, that within 10 km of the ocean, is home for 22 per cent of the region's *urban* inhabitants.[4] Most of that coastal zone is separated from the interior by rugged mountains. With the exception of one gravel road in the north (from Magadan to Yakutsk), access inland is provided only through two portals of the extreme south: those through Vladivostok–Wrangel Bay and Vanino–Sovetskaya Gavan'. From the Chukchi Peninsula to the Chinese–Korean border, the north–south communication is solely by water. Ocean transport represents the exclusive link between Sakhalin, Kamchatka and most of Magadan Oblast and the rest of the country (*Pravda*, 21 May 1986, p. 3). Maritime transport for the Soviet Far East, therefore, assumes exceptional import-ance. Altogether twenty-four domestic harbours are involved in intra-regional marine shipment, with only three of these (Nakhodka, Vostoch-nyy and Vladivostok) engaged in international trade or trade with Soviet ports in the European USSR (Savin 1984, p. 64 and Bartkova 1982, p. 145). During the middle and late 1970s, the proportion of the transport-communication sector in the total employment in Kamchatka and Sak-halin Oblasts reached 14.6 and 13.6 per cent respectively. (See Tables 11.3 and 11.4 later.) These shares surpassed the national share of this sector by 4 and 3 percentage points (*Nar. khoz. SSSR v 1975 g.*, pp. 532–3 and *Nar. khoz. SSSR v 1978 g.*, pp. 366–7).

As a fishing, forestry and non-ferrous metal frontier, the Soviet Far East calls to mind the province of British Columbia, Canada, although the size and population of the latter are very much smaller and climat-ically the Maritime Provinces are a better analogue. The importance of coastal transport, the dichotomy and weak transport links between the littoral and the interior also calls to mind the coastal provinces of Canada, especially British Columbia. However, the latter is strongly integrated with the whole North American market and, indeed, with the world market as a whole. In the mid–1980s, the ratio of exports to GDP (gross domestic product) for British Columbia reached 22–3 per cent, half of the exports going to the US and about a quarter to Japan (North 1988). By contrast, the participation of the Soviet Far East in the world econ-omy remains very modest and all but a miniscule portion of this exchange (less than 2 per cent) is controlled from Moscow.

On the other hand, the dependency of the Far East on far-away centres of the USSR is very strong, even though remoteness makes economic integration into the national mainstream much more difficult than for regions farther west. Every rouble of gross economic output in the Far East requires the importation of 30 kopeks of inputs (Klin *et al.* 1982, pp. 13–32). Almost all of this originates in regions west of Lake Baykal, much of it from the Urals and the European USSR, 6,000 to 8,000 km away by rail. (For the Far Eastern North and Sakhalin, further trans-shipment by river and coastal freighter must be added.) The remoteness of the region from the country's major centres of population and industry have so far prevented the generation of major export flows to pay for the high cost of development and the expense of naval and military installation. Inshipment into the Far East exceeds outshipment from it 1.8 times by value and 2.5 times by volume. The dominance of incoming over outgoing freight in tonnage is especially striking, for it is in stark contrast with Siberian regions further west (Kolesov 1982, pp. 95 and 100). Clearly, the Far East is a resource frontier so removed from markets that it can ship out only a restricted volume of bulk cargo. At the same time, the early stage of its economic development and the prevalence of military installations are responsible for very large volumes of incoming freight.

Specific examples of this 'parasitic' relationship between the Far East and the rest of the USSR abound. Approximately 95 per cent of all products of ferrous metallurgy, 80 per cent of those of the chemical industry, 70 to 80 per cent of those of light industry, and over half of those of machine building (by another source, 70 per cent) must be shipped in from elsewhere, chiefly from the European USSR (Minakir *et al.* 1986, pp. 151–2; Chichkanov 1985, pp. 98–9; Chichkanov and Minakir 1984, p. 89 and Klin 1982, pp. 13–32). Four-fifths of all metal-cutting machine tools used, for example, is railed in from the European provinces, 9 per cent from the Urals. Of rolled metal alone, 2.1 million tons are railed in each year at a cost of 32 to 35 million roubles, and even the reconstruction of the old Komsomol'sk steel mill and the building of a new plant will not satisfy more than 50 to 60 per cent of demand (*Ekon. gazeta*, July 1984, p. 9). This subordinate, dependent relationship with the rest of the country is shown even more clearly by the huge volume of interregional subsidy flowing into the Far East and Transbaykalia. Through the vast area east of Lake Baykal, a full third of the regional income utilized during the mid–1970s was subsidy from other regions of the country (Dienes 1982, pp. 214–23).

This subsidy does not stem from the underpricing of primary products, which has indeed been a problem in Soviet national and regional accounting. At any rate, high-value precious metals, diamonds and fish products, which are not under-priced, represent a very large part of the output of primary industries in the Far East. By contrast, inputs, *except labour*, are more widely under-priced. Capital for Far Eastern investment carries a lower putative interest charge, the normative period for construction is longer than the Soviet average (Minakir *et al.* 1986, p. 167; Aparin and Krinitskaya 1979, p. 50), investment goods and construction materials are made available at low prices and a large portion of such goods are shipped into the Far East from the Urals and the European USSR. Two Soviet economists declared forcefully a few years ago that the interregional subsidies to the Trans-Ural provinces of the country are real and not an illusion created by the anomalies of the Soviet price system. Rather, the root cause of these subsidies is found in the raw, undeveloped state of this territory; the reason for them in national priorities associated; with its development (Bakhrakh and Mil'ner 1984, pp. 20 and 22–3).

As expected from the underdeveloped and imbalanced economy of the Far East, extra-regional linkages with the rest of the country are more important than economic links among the provinces of the region, which do not complement one another significantly and are able to provide for each other only a part of regional inputs and final demand. The economic links of the north east (Magadan Oblast and the Yakut ASSR) with the rest of the Far East are especially weak and one-sided. The north east receives only a small fraction of its inputs not produced locally from elsewhere in the Far East. It also ships virtually all of its products (overwhelmingly consisting of gold, diamonds, tin and fur) to the European USSR and, indirectly, for export. Sakhalin and Kamchatka Oblasts, however, rely on the rest of the Far East (principally on the extreme south of the mainland) for inputs almost as heavily as on provinces outside the Far East. They similarly distribute their output both inside and outside the region (Mikheyeva 1982, p. 53 and 1983, p. 79).

The economic indicators for the development of the Far East are deteriorating sharply, far more rapidly than for the USSR as a whole. According to the calculations of Soviet scholars, the growth of factor inputs during the 1960–80 period accounted for almost four-fifths of the increment in the net output of the region's industry; growth in factor productivity contributed less than 21 per cent. The corresponding ratio for Soviet industry as a whole was 53 and 47 per cent (Table 11.1). Labour productivity has been growing significantly slower than in the Russian Republic as a whole (according to official data until 1975 and partial evidence since). Capital productivity in the Far East during the 1970s declined with notable rapidity and at an increasing rate (Table 11.2). For agriculture and construction, the contribution of factor productivity to output growth appeared even lower than for industry, being actually negative for construction, with factor inputs rising somewhat

Table 11.1 The share of extensive and intensive factors in the growth of industrial output in the Soviet Far East, 1960–80

Regions and country	Percent of total increment in final (net) industrial product attributable to	
	Growth of factor inputs	Improvement in factor productivity
	%	%
Far East	79.2	20.8
Southern zone	72.3	27.7
Northern zone	69.7	30.3
USSR	53.2	46.8

Source: Minakir, P.A., Ekonomicheskoye razvitiye regiona: programnyi podkhod. Moscow: Nauka, 1983, pp. 185–91.

Note: The author assumes a modified Cobb-Douglas production function. Final product in the industry of the Far East and its zones was estimated as gross output – material costs + turnover tax, available by industrial branches for the Russian Republic, and corrected according to the branch structure characteristics of the Far East.

Table 11.2 Dynamics of industrial growth in the Far East by time period

Factor	1960–65	1966–70	1971–5	1976–80
	%	%	%	%
Percentage of total increment attributable to growth in factor productivity	35.2	32.5	2.6	−50.5*
Labour productivity	36.2	36.0	33.9	27.9
Productivity of capital	−1.0	−3.5	−36.5	−78.4
Percentage of total increment attributable to growth of factor inputs	64.8	67.5	97.4	150.5*
Growth in employment	35.8	23.2	25.6	18.7
Growth in value of fixed assets	29.0	44.3	71.8	131.8

Source: Minakir, P.A., *Ekonomicheskoye razvitiye regiona: programnyi podkhod*. Moscow: Nauka, 1983, p. 192. (Reprinted with permission of V.H. Winston and Sons, Inc. from *Soviet Economy*, April–June 1985, p. 155.)

Note: *Combined inputs of capital and labour growing some 50 per cent faster than industrial output.

faster than the value of production in this branch (Minakir *et al.* 1982, pp. 40–6). In the Far East, the time lag before production factors yield any return is longer than elsewhere, and, still more serious, capital and labour are used with lower coefficients of intensity (Minakir 1983, pp. 194–5).

In the 1970s, industrial growth in the Far East slowed even more drastically than in the USSR as a whole. If, during the 1960s, the volume of industrial output in the Far East increased at a slightly faster annual rate than the average for the Soviet Union the following decade and a half saw the reversal of that relationship. From 1970 through to 1986, Soviet industrial production rose 224 per cent, but in the Far East only 210 per cent. The rate in the region remained below the Soviet mean both during the 1970s and in the first six years of the 1980s as well (*Nar. khoz. SSSR za 70 let*, p. 135). That drastic slow-down was due mainly to a sharp deceleration in the expansion of the three principal industries of the Far East. The growth rate of the fishing industry and mining each dropped by one half (from an average yearly increase of 11 per cent and 12.8 per cent respectively during the 1960s to 5.5 per cent and 6 per cent during the 1970s), that of non-ferrous metallurgy declined from 7.5 per cent to 5 per cent (Minakir 1983, pp. 66–7).

Much of the slow-down in the expansion of these principal branches was due to deteriorating resource conditions and quality, aggravated by inadequate geological prospecting and preparation. Soviet sources unanimously play up the presence of huge mineral riches in the Far East. It is far less frequently mentioned that, as a rule, the minerals of the region have very complex geochemical composition (Minakir *et al.* 1986, p. 37 and Aganbegian *et al.* 1984, pp. 12 and 71). This would complicate exploitation anywhere, but in a sparsely populated vast territory, where supporting infrastructure is very undeveloped, the effect is particularly adverse. In addition, hasty and inadequate exploration has led to errors and miscalculations in the projections and technical parameters of mining, enrichment and smelting facilities (Minakir *et al.* 1986, p. 37). Similarly, a sharply worsening quality of fish catch characterizes the coastal fishing industry. The Ministry of Fisheries cans a high percentage of this inferior catch in its Far Eastern plants because cans fetch a higher price. Yet they cannot be marketed and the equivalent of 450 million 'standard sized jars', unsellable but worth 300 million roubles, were delivered in 1985 alone (*Literaturnaya gazeta*, 19 November 1986, p. 19). Coastal fisheries still account for 94 per cent of all the catch, as opposed to the 6 per cent that originates from the open ocean. Though the industry hopes to raise the latter portion to 15–20 per cent by the end of the 1990s, coastal fishing, with its deteriorating return, will continue to contribute the lion's share of the haul well into the next century (Minakir *et al.* 1986, pp. 43 and 45).

For some years now, the Far East has been struggling with a 'chronic lag in its fuel and power complex' and a 'palpable energy deficit', emphasized by Gorbachev, Aganbegian and others (*Pravda*, 11 November 1985; 29 July 1986 and Aganbegian, 1984, p. 18). Per capita power consumption here fell below the Soviet average in the early 1980s. Conditions must have worsened since, because plans for capacity expansion during 1984–5 were fulfilled only 50 per cent (*Pravda*, 10 October 1982, p. 2 and *Ekon. gazeta*, no. 41, October, 1987, p. 7). A million tons of coal and some 13 million tons of fuel oil (87 per cent of total consumption) are railed in the Far East each year, together with other refinery fractions, the oil products coming from west of Lake Baykal (*Ekon. gazeta*, no. 17, April, 1987, p. 15 and no. 41, October, 1987, p. 7).

In fact, before 1990, the Far East will not be able to guarantee *any* substantial growth in its principal industries of specialization. The necessity of moving to new mineral deposits in the north and the need to reconstruct thoroughly the technology of these branches restrict the possibility of expansion. Such expansion, feasible only in the longer term, will require 15 per cent more capital investment for a unit increase in output than the average for the USSR. This is the assumption of the basic central variant of an interregional programming model (OMMM), constructed to analyse the contribution of economic regions to national economic growth. Even this assumption relies on a complementary growth of labour productivity above the Soviet mean, a net in-migration of labour and an expansion of social infrastructure and services at such a rate that per capita it would put the Far East at 120 per cent of the average Soviet level by the end of the century (Chichkanov and Minakir 1984, pp. 80–82).

The BAM programme

Soviet plans called for the completion of the Baykal–Amur Railway (BAM) by the 7 November 1984, anniversary of the Revolution, a year ahead of schedule. The laying of the tracks to achieve such a feat was apparently accomplished, thanks in part to the crucial role of military construction crews which built the entire eastern half of the railway (i.e., east of Tynda) (*Stroitel'naya gazeta*, 26 July 1987, p. 1). However, two major tunnels will not be ready for some time. Complex geological conditions, combined with slipshod work, have led to 'intolerable delay in the work connected with the opening of train traffic', in the words of a special investigative commission, headed by then Politbureau member Aliyev in 1984 (*Sovetskaya Rossiya*, 3 June 1984 and *Gudok*, 6 March 1984, p. 1). Conditions have not improved much since that time. As of November 1987, the 15.3 km North Muya tunnel is barely half constructed, and its completion date is now put off till 1992 (*Pravda*, 27 October 1987, p. 3). Meanwhile, a 28 km bypass has been laid, which is so steep – a 40 m change in altitude for every km – that (in the words of a journalist) only slalom skiers can negotiate it and rolling stock must be restricted to a maximum speed of 15 km per hour. Another 52 km bypass did not help matters much and traffic remains negligible owing to high cost and tariff, much above that of the old Trans-Siberian Railway. (*Stroitel'naya gazeta*, 11 August 1987, p. 1 and *Pravda*, 11 June 1987, p. 2).

Even more serious is the rapid deterioration of facilities and, in numerous places, even of the rails already built. Because of complex geological and engineering conditions, inadequate preliminary studies and violations of standards, there are today over seventy construction objects at seven major BAM stations alone 'which now cannot be used even for temporary exploitation. The number one problem already today is not how to proceed with the construction of the railway or what it will carry, etc., but how to maintain in usable shape what has been so far built' (*Izvestiya*, 19 August 1987, p. 2. Also *Pravda*, 27 October 1987). Even the rails themselves are being damaged by rust, years before normal exploitation will begin (*Sovetskaya Rossiya*, 1987, summer, quoted in *NRS*, 2 August 1987, p. 8 and *Sots. industriya*, 17 June 1987, p. 2). Similarly, on some sections of the Little BAM, the transverse north–south section between the Trans-Siberian and the South Yakutian coal basin (which, in contrast to the BAM itself, is heavily used), low-quality, improperly hardened R–50 rails and light-rail types were laid to save money and speed construction. They have now deteriorated, causing wrecks and requiring thorough repairs and upgrading. Soon the entire line will have to be extensively upgraded and overhauled. From Tynda to BAM station on the Trans-Siberian, trains take eight hours to negotiate the 187 km section (Al'ter 1987, p. 41 and *Pravda*, 11 June 1987, p. 2).

Preparatory work on extending the north–south railway (Little BAM) towards Yakutsk has also started and the first half of this line to Tommot (on the Aldan river) is among the projects approved for the 12th Five Year Plan (*Materialy* 1986, p. 320). This new line, named AYAM (Amur–Yakutsk Mainline), will not, however, help in the development of the BAM zone. Indeed, by the premature transfer of construction brigades and equipment from that east–west line to the new north–south railway, it will contribute to the already drastic scaling-down and postponement of projects along the BAM and may even delay the full completion of that railway itself. Indeed, recently a Soviet expert severely criticized the establishment of a separate construction trust for AYAM. He foresees the repetition of the same errors committed on BAM, thus compounding problems on both projects (*Ekon. gazeta*, no. 29, July, 1984, p. 9). In addition, building organizations from the Kuzbas and the Kansk-Achinsk fuel-energy complex west of Lake Baykal are reportedly raiding construction brigades both on BAM and AYAM. These raiders offer the same pay and much better living conditions in their more established, climatically less extreme areas (*Izvestiya*, 26 December 1986, p. 2).

The effect of the new AYAM railway, when and if it is completed a decade or more hence, will be felt in central Yakutia, chiefly by reducing sharply the cost of supplies *into* the region. While shipment out of Yakutia is projected to grow between 1985 and 2000 by 80 per cent, shipment into that ASSR is expected to increase by almost two and a half times (from 6.4 million to 15.3 million tons) (Mote 1987a, p. 20). AYAM will not alter appreciably the economic structure of the Yakut ASSR or the north east of Siberia in general. That economic structure is expected to remain extremely narrow, based overwhelmingly on the extraction and smelting of precious and strategic metals and tin. Excepting the extreme south-west corner of Yakutia, not even petroleum production is envisaged from the basins of the North-east (Yakut ASSR and Magadan Oblast) in this century. Recent works designate most of this territory as 'reserve areas' for the more distant future (Robinson 1985, pp. 98–106 and Privalovskaya 1984, pp. 319–27).

As I wrote in 1985 (Dienes April-June, 1985, p. 169), the larger BAM 'programme' for the development of resources along the railway is at a virtual standstill. In only one of the TPCs (territorial production complexes), South Yakutia, is work moving ahead and, so far, only coal mining and electric power development is a reality even here. Aside from that mining and power centre, not a single new enterprise appeared along the BAM, according to a July 1987 report (*Sotsialicheskaya industriya*, 17 July 1987, p. 2). The guidelines of the 12th Five Year Plan, published in 1986, were, indeed, very brief concerning the development of the BAM zone, noting the *start* of construction of only one new resource project, that of the Seligdar apatite concentrator, thus presumably approving the opening of the mine itself. (The huge Far East produces no fertilizers at all and its soils are in dire need of phosphorus). Only preparatory work on creating a ferrous metallurgical base using local coking coal and iron ore is endorsed, with not even the location of the plant decided yet (*Materialy* 1986, p. 320). None of the other resource projects proposed since 1975 for the BAM zone, e.g., Udokan copper, Kholodnaya lead, Molodezhnoye asbestos has been approved. With respect to Udokan, among the two to three largest copper deposits in the world, the Minister of Non-Ferrous Metallurgy recently declared that its exploitation today is not economically feasible (*nevygodno. Stroitel'naya gazeta*, 12 May 1987, p. 2). The preliminary draft of the 12th Five Year Plan also envisaged the start of construction on a new underground mine in the South Yakutian coal basin, with the Coal Ministry asking for 1,000 million roubles of outlay for the mine, but the project was dropped from the final version (Shabad 1986, p. 271).

Given the postponement of major resource projects envisaged earlier, delays and decay on the BAM itself, the economic contribution of the BAM zone even by the end of the century must be questioned. Today, the railway essentially serves only workers on its own construction (Al'ter 1987, p. 40). Wood and South Yakutian coal are the sole important commodities carried out of the BAM zone. Ironically, even that coal to date moves to the Pacific not through the BAM proper but south (via the Little BAM) to the Trans-Siberian, then through that old railway. It seems that wood and coal will remain the only major commodities shipped *out* for at least another decade. The much heralded 'gateway to Siberia', which was supposed to carry 35 million tons of freight per year per kilometre *by* 1983–5, and 10 million tons without Tyumen' oil, hauls less than one million tons today (*Izvestiya*, 19 August 1987, p. 2 and 21 August 1987, p. 2). At Komsomol'sk, on the eastern end, fifty-five to sixty trains are assembled and dispatched for a westward journey each day. Only four to five of these move along the BAM, although the eastern half of the railway is supposedly operational. 'Nothing to haul', says the chief of the marshalling division, so the rest of the trains use the Trans-Siberian (*Izvestiya*, 19 August 1987, p. 2). The Ministry of Transportation subsidizes the operation of BAM, such as it is, to the tune of 150 million roubles each year (Al'ter 1987, p. 40). According to Victor Mote, Academician Aganbegyan 'had concluded as early as 1983 that the new railway would not be able to carry more than 35 million ton-kilometers per kilometer *before the end of the century*' (emphasis added. Mote 1987b, p. 379). By now, even these projections must be in doubt.

A new Far Eastern programme: myth or reality?

In mid–1986, as part of his major Vladivostok address, Gorbachev high-lighted the slowdown in the development of the Far East, the region's narrow, very unbalanced economic profile and the chronic shortage of social services and food, almost half of which is shipped in from the outside. He seemed to promise a long-term complex programme for the territory 'within the framework of a comprehensive government policy for regional development' (*Ekon. gazeta*, no. 32, August, 1986, pp. 2–3). The idea of balance and comprehensiveness underlined the portion of the speech devoted to the economy of the region. 'The Far East must no longer be regarded solely as a raw material base.' It must create 'complete production cycles . . . for diverse finished products'. The development of infrastructure must be a priority task and the development of an up-to-date construction industry must be accelerated (*Ekon. gazeta*, no. 32, August, 1986, p. 2). A year later, in July 1987, the Politburo approved the 'Long-term Programme for the Comprehensive Development of the Far Eastern Region and Transbaykalia till the year 2000 (*Pravda*, 24 July 1987, p. 1).

In the following month, the first deputy chairman of Gosplan made public a very ambitious blueprint for the area's development. The striking figures revealed by Gosplan's deputy chief, however, were reported in the form of a newspaper interview (*Pravda*, 26 August 1987, p. 2), with not one of them appearing in the brief notice which announced Politburo approval of the Long-term State Programme for the region. (The same Politburo session also dealt with three other totally unrelated issues. *Pravda*, 24 July 1987, p. 1). The only concrete information ever to emerge about the Programme was reported in the form of a newspaper interview with the First Deputy Chairman of Gosplan (*Pravda*, 26 August 1987, p. 2), with some of the figures repeated in *Stroitel'naya gazeta* (5 April 1988, p. 1) and in the form of a brief article by the economist N. Singur in Gosplan's own journal, *Planovoye khozyaystvo*, no. 3, 1988. Given the long lead times for publications in Soviet periodicals even for short notes, the two commentaries must have been produced about the same time. They were very similar and clearly put out by Gosplan's bureaucracy. *Stroitel'naya gazeta* (5 April 1988, p. 1) refers to the Far Eastern Programme as a detailed document in which the volume, time and places of specific projects are designated. Yet it also speaks of wide-spread opposition to it within Gosplan, Gosstroy (State Construction Committee) and ministries, where many regard it as a threat to intensification in the country's European heartland. In view of the huge investment figures, the very ambitious nature of the projections (see below) and the admitted existence of such opposition, the possibility cannot be ruled out that these figures do not represent firm official commitment.

They may be floated in part to gauge Japanese reaction and increase Japanese willingness towards economic participation. The *Pravda* interview with Gosplan's deputy chairman appeared exactly a year after Gorbachev's Vladivostok address, in which he invited, indeed pressed, the Japanese for joint investment in the Far East. The deputy chief of Gosplan uses the verb 'proposed' for the huge figures on capital investment during the remainder of this century (*Ekon. gazeta*, no. 32, August, 1986, p. 4 and *Pravda*, 26 August 1987, p. 2).

How ambitious and realistic is the proposed programme for the Far East? The *Pravda* interview (26 August 1987, p. 2) states that cumulatively a sum of 198,000 million roubles of centralized investment is proposed for the Far East through to the year 2000 and an additional 34 billion for the two Transbaykalian provinces of Chita Oblast and the Buryat ASSR. Thirty per cent of this huge outlay is to flow into the development of the social-cultural infrastructure, which would guarantee each family separate quarters by the end of the century. The lag in nurseries and schools, medical and cultural facilities is to be eliminated. The Far East and Transbaykalia are to become fully self-sufficient in the supply of pork, milk and eggs, potatoes and some vegetables, as well as in electricity and fuel. Coal extraction would reach 82 to 85 million tons, oil extraction 6 to 8 million tons (a 3.1- to 3.8-fold growth), electricity generation would rise 2.6 times, and gas production 7.2 to 9.3 times. Very ambitious growth rates are presented for the cellulose, pulp and paper, and wood products industries. Overall, the value of industrial output is projected to increase 2.4 to 2.5 times. A key aspect of the programme is the transformation of the Far East from a basic supplier of raw materials to a zone of significant processing. To accomplish this, the machine-building and metallurgical branches would be overhauled and re-equipped and the construction material industry would need to expand its output significantly. In the BAM zone, production would be multiplied 5.6 times. Such a move towards a more balanced, comprehensive industrial structure would, of course, require significant in-migration. The population of the Far East as a whole would increase by 20 per cent (Singur 1988, p. 97) and that of the BAM zone by 70 per cent (*Pravda*, 26 August 1987, p. 2).

If these figures represent official targets, the coherence of the national planning process itself must be questioned. In the first half of the 1980s, the volume of capital allocation to the Far East seems to have been about 33,000 to 34,000 million in pre–1982 roubles. This suggests an *annual* average of around 8,000 million roubles, computed to that older price base, during the first two to three years of the 12th Five Year Plan (1986–90). A 1987 source, however, states that during the 1981–5 Five Year Plan investment allocation to the Far East decreased *absolutely* by 6,000 million roubles when compared to the 1976–80 period (*Stroitel'naya gazeta*, 5 April 1988, p. 2). If this is true, average yearly investment in the first years of the 12th Five Year Plan did not even reach 7,000 million in pre–1982 prices.[5] In post–1982 prices, investment in the Far East may have averaged 9.05 to 9.1 billion roubles in the last few years, but only 8 to 8.05 billion if the claim of the huge absolute decline of investment in the first half of the 1980s is accepted. To invest a sum of 198 billion roubles by the year 2000 would require an annual growth of well over 9 per cent, if a constant exponential rate is assumed. If the plan is 'frontloaded', and half the volume, for example, is allocated by 1995 or before, far higher rates of increase would be necessary for the rest of the 1980s and the early part of the forthcoming decade. Yet in the first three years of Gorbachev's stewardship (1985 to the end of 1987), the annual growth of investment in the national economy as a whole averaged no more than 5.3 per cent and the 1988 plan calls for a 3.6 per cent rise only (*Nar. khoz. SSSR za 70 let*, p. 323; *Nar. khoz. SSSR v 1985 g.*, p. 365; *Ekon. gazeta*, no. 4, 1987, p. 18 and no. 5, p. 2). As for the Far East (as indicated in note 5), capital allocation to the region rose by little more than 4 per cent per annum as an average rate through the 1970s, more slowly than that during the second half of that decade and more slowly still during the first half of the 1980s.

Gosplan's Deputy Chief projects a 70 per cent rise in the population of the BAM zone. Yet the housing shortage here is so severe that recently per capita living space in Tynda, the 'capital' and presumably best provided settlement of the zone, was reported to be a mere third of the Soviet norm (Ponkratova 1987). Rapid improvement cannot be expected when 60 per cent of the construction labour in the Far East turns over every year (Singur 1988, p. 95). The anticipated population increase of 20 per cent for the Far East as a whole seems more realistic, although in the region today average housing space per capita falls well below the mean for both the USSR and the Russian Republic (Singur 1988, p. 96).

The Central Committee of the CPSU and the USSR Council of Ministries recently published a resolution 'On ensuring the effective employment of the population, improving the system of job placement and enhancing the social guarantees for working people' (*Pravda*, 19 January 1988). As a result of Gorbachev's *perestroyka*, the number of unemployed, euphemistically called 'redundancies', could reach 16 million, according to official estimates. The resolution specifically mentions encouraging the unemployed to move to labour-deficit areas in the east and the north. Yet the extreme housing shortage and unsatisfactory living conditions in the region, even if improved, will remain a grave obstacle to such migration.

The issue of a broad programme for the harmonious development of the Far East also must be put into a larger policy context. The objective of balanced, more diversified development east of the Urals has appeared before but 'always fell into the background when faced with the immediate need for . . . a raw material, a metal, or fuel, whether it was Kolyma gold during the war years or Tyumen' oil and gas in our day' (Rumer 1984, pp. 134–5). At the onset of the Gorbachev era, the expansion of machine building, infrastructure, health and social services throughout the USSR were declared as key pillars in the new policy of *perestroyka*. What Gorbachev said in Vladivostok was declared in Tyumen' and Novosibirsk about Siberia as well. At a major conference in the latter city, July 1985, the Vice-President of the Soviet Academy of Sciences stated that 'Siberia can no longer obtain its advantage . . . through the extractive branches . . . Today's strategy envisages a shift in the centre of gravity towards the processing branches.' Politburo member V. Vorotnikov echoed that sentiment, pointing to the need for a harmonious development and concern for all spheres of the economy (*Sots. industriya*, 20 July 1985, p. 3 and *Pravda*, 19 July 1985, p. 2).

The creation of a more balanced economic structure in the Far East, with some elements of autarchy, is indeed a desirable direction for population stability and economic health, if the process is allowed to take place in a relatively autonomous fashion. Such a development strategy would need to emphasize the satisfaction of local demand, through local linkages and initiative. However, the autonomous evolution of a reasonably diversified economy in an outlying region is a very slow process in any country. On the other hand, an accelerated programme on the basis of autarchic principles, enforced and financed from outside power centres, is always very costly and fraught with its own dangers. Nor is it congruent with current economic reforms and the framework of economic management provided by the adoption of the new Enterprise Law (on 30 June 1987).

The new law requires enterprises to become self-financing, establishes a series of user fees for labour and social-communal services, for land and natural resources (amounting to rent payment) and sets a uniform capital charge for all firms irrespective of location, at least within the same sector (*Ekon. gazeta*, no. 27, July 1987, pp. 10–12). While the wealth in natural resources, resulting in consequent low user fees, will work to the benefit of the Far East, much higher payment for labour and services and the loss of preferential capital charges will more than counterbalance this gain. Even the advantage in natural resources (hence, lower rent payments) holds, as a rule, only *vis-à-vis* the European USSR but not compared to Siberian regions west of Lake Baykal. For some metals, it does not hold with respect to Kazakhstan and Central Asia either.

For most economic activities, the required payment for labour and services and the uniform capital charge will disadvantage Far Eastern enterprises compared to those in any other Soviet region. Those north of the most clement border zone will be especially unfavourably affected. Indeed, the Director of the Institute of Economics and Organizations of Industrial Production (IEOPP) of the Soviet Academy of Sciences, A. Granberg, stated recently that, in his opinion, the general impact of the new laws and economic mechanism on the BAM zone will be detrimental. In particular, industries relying even moderately on inputs shipped in from elsewhere cannot pay for these inputs plus the much higher wages prevalent in the Far East and modernize and renovate from enterprise funds at the same time. In comparison with better located regions, the geographic burden is too great. Granberg specifically notes that machine-building firms, which had been projected widely in the BAM zone, would be unable to sustain themselves. Even the plans of the 1970s for resource exploitation are outdated under the new conditions (*Ekon. gazeta*, no. 29, July 1987, p. 8. See also Bond 1987, pp. 511–14). A whole range of general industries and consumer branches must be equally affected.

Untrammelled cross-border trade with Heilongjiang province of China (Manchuria), fully in the purview of any local enterprise on both sides of the border and actively encouraged by both Moscow and Peking, could make Far Eastern enterprises much more viable and competitive under the new Enterprise Law. Yet, despite the easing of controls on foreign trade, neither government is prepared to go this far. Local trade between the Soviet Far East and Manchuria more than doubled from 1984 to 1987, but still accounts for less than 2 per cent of centralized Sino–Soviet trade and for a trifle against what each country sells to the rest of the world (*Financial Times*, 18 December 1987, p. xiv and *Vneshnyaya torgovlya SSSR*, no. 12, 1987, *Supplement*). On the Chinese side there are also memories of Russian and Soviet exploitation, particularly by Soviet troops in 1945. Cross-border barter trade, with its cumbersome restrictions and *modus operandi*, 'is probably destined to remain at the margin' of the USSR-China trade (*Financial Times*, 18 December 1987, p. xiv). It will not help much in the creation of a more balanced economic structure in the Soviet Far East.

So far *perestroyka* has not been sufficient to free enterprises from the petty tutelage of Moscow ministries, even in small-scale, local industries, to respond flexibly to regional needs. At the very end of 1987, the Yakut ASSR's Minister of Local Industry revealed that each year whole contingents of officials and specialists from Yakutia must fly to Moscow to submit their plans for approval to the RSFSR Ministry of Local Industry and the Russian Republic's Gosplan. Though the plans have been balanced and cleared with regional authorities, these superior organs in Moscow barely look at the figures. They simply hand down their own plan, which ignores local constraints and circumstances. Claims of co-ordination between regional and national plans is simply a farce (*Izvestiya*, 24 December 1987, p. 2). Under Gorbachev's reform package, ministries and regional bodies are both 'held responsible for production results in their respective areas and for ensuring that subordinate firms act "properly" '. Simultaneously they have been enjoined 'not to interfere in enterprise decision-making'. Given this inconsistency (among others in the package), the need of the centre will surely continue to prevail. Indeed, with the confusing new arrangements of many normatives, firms most likely will 'be eager for ministerial aid . . . They will seek to have as much of their production as possible classified as "state orders" backed up by rational inputs' (Schroeder 1987, pp. 234–5). Under these conditions, enterprises cannot be responsive to local demand and promote balanced development.

Current military deployment and strategic prospects under Gorbachev

Only a brief sketch of military capabilities needs to be provided here on both defensive and offensive capacities. The region plays a pivotal geographic role in that defensive arc of large phased array radar (LPAR) for early warning and ballistic missile tracking. Two of the six LPARs are in the Far East and, according to one source, the seventh is under construction in Kamchatka. MIG–31 *Foxhound* interceptors with full look-down/shoot-down capabilities operate in the region, eighteen of the latter identified on Sakhalin (Tan Eng Bok 1986, p. 45). Soviet offensive capacity in the region has grown enormously since 1965, though unevenly and in distinct phases. Today roughly one-third of all the country's offensive forces are deployed in the Far East. At the same time, this build-up is characterized by a steady improvement in quality, the development of air mobile capacity (especially in the 1980s) and 'a reduction in the time lag for the delivery of new equipment as compared to the European front. Traditionally, new equipment in the Far Eastern region appeared with a time lag of between three and six years, sometimes even a decade. In some cases now it does not exceed a year' (Tan Eng Bok 1986, pp. 45 and 50). The mix of factors driving the build-up has also been shifting, changing the composition and character of Soviet Far Eastern forces and their blend of mission. The initial motivation to respond to Chinese territorial claims, ideological challenge and the violence of the Cultural Revolution has broadened to include multi-faceted issues in the wider Asian and Pacific theatre. These include both safeguarding Vietnam from Chinese force and profiting from its strategic position, in order to maintain Soviet strength against the worst-case possibility of an eventual Chinese–US–Japanese military co-operation and respond to new global opportunities for projecting Soviet power from the Far East (Gelman 1987, p. 181).

Since the mid–1960s, the Soviets have put about a fourth of all their ground forces and somewhat more than a quarter of their ground-based airforce into Asia, chiefly along the Pacific coast and the Chinese border (Ellison 1982, pp. 200–6 and *The Christian Science Monitor*, 17 February 1983, p. 9). Thirty-nine divisions, with 360,000 men, are stationed east of Lake Baykal, although only half of these forces are at Categories 1 and 2 readiness. (Respectively, 75–100 per cent and 50–75 per cent of wartime strength: Ha 1987, p. 142). The number of aircraft of all types in the Far East reaches 2,700, with 60–70 per cent of them, perhaps more, being third-generation models. In times of emergency, twice as many could be accommodated by the region's 140 airfields, more than fifty of which have had their runways doubled or even tripled in recent years, according to Landsat photographs. Since 1981, the number of SS–20 missiles has more than doubled, reaching 171 by the end of 1985 (Tan Eng Bok 1986, pp. 47 and 50). According to the new US–Soviet agreements, signed after tortuous negotiations in December 1986, all Soviet SS–20s, including those deployed in Asia, will now have to be destroyed.

Even more significant from a geostrategic point of view has been the build-up of maritime forces. With 825 vessels (860 according to the US Assistant Secretary for Defense), including 135 submarines, the Soviet Pacific fleet is the largest of the four Soviet fleets. Since 1965, the number of *major* surface combatants based in the Far East rose from 50 to 90 (Gelman 1986, p. 50 and *Far Eastern Economic Review*, 4 August 1988, p. 28). The share of all Pacific blue-water and coastal forces in the Soviet naval arsenal is more than twice as large as the 20 per cent share that the Pacific *commercial* fleet represents in all Soviet civilian tonnage (Ivanov and Malakhovskiy 1983, p. 243). Two of the USSR's four aircraft carriers for combat planes with vertical take-off capabilities are also assigned to the region. Almost a third of all submarines and 35 to 40 per cent of all ballistic missile submarines are based here, three-quarters of the latter at ice-free Petropavlovsk-Kamchatskiy (*Soviet Military Power*, 1987, pp. 8–9, 17–18, 68–9 and *Jane's Defence Weekly*, 14 April 1984, p. 560). This huge base may be half the size of Murmansk-Severomorsk on the Kola Peninsula, the biggest in the USSR. Petropavlovsk is the largest city in the entire Siberian north and, with 252,000 persons in January 1987, it accounts for almost three-fifths of the entire population of Kamchatka and a fifth of the urban population on the entire Pacific coast (*Nar. khoz. SSSR za 70 let*, 1987, pp. 390 and 399 and footnote 4). Altogether, ten major naval bases line the Pacific littoral from Vladivostok to Anadyr', near the Bering Strait. Komsomol'sk on the Amur river, with the only steel plants in the Far East, has been developed into one of the largest naval construction centres in the world (*Jane's Defence Weekly*, 14 April 1984, p. 560 and Tan Eng Bok 1986,

p. 51). Assembly for major vessels, however, must take place on the coast.

On Kamchatka and Sakhalin, with their exposed positions, small populations and narrow economic bases, the employment structures clearly reflect the prominence of military-related activities. Tables 11.3 and 11.4 highlight a large residual category in both provinces, with significantly rising shares of total employment over the years. The sector, 'other branches', in Kamchatka obviously represents activities overwhelmingly connected to the military. In Sakhalin Oblast, the large employment discrepancy between the 1975 and 1978 data also points in the same direction. So does the incongruous fact that on this remote island the share of the labour force in the 'non-material' production sphere of the economy matches that for the country as a whole (*Nar. khoz. SSSR v 1978 g.*, p. 364). While Sakhalin lacks the huge naval concentration of Kamchatka, it hosts six aviation and two army bases with the Fifteenth

Table 11.3 Workers and employees in the economy of Sakhalin Oblast

Economic sectors	1975 ('000s)	%	1978 ('000s)	%
Total for Oblast	336.6	100.0	350.2	100.0
Industry	105.8	31.4	110.4	31.5
Agriculture	17.2	5.1	18.4	5.3
Construction	37.4	11.1	39.3	11.2
Transport and communication	47.2	14.0	47.6	13.6
Services related to the material production sphere	38.5	11.4	42.4	12.1
Non-material production sphere	74.4	22.1	88.9	25.4
Discrepancy	16.1	4.8	3.2	0.9

Source: Computed from Zykin, B.N. (1981) *Effektivnost' regional'noi ekonomiki*, Moscow: Nauka, pp. 52–3.

Table 11.4 Workers and employees in the economy of Kamchatka Oblast (as percentage of total)

Economic sectors	1950	1970	1975
Total for Oblast	100.0	100.0	100.0
Industry (industrial production personnel)	45.9	27.6	27.5
State farms and agricultural production enterprises	4.7	5.1	5.1
Transport	9.5	13.2	14.6
Construction	2.9	10.2	7.6
Retail trade, public catering, procurement, and supply management	9.4	11.3	11.0
Housing and municipal economy	3.9	4.0	4.1
Health and social services	5.1	5.8	5.7
Science (including supporting personnel)	1.8	3.2	3.4
Government and Party organization	4.8	2.8	2.9
Other branches	12.0	16.8	18.1

Source: *Narodnoe khoziaistvo Kamchatskoi Oblasti 1971–1975 gg.*, (1977) Petropavlovsk-Kamchatskii, p. 84.

Army headquartered at Yuzhno-Sakhalinsk (*Jane's Defence Weekly*, 14 April 1984, pp. 560–1). In addition, the oblast is saturated with early warning and intelligence-gathering electronic gear and since 1981 has been the location of an air-command centre that co-ordinates all air and naval air missions in the Soviet Far East (Tan Eng Bok 1986, p. 47). This clearly requires a large number of supporting service personnel.

As stated earlier, Soviet *de facto* control of Cam Ranh naval base places the USSR very near the Malacca Strait, at the doorstep of the Indian Ocean. Now she also flanks China from the south. The growth of the Soviet presence at Cam Ranh has been spectacular, and thirty Soviet warships are permanently stationed there (Tan Eng Bok 1986, p. 52). The utility of this naval base, however, has an even wider significance. As regional interdependence between the Far Eastern periphery and the European USSR increased throughout the Brezhnev era, a major decision was apparently made to establish a regular, extraterritorial sea line of communication linking the two by means of a 12,000 nautical mile route via the Black Sea, Suez Canal, Indian Ocean and South China Sea. By 1984, the USSR was the world's largest user of the Suez Canal among nation-states in numbers of transits per year. The German Intelligence Service (BND) estimated at the end of the 1970s that four-fifths of the Soviet transcontinental freight, i.e. that moving all the way to the Pacific coast, travelled via this southern sea route rather than rail (Westwood 1985, p. 47; see also Trainor 1986, p. 48). This estimate, however, is so high that it may refer only to specific categories of freight. From Vladivostok and Nakhodka, Cam Ranh is located one-third of the way along this route. Shipping between domestic ports comprised 34 to 37 per cent of all shipping by the Soviet merchant marine during 1980–3 (Levenson 1984, p. 7). While most domestic maritime transport is still within the same sea, inter-basin transport was expected to grow significantly during the 1970s and 1980s (Nikol'skiy *et al.* 1973, Fig. 19, p. 82 and Danilov, 1977, p. 300). Marine links between the Pacific ports and the Black Sea comprise an overwhelming share of such inter-basin shipments (Levenson 1984, pp. 24 and 26, Table 9b).

Hanoi is in dire need of Soviet aid. Vietnam's economy suffered an absolute decline in 1987 and its government is promising a return to the 1985 living standard only by the end of this decade (*Heti Világgazdaság*, 16 January 1988, pp. 12–13). Soviet leverage on Vietnam for Cam Ranh is, therefore, considerable. Yet, the importance of Cam Ranh for the USSR has become so great that Vietnam today enjoys a significant reverse leverage on the USSR. Notwithstanding the value of Cam Ranh, however, and the still unresolved issues between Moscow and the East Asian powers, the Far Eastern theatre is now feeling the impact of Gorbachev's diplomatic initiatives. In his 1986 Vladivostok speech, the General Secretary offered China greater economic co-operation and revealed that talks about 'substantial' troop withdrawal from Mongolia were underway (*Ekon. gazeta*, no. 32, August 1986, pp. 4–5). Two years later he proposed to withdraw Soviet forces from bases in Vietnam in exchange for the removal of US bases from the Philippines (AP Report, *Lawrence Journal World*, 16 September 1988, p. 1). With Vietnam's disengagement from Cambodia and economic normalization with China both in process, Moscow will almost certainly come out with dramatic initiatives in the Far Eastern theatre. Major troop reduction along the Chinese border will likely be offered, the Cam Ranh base may be given up and the modernization of some obsolescent naval forces along the Pacific coast may now be postponed or foregone. It is my belief that these anticipated initiatives will substantially affect Soviet position *vis-à-vis* China but not *vis-à-vis* Japan. And they will not alter the crucial importance of the Pacific coast and the Okhotsk Sea in Soviet global strategy and superpower relations.

During the past decade, the Soviet military build-up in the Far East seems to have been driven by superpower competition and strategic considerations which now embrace three of the world's largest oceans and reach well beyond the immediate East Asian theatre. With the key role played by the nuclear-ballistic-missile-carrying submarine in the global superpower strategic stand-off, even Soviet domestic waters behind historic chokepoints can become sanctuaries rather than traps. The Sea of Okhotsk has, in fact, been used as an SLBM (submarine-launched ballistic missile) test range, with many such missiles fired from it during the last decade. In addition, over most of the Okhotsk Sea large submarines are not cramped in winter between a shallow sea bottom and very thick pack-ice, a problem which severely limits their counterparts in the Barents Sea to a mere third of that basin (Bergesen *et al.* 1987, pp. 58–87). Sanctuaries, however, have to be defended. Soviet determination to protect the approaches to the Okhotsk and Japan Seas, even at the price of permanently alienating Japan, is proved by the militarization of the Southern Kurils, which the Japanese claim and consider to be a historic part of their 'homeland'. These islands have been garrisoned with more than 14,000 men, and in 1986 reportedly received SSC–1 rockets, with a 450-kilometre range (Ha 1987, p. 96 and *Sankei Shimbun*, quoted in NRS, February 1986, p. 2). The 1987 arms control treaty does not require their elimination. Perhaps forty MIG–23s and a dozen helicopter gunships are also stationed there now (Tan Eng Bok 1986, p. 78 and *Japan Times*, 23 October 1987, p. 3).

Given the new geostrategic role of the Far Eastern littoral, the high military profile on the Pacific must be regarded as permanent, as long as the USSR aspires to superpower status. US–Soviet progress in arms control will not change that. Nor will Gorbachev's success or failure at reforms on the home front and diplomacy abroad. Gorbachev may downgrade the military complex in the relative ordering of Soviet priorities. *Pari passu* this would affect the Far East also. Already in 1987 changes were reported in the top Far Eastern command – from a general to a lieutenant-general for land-based forces and from an admiral to a vice-admiral for the Pacific navy. Nor are the new commanders members or alternate members of the CPSU's Central Committee, while the outgoing commanders were candidates. General Postnikov, commander of the Transbaykal MD, however, remains a full member (*Foreign Report*, 2 April 1987). What seems clear, however, is 'that a second major military arena . . . has become a fixture in Soviet military planning' (Gelman 1987, p. 214). In particular, the southern coast of the mainland, Kamchatka and Sakhalin fulfil the same strategic function on the Northern Pacific as the Kola Peninsula does on the Northern Atlantic Ocean. Tactics have changed. The number of naval passages from Soviet Pacific harbours and large-scale blue-water exercises have decreased significantly since 1986. By some accounts, naval activity was down by 50 per cent. Moscow may also forgo the modernization of some aging assets. Yet the capabilities of the Pacific fleet are unlikely to change greatly in the future (*Far Eastern Economic Review*, 4 August 1988, p. 88).

The peculiar position of the USSR in the Far East will effectively circumscribe and channel Moscow's options, despite all the diplomatic and public-relations skills of the Gorbachev team. The USSR will not reap any reward of 'Finlandization' in China for a skilful peace offensive and arms-control diplomacy. Normalization is mutually beneficial to both countries. However, the political and economic windfalls already manifest from the new Gorbachev policy in Central Europe cannot be duplicated in Beijing. In addition, in Europe Moscow is in command of all the Warsaw Pact forces and exercises ultimate control over the foreign policy of all states in the Pact. The Soviet Union wields no such control over Vietnam and even less over North Korea. The Korean Peninsula, in particular, will continue to present dangers as much as opportunities for Moscow, just as for Washington, Tokyo and Beijing.

Real concessions to Japan could be clearly rewarding to Moscow. Indeed, during recent years Japanese experts have often raised the question of why the USSR continues to pursue such a self-defeating policy towards their country. A well-known Japanese specialist sees on the Soviet side a combination of the following: a misperception about military strength as an effective source of international influence; an out-of-date image of today's increasingly self-confident, even self-assertive Japan, as a country that must feel weak and vulnerable because of lack of natural resources and large armed forces; a preoccupation with superpower relations and global strategy; and an institutional structure in which Japanologists have had little influence in the making of foreign policy (Kimura 1986, pp. 114–18 and 1984, pp. 29–34). A reassessment of

Japan's role 'as an important, independent source of political and economic power in the world, particularly in the Asian–Pacific region' is underway in Moscow. The 'Japanese people's unique capacity to creatively master technology' is held up as an example in the current drive for *perestroyka*. In his Vladivostok speech Gorbachev referred to Japan as a 'power of paramount importance'. Moscow's most recent policy towards Japan is thus a mixture of these two perceptions, but the Kremlin is still intent to improve relations with Tokyo essentially on its own terms (Kimura 1987b, pp. 82–4 and 88 and *Ekon. gazeta*, no. 32, August 1986, p. 4).

As I have already shown, Japan has relatively little to gain economically by plunging into resource and infrastructure development in Siberia and is reluctant to do so. Japan's cumulative trade surplus with the USSR from 1980 to September 1987 has already surpassed $13,000 million and Soviet imports from that country have exceeded exports to it in every year by a wide margin. The Soviet Far East cannot compete with other resource frontiers on the Japanese market, Moscow's hard currency shortages will further constrain economic ties and 'Gorbachev's attitude on East–West trade is far from unequivocally positive' (Becker 1987, p. 7 and *Vnyeshnyaya torgovlya SSSR*, no. 12, 1987. Supplement).

To improve relations with Tokyo substantially, the enticement from Moscow would have to be military and territorial. Fundamental concessions here, however, are very unlikely, for the strategic–military significance of the Pacific coast goes well beyond Japan. Indeed, recent polls in that country have concluded that 70 per cent of the Japanese public do not expect the Soviets to return the Southern Kurils. Moreover, among younger age-groups there has been a progressive erosion of sentiment for the retro cession of the islands (only 21 per cent of those in their thirties and 10 per cent of those in their twenties professed to be interested in the issue). These polls seem to confirm that 'the longer [the Soviets] wait, the less interested and less emotional the Japanese will become on the islands'. Moscow knows that and therefore resists any change in its policy on this issue. Apparently, it is willing to accept 'the continuation of the chilling relations . . . between the two countries' (Hasegawa 1987, pp. 56–7).

Significantly, a recent survey by the Tokyo Broadcasting System also reported a dramatic rise (up to 55 of every 100 Japanese polled) in the percentage of those who view US–Japanese relations as 'unfriendly', a perception surely impelled by economic circumstances. Some observers even speak of a 'nascent Japanese Gaullism' (*Insight*, 18 January 1988, p. 29). Increased strain between Tokyo and Washington will most likely serve as a further argument for Moscow that they need not make any real concession on the territorial question or even the demilitarization of the Southern Kurils. In time, with the inevitable divergence between

the US and Tokyo on certain strategic issues and a new Japanese generation not emotionally attached to the Northern Territories, Moscow's
fait accompli on the Southern Kurils will extract from the Soviets a
progressively smaller political price in their relations to Japan. Indeed,
at the latest Japan–Soviet Working Level Consultation, Soviet Deputy
Foreign Minister Igor' Rogachev dismissed the Japanese position on
the territorial problem outright. Both Rogachev and Foreign Minister
Shevernadze struck a hard line even about the militarization of these
islands and of the Far Eastern coast in general. These are internal Soviet
matters unrelated to improvements of relations between the two powers
(*Japan Times*, 18 November 1987, p. 1 and 20 November 1987, p. 1 and
FBIS, 28 September and 18 November 1987, quoted in *SUPAR Report*,
no. 4, February 1988, pp. 5, 10, and 11).

Moscow is concerned about increased Japanese military spending, now
over the psychological barrier of 1 per cent of the GNP and approved
to be close to $30,000 million in 1988 (*Insight*, 18 January 1988, p. 29).
Yet, 'the Gorbachev government has not budged from its basic stance
that the problem [of the ownership of the islands] has already been
solved'. Nor can anyone so far discern any change in that government's
attitude and policy from those of its predecessors with regard to the
military build-up on the Kurils. It may very well be that Gorbachev
thinks that 'there is no need, or even perhaps no way, to change the
substance or goals of Soviet policy toward Japan' but greater attention
to diplomatic niceties and style 'might create the appearance of substantive change and enable the Soviets to deal more effectively with the
Japanese' (Kimura 1987a, p. 9).

Conclusion

The role of the Far East in the Soviet regional system is conditioned by
its remoteness from the controlling power centre, by its singularly harsh
physical environment, and by its exceptional strategic importance. The
region is weakly integrated into the mainstream of the Soviet economy
and its industrial structure remains extremely narrow and unbalanced.
At the same time, the Far East is tied to the distant heartland of the
country in a strongly dependent, parasitic relationship. Such dependence
is further enhanced by the military-strategic importance of the region,
especially of its coast, to Moscow. The recent efforts by the Soviet
government to open up its vast natural riches and accelerate regional
growth via resource exports are proving largely unsuccessful despite the
construction of the BAM. In a period of severe capital and labour
constraints, national economic priorities are shifting to West Siberia,
which is far more accessible to the country's economic core. Concur-
rently, reduced world demand for energy and raw materials, combined

with the high costs and often poor quality of Soviet natural resources, make them increasingly less attractive to countries of the Pacific basin, especially Japan.

In mid 1987, the Politburo approved a Long-term Programme for the Comprehensive Development of the Far East. Yet the current status of the programme is very uncertain. A set of strikingly ambitious figures did appear in the form of an interview by Gosplan's deputy chief but widespread opposition to the programme has also been reported in the highest government agencies. As this chapter has shown, the target figures aired at that interview plus the huge investment increases called for seem wholly unrealistic. To date, the yearly construction goals are not being met (*Stroitel'naya gazeta*, 5 April 1988, pp. 1–2).

Strategic and security concerns about the huge and remote Far East go back to the nineteenth century. In the Soviet era, crucial developments in political and military relations with China and Japan have heightened these concerns. More recently still, the increasingly global ocean policy of the USSR has sharply raised the importance of the Pacific coast and its approaches for the Moscow leadership. Only in part has the evolution of the USSR from a mere great power to a superpower been based on the attainment of nuclear parity with the US. The other pillar of this transformation has been the achievement of a truly global reach, an assured sustained presence on all three of the world's largest oceans. The huge Pacific military build-up and Soviet *de facto* control of Cam Ranh Bay near the portal of the Indian Ocean are driven by this geostrategic development. Whatever normalization and arms reduction may be achieved under Gorbachev, the USSR will not retreat from being a full-fledged Pacific power. Normalization with China will proceed slowly. Better relations with Japan will be obstructed by Soviet refusal to make substantive concessions on the Southern Kurils, their demilitarization and significant military reduction along the Pacific coast. A notable reduction in the Soviet naval/military presence on the Pacific would mean a contraction of its global reach, hence a shrinking of its superpower status.

Given this geostrategic reality, continued high priority will be accorded by Moscow to the Far Eastern *coastal* provinces in the years ahead. This priority will remain independent of the general development strategy employed and of Gorbachev's overall policy of restructuring. While the optimistic targets of the Long-term Far Eastern Programme, quixotically presented only in the form of a *Pravda* interview, appear chimerical, the Soviet leadership will continue to subsidize the Far East in a huge way. Moscow will continue to regard the region as of critical strategic value, essential for its superpower status. Because of vast natural riches, it will also view it as a *potential* economic asset. Yet in the short, even medium term, economic development here will be very slow and its nature extremely unbalanced. New production centres, even of resource extraction, will be very few. The important position of rivalling other nations on the vast Pacific, a position that Russia has hoped to gain since the nineteenth century has so far become a reality only in the military sense. I do not expect it to change in the forthcoming decades.

Notes

*A small portion of this chapter is based on the author's article that appeared under a similar title in *Soviet Economy*, vol. 1, no. 2 (April–June), 1985. We would like to thank the publisher for permission to use this material.

1. The largest is Cuba.
2. In 1980, the Soviet Far East accounted for a mere 2.8 per cent of Soviet national income, i.e. a little less than 13,000 million roubles in current prices, according to official calculations. Mozhin 1980, p. 78 and *Nar. khoz. SSSR za 70 let*, p. 122. Only a fifth of Soviet exports to the Asian–Pacific region originated in the Far East (Shlyk 1981, p. 35) amounting to a mere 1,350 million roubles at that time. (The Asian–Pacific region, as defined here, excludes the Americas). Even at the official exchange rate, this export would represent only a small fraction of the exports of such Pacific economies as Taiwan, South Korea, Singapore or Malaysia (Bradshaw 1988).
3. The Far Eastern North and Transbaykalia, however, are very important producers of gold. Gold sales are not included in Soviet exports but are vital to the country.
4. In 1981, this narrow coastal zone of the Far East accounted for 9.1 per cent of the urban population found along all Soviet seashores. Since at that time 10.85 per cent of the Soviet urban population lived at no more than 10 kilometres from a sea, the urban population of the immediate Pacific littoral totalled 1.7 million people. By 1987, this figure was surely over 1.8 million, 70 per cent of whom lived south of the Amur estuary and along the coast of Sakhalin (Vas'kov 1986, pp. 77 and 81 and *Nar. khoz. SSSR 1922–1982*, p. 15).
5. During the 1960s, the average yearly increase of capital investment in the economy of the Far East reached 8.3 per cent per annum. During the 1970s, this rate declined to a little over 4 per cent. For 1970 and 1975, the official figures for investment were 3,673,000 and 5,445,000 roubles respectively (Minakir 1983, p. 66; Chichkanov and Minakir 1984, p. 84 and *Nar. khoz. RSFSR v 1975 g.*, p. 329).

Bibliography

Aganbegyan, A.G. and Mozhin, A.A. and V.P. (eds) (1984) *BAM. Stroitel'stvo. Khozyaystvennoye osvoyeniye*, Moscow: Ekonomika.

Al'ter, Igor' (1987) 'BAM bez litavr', *Yunost'*, no. 10, pp. 40–42.

Aparin, I.L. and Krinitskaya, M.Ye. (1979) *Industrial'naya baza stroitel'stva Severnoy zony*, Leningrad: Stroyizdat.

Bakhrakh, M. and Mil'ner, G. (1984) 'Proizvodstvo chistogo produkta i ispol'zovaniye natsional'nogo dokhoda po regionam v RSFSR', *Vestnik Statistiki*, no. 6, pp. 14–23.

Baldwin, Godfrey S. (1979) *Population Projections by Age and Sex: For the Republics and Major Economic Regions of the USSR. 1970 to 2000*, Washington, D.C.: Foreign Demographic Analysis Division, Series P–91, no. 26, September.

Bartkova, I.I. (1981) 'Portovye goroda Dal'nego Vostoka', in P.Ya. Baklanov and V.N. Bugromenko (eds) *Territorial'no-khozyaystvennyye struktury Dal'nego Vostoka*. Vladivostok, pp. 142–51.

Becker, Abraham S. (1987) *US–Soviet Trade in the 1980s*, Santa Monica, Ca: Rand Corporation, November.

Bergesen, Helge, Ole, Moe Arid and Ostreny, Willy (1987) *Soviet Oil and Security Interests in the Barents Sea*, London: St. Martin's Press.

Bond, Andrew R. (1987) 'Spatial Dimensions of Gorbachev's Economic Strategy', *Soviet Geography*, September, pp. 490–523.

Bradshaw, Michael J. (1988) 'Soviet–Asian Pacific Trade and the Regional Development of the Soviet Far East', *Soviet Geography*, April.

Chichkanov, V.P. and Minakir, P.A. (1984) *Analiz i prognozirovaniye ekonomiki regiona*, Moscow: Nauka.

—— (1985) 'Problemy i perspektivy razvitiya proizvoditel'nykh sil Dal'nego Vostoka', *Kommunist*, no. 16, pp. 93–103.

Danilov, S.K. (1977) *Ekonomicheskaya Geografiya Transporta SSSR*, Moscow: Transport.

Dienes, Leslie (1982) 'The Development of Siberian Regions: Economic Profiles, Income Flows and Strategies for Growth', *Soviet Geography*, April, pp. 205–44.

—— (April–June, 1985) 'Economic and Strategic Position of the Soviet Far East', *Soviet Economy*, vol. 1, no. 2, April–June, pp. 146–76.

—— (September, 1985) 'Soviet–Japanese Economic Relations: Are They Beginning to Fade?' *Soviet Geography*, September, pp. 509–25.

Ekonomicheskaya gazeta. Weekly.

Ellison, Herbert J. (ed.) (1982) *The Sino–Soviet Conflict: A Global Perspective*, Seattle: University of Washington Press.

Foreign Broadcasting Information Service.

Foreign Report London: Economist Newspapers Ltd.

Gelman, Harry (1986) 'The Soviet Far Eastern Military Buildup: Motives and Prospects', in Richard H. Solomon and Masataka Kosaka, *The Soviet Far East Military Buildup*, Dover, Mass.: Auburn House Publishing Co., pp. 40–55.

—— (1987) 'The Siberian Military Buildup and the Sino–Soviet–US Triangle', in Rodger Swearingen (ed.) *Siberia and the Soviet Far East. Strategic Dimension in Multinational Perspective*, Stanford: Hoover Institution Press, pp. 179–225.

Ha, Joseph M. (1987) 'Soviet Policy in the Asian–Pacific Region: Primary and Secondary Relationships', *Acta Slavica Iaponica*, Tomus V, pp. 92–110.

Hasegawa, Tsuyoshi (1987) 'Japanese Perceptions of the Soviet Union: 1960–1985', *Acta Slavica Iaponica*, Tomus V, 1987, pp. 37–70.

Hauner, Milan (1985) 'Seizing the Third Parallel: Geopolitics and the Soviet Advance into Central Asia', *Orbis*, Spring, pp. 5–31.

Heti Világgazdaság. Budapest. Weekly.

International Institute of the Steel Industry. October 1987 Session. Washington, D.C. Quoted in *Heti Világgazdaság*, Budapest. 21 November 1987, pp. 13–15.

International Institute of Strategic Studies (1986) *The Military Balance*, London.

Interflo/Report. East–West Trade News Monitor.

Ivanov, V.I. and Malakhovskiy, K.V. (eds.) (1983) *Tikho-okeanskiy regionalism: kontseptsii i real'nost'*, Moscow: Nauka.

Jane's Defence Weekly, London: Jane's Publishing Co.

Keizai Koho Center (1984) *Japan 1984. An International Comparison*, Tokyo.

Kimura, Hiroshi (1984) 'Soviet Policy Towards Japan', *Washington Quarterly*, vol. 7, no. 4, Summer, pp. 21–37.

—— (1986) 'The Soviet Military Buildup: Its Impact on Japan and Its Aims', in Richard H. Solomon and Masataka Kosaka, *The Soviet Far East Military Buildup*, Dover, Mass: Auburn House Publishing Co., pp. 106–22.

—— (1987a) 'Soviet Focus on the Pacific', *Problems of Communism*, May–June, pp. 1–16.

—— (1987b) 'Basic Determinants of Soviet–Japanese Relations: Background, Framework, Perceptions, and Issues', *Acta Slavica Iaponica*, Tomus V, pp. 71–92.

Klin, P.M., *et al.* (1982) 'Mezhotraslevyye svyazi i proportsii v narodnom khozyaystve Dal'nego Vostoka', in *Retrospektivnyy analiz ekonomii Dal'nego Vostoka*, Vladivostok, pp. 13–32. Abstracted in *Referativnyy zhurnal. Geografiya*, no. 11, E209.

Kolesov, L.I. (1982) *Mezhotraslevyye problemy razvitiya transportnoy sistemy Sibiri i Dal'nego Vostoka*, Novosibirsk: Nauka.

Levenson, Marcia (1984) *Soviet Maritime Freight Transport*, Washington, D.C.: Wharton Econometric Forecasting Associates.

Materialy XXVII s"yezda Kommunisticheskoy Partii Sovetskogo Soyuza (1986) Moscow: Politizdat.

Mikheyeva, N.N. (1982) 'Uchet produktov nedopol'nyayushchego vvoza v analize mezhregional'nykh i mezhotraslevykh svyazey', *Izvestiya SO AN SSSR, Seriya obshchestvennykh nauk*, no. 6, pp. 52–8.

—— (1983) 'Analiz mezhotraslevykh i mezhregional'nykh vzaimodeystvii na osnove informatsii otchetnykh mezhotraslevykh balansov Dal'nego Vostoka', *Izvestiya SO AN SSSR, Seriya obshchestvennykh nauk*, no. 6, pp. 76–9.

Minakir, P.A. (1983) *Ekonomicheskoye razvitiye regiona: programmnyy podkhod*, Moscow: Nauka.

——, Mashteler, T.M. and Prokapalo, O.M. (1982) 'Predplanovoye issledovaniye faktorov regional'nogo ekonomicheskogo rosta', *Izvestiya SO AN SSSR, Seriya obshchestvennykh nauk*, no. 1, pp. 40–6.

——, Renzin, O. and Chichkanov, V. (1986) *Ekonomika Dal'nego Vostoka*, Khabarovsk: Khabarovskoye Knizhnoye izdatel'stvo.

Mote, Victor (1987a) 'The Amur–Yakutsk Mainline: A Soviet Concept or Reality?' *The Professional Geographer*, vol. 39, no. 1 (February), pp. 13–23.

—— (1987b) 'The BAM and the Pyramids of Power', in U.S. Congress, Joint Economic Committee, *Gorbachev's Economic Plans*, vol. 2, Washington, D.C.: U.S. Government Printing Office.

Mozhin, V.P. (ed.) (1980) *Ekonomicheskoye razvitiye Sibiri i Dal'nego Vostoka*, Moscow: Mysl.

Narodnoye khozyaystvo SSSR v 19xx godu and *Narodnoye khozyaystvo SSSR za 19xx let*, Moscow: Statistika, 19xx. Annual.

Narodnoye khozyaystvo RSFSR v 19xx godu, Moscow: Statistika, Annual.

Nikol'skiy, I.V., Tonyayev, V.I. and Krasheninnikov, V.G. (1975) *Geografiya vodnogo transporta SSSR*, Moscow: Transport.

North, Robert, University of British Columbia. Personal communication, 1988.

Novoye Russkoye Slovo, New York.

Platt's Oilgram News.

Ponkratova, L.A. (1987) 'Osobennosti sfery obsluzhivaniya v gorodakh Amursky oblasti,' in *Ekonomicheskiye i sotsial'no-geograficheskiye aspekty urbanizatsiya*, Moscow, pp. 28–37. Quoted in *Referativnyy zhurnal. Geografiya*, no. 11E306.

Privalovskaya, G.A. (1984) 'Regional Development and Natural Resources of the USSR', *Geoforum*, vol. 15, pp. 39–48.

Proskuryakova, A.G. (1982) 'Osobennosti i tendentsii formirovaniya i razvitiya regional'nogo transporta Dal'nego Vostoka', in *Regional'nyye ekonomicheskiye problemy razvitiya transportnoy sistemy Dal'nego Vostoka*, Vladivostok, pp. 9–19. Abstracted in *Referativnyy zhurnal. Geografiya*, no. 8, 1983, E178.

Robinson, B.V. (1985) 'Economic Geographic Assessment of the Oil Resources in East Siberia and the Yakut ASSR', *Soviet Geography*, vol. 26, no. 2.

Rumer, Boris Z. (1984) *Investment and Reindustrialization in the Soviet Economy*, Boulder: Westview Press.

Savin, N.I. (1984) 'Morskoy transport v narodnokhozyaystvennom komplekse Dal'nego Vostoka', in V.N. Bugromenko (ed.) *Territorial'nye aspekty razvitiya transportnoy infrastruktury*, Vladivostok.

Schroeder, Gertrude E. (1987) 'Anatomy of Gorbachev's Reforms', *Soviet Economy*, vol. 3, no. 3 (July–September), pp. 219–41.

Shabad, Theodore (1986) 'News Notes', *Soviet Geography*, April, pp. 248–79.

Shlyk, N.L. (1981) 'Eksportnaya spetsializatsiya Dal'nego Vostika: osnovnyye napravlenya razvitiya', *Izvestiya S.O. AN SSSR ser. obshch. nauk*, vol. 6, pp. 34–7.

Singur, N. (1988) 'Dal'niy Vostok: kompleksnoye razvitiye proizvoditel'nykh sil', *Planovoye khozyaystvo*, no. 3, pp. 94–8.

Slavin, S.V. (1982) *Osvoyeniye Severa Sovetskogo Soyuza*, Moscow: Nauka.

Soviet Military Power (1987) Washington, DC: US Government Printing Office.

Soviet Union Pacific Area Report, University of Hawaii at Manoa.

Tan Eng Bok, Georges (1986) *The USSR in East Asia. The Changing Soviet Position and Implications for the West*, Paris: The Atlantic Institute for International Affairs.

Trainor, Bernard, 'Southeast Asia Ten Years After the Fall of Saigon', *Naval War College Review*, vol. 39, no. 4, Autumn 1986, p. 48.

Treyvish, A.I. (1982) 'Rol' ekonomiko-geograficheskogo polozheniya Dal'nego Vostoka v formirovanii ego territorial'no-khozyaystvennoy struktury', in P.Ya. Baklano and V.N. Bugromenko (eds) *Territorial'no-khozyaystvennyye struktury Dal'nego Vostoka*, Vladivostok, pp. 104–18.

Vas'kov, S.T. (1986) 'Primorskiye ekonomicheskiye zony v sisteme intensivnogo territorial'nogo razvitiya', *Izvestiya AN SSSR, Seriya geograficheskaya*, no. 2, pp. 77 and 81.

Vneshnyaya torgovlya SSSR. Monthly.

Westwood, James T. (1985) 'Soviet Maritime Strategy and Transport', *Naval War College Review*, vol. 38, no. 6/Sequences 312, November–December, pp. 42–9.

Zykin, B.N. (1981) *Effektivnost' regional'noy ekonomiki*, Moscow: Nauka.

Chapter twelve

Conclusions and recent developments[1]
Elisa B. Miller and Allan Rodgers

While it would be repetitive to summarize the contributions offered by
the authors of this book, some of their conclusions merit re-emphasis,
at the risk of over-simplifying their arguments.

Philip Pryde and Victor Mote argue that the vast majority of the Far
East will never be developed because of environmental constraints, yet
the portion that is amenable to development has considerable potential.
They stress that the southernmost part of the region will play a considerably larger role in the future regional economy of the Pacific Rim and
that no natural constraints other than sheer distance can preclude that
growth. Of course, much depends on the willingness of Soviet planners
to accept the higher development costs of expanding the economic base
of this more favoured region.

Ann Helgeson, whose thrust was on labour force and the social infrastructure, focuses on the increased possibilities for the use of foreign
labour in the Far East. Those workers would presumably come from
China, Korea and possibly Vietnam; so far, however, the numbers
involved are minuscule. Internal migration is the other labour source
which may ease shortages in the region. But to attract migrants, there
is the need for higher wage incentives and the need for improvement in
living conditions that must materialize if new labour resources are to
become available, otherwise development prospects are dim.

Craig ZumBrunnen's chapter on resources, he argues, was handicapped
by the lack of detailed data on reserves, production and costs. Yet his
conclusions add an important dimension to this study. First, as all con-
cede, the energy and mineral resource endowment of the Far East is
vast, though at the same time, throughout much of the region environmental factors pose very significant physical, transportation and economic
obstacles to rational resource development. To these must be added the
problems of Soviet institutional, personnel and infrastructure difficulties
to enhanced Far Eastern trade and co-operative resource development
in the Far East.

The forest and fishing industries of the Far East are the focus of

Brenton Barr's chapter. He argues that the local fishing industry has a national importance and momentum which guarantee it a strong future, whereas the region's forestry and wood-processing industries are likely to remain relatively underdeveloped unless the European-Uralian forests become irrevocably degraded.

In his chapter on the Southern Yakutian Territorial Production Complex, Victor Mote argues that further development of this region has been the subject of considerable debate within Gosplan. Like others, he feels that enough evidence exists that investment priorities will inevitably go to Soviet Europe. He also argues that Yakutia's extractive industries and ultimately her proposed 'smoke-stack' industries are not the wave of the future in the Soviet Union. Thus it follows that they will not receive investment priority. Even the question of the completion of the AYAM is still a subject of considerable controversy. Thus in the overview it would appear that the bright future originally forecast for this sub-region will not come to pass. While abandonment appears highly unlikely, development prospects will surely be scaled back.

Robert North's study of transportation in the Far East argues that the standard Soviet attitude towards transportation development in a frontier region differs from that in the west, particularly that of northern Canada. Transportation is not seen as a leader of economic development; there has been a consistent underestimation of the stimulating effects of transport investment. Thus the Little BAM is cited as a facility which was soon swamped by growing traffic. In contrast, he notes the BAM is a rare case of heavy investment ahead of demand and its outcome has lagged far behind expectations. He argues that the impact of *perestroyka* in the Far East is likely to be negative because national sectoral organizations may try to minimize their investments, concentrating on other parts of the Soviet Union where returns come more readily rather than in a remote, environmentally hostile region like the bulk of the Far East.

Allan Rodgers is pessimistic about the outcome of the Gorbachev programme, but he does anticipate a modest increase in interregional commodity flows in the coming decade. There would be increased interaction with European Russia and a consistent decline of flows with Siberia. There may also be a reduction of the disparity between inbound and outbound hauls, but the former will continue to predominate.

Michael Bradshaw's focus is on Soviet Far Eastern trade. In the 1970s the region benefitted from a basic complementarity with the resource-poor industrial economy of Japan. As we enter the 1990s highly competitive resource markets and the changing structure of the Japanese economy have undermined the complementarity that had previously promoted trade. In the future he believes that there is little chance of large-scale resource exports to the Pacific Rim. Like Dienes and other contributors to this book he fundamentally believes that Gorbachev's policies do not favour remote-area resource projects. In his opinion foreign trade will play a more modest role in the economy of the Far East in the future than it did in the past.

To this point, we have not directly discussed the Dienes contribution. We choose to do so now because it is so central to the major themes addressed in this volume. That is, what are the prospects for this immense remote region? Fundamentally, Dienes treats the viability of the Gorbachev and Gosplan's 'Long-term Programme for the Comprehensive Development of the Far Eastern Region and Transbaykalia till the Year 2000'. Dienes argues that, given the investments and achievements of the 1970s and 1980s, the proposed investment programme and its targets seem wholly unrealistic and chimerical. Yet the Soviet leadership will continue to subsidize the Far East, perhaps at greater levels than in the past. Moscow, he argues, will continue to regard the region as one of critical strategic value; because of vast natural riches it will also view it as a potential economic asset. Yet its development will be very slow and extremely unbalanced.

Recent developments

What is the future of the Soviet Far Eastern economic region and especially its role in the Asia–Pacific region? Three areas where Soviet initiatives have been evident help us to assess this question: first, the long-term programme for the development of the region to the year 2000; second, national-level policy enactments concerning the region; third, economic reform in both domestic and foreign economic relations as they have an impact on the region.

Long-term programme for the development of the region to the year 2000[2]

The long-term programme for the comprehensive development of the Far Eastern region and Transbaykalia to the year 2000 – with a price tag of 0.2 billion roubles – was developed in the mid-1980s, and approved by the Central Committee in the summer of 1987 (*Pravda*, 25 July 1987). To some observers, this was a victory for those who were trying to draw attention to the region's importance, and to the region's plight. This programme is a set of projects which would require investment funds 2.5 times greater than what had been invested in the previous fifteen years and implies a commitment on the part of Moscow to attend to the region, a commitment that has heretofore been lacking (*Pravda*, 26 July 1987).

However, in his Krasnoyarsk speech of September 1988, Gorbachev indicated that the long-term programme for the development of the region, as passed in 1987, is now seen to have basic insufficiencies;

that the plan needs, and indeed is undergoing, 'correction' (*Pravda*, 18 September 1988). This is not surprising. First, the plan was developed before Gorbachev came to power and as a result does not reflect – in mechanism, nor in policy initiatives – the substantial policy changes of economic reform. Furthermore, while some applauded the plan, most observers saw it as stillborn: a 'wish list' of projects, with a price tag but without budget allocations, without real ministerial commitment, without any co-ordinating authority to enforce the programme, and, perhaps most importantly, without a sufficient conceptual basis – the programme did not go far enough to address the real ills of the region's economy.

For the authors–architects of the plan, the programme was meant to implement a radical restructuring of the Soviet Far Eastern economy: less emphasis on extensive growth of the traditional sectors of specialization (fish, non-ferrous metals and forest products), and more emphasis on the development of sectors which could provide the basis for dynamic, intensive growth of the region. Emphasis would be placed on the construction sector (a bottleneck), on social infrastructure (housing, health and recreation) to meet the needs of a stable labour force, on the development of the chemical industry, and on the reorientation and development of the machine-building sector to meet the need for technical innovation and reorientation of the economy from raw-material production to raw-material processing (Miller 1988a).

From the short statement of Gorbachev in Krasnoyarsk, and from interviews with the major architects of the plan, the impression received is that the plan, as it was approved, does not succeed in its purposes: it establishes neither a programme nor a mechanism for solving the long-term stagnation of the Soviet Far East economic region (Miller 1988a). But, in addition, the plan did not envisage the major economic reform initiatives that have taken place since Gorbachev. Thus, besides questioning the value of the plan as it was approved in 1987, new questions now must be addressed. In particular: a) the role of foreign investment capital, how to use this factor either as an addition to, or in lieu of, central investment funds, and b) how will the plan work under the new economic mechanism where enterprises will make decisions more autonomously and, unless there is a co-ordinating mechanism, not necessarily in tandem with the long-range goals for economic restructuring in the Soviet Far East.

Nevertheless, even if we do call the plan as approved in 1987 a 'paper tiger' in need of revision, the basic thrust behind it – the need for dynamic growth planning – is correct. And, if planners persevere, it could be a much-needed instrument even as reform and decentralization proceed. Indeed, it is just when enterprise managers have autonomy but an irrational price system that the need for project evaluation and investment choice *and* a set of priorities or plans for the region becomes most relevant. Therefore, in addition to setting guidelines for the allocation of central investment capital, the plan can serve a useful and positive purpose as a guide for investment choice and project evaluation as regards direct foreign investment. This function grows in importance as we approach a period of active direct foreign investment, rivalling the era of the 1920s in the Soviet Far East, where British, French and Japanese concessions were active (in gold, silver, non-ferrous ores and metals, coal and oil) and decision making was more decentralized (Koltsov 1979).

Special national-level policy enactments as they concern the Soviet Far East

In Gorbachev's Vladivostok speech (*Ekonomicheskaya gazeta*, 26 July 1987) much emphasis was made on the role of the USSR in the Asia Pacific region – with suggestions for directions of development for the Soviet Far East and areas of future policy. In Spring 1987, the USSR Council of Ministers created SOVNAPEC, the Soviet National Committee for Co-operation with Asia Pacific Countries (see 'Soveshanie po problemam i perspektivam razvitiia vneshne-ekonomicheskikh sviazei Dalnevostochnogo Ekonomicheskogo Regiona' in *Problemy Dalnego Vostoka*, no. 4).

SOVNAPEC is closely tied to the important advisory institutions in Moscow (Primakov, its first head, was Director of the prestigious Institute of International Economy and World Relations, IMEMO). It is tied, as well, to other centres of power in Moscow (its first deputy, Igor Khotsialov, is head of a department at the State Commission on Foreign Economic Relations). In addition, SOVNAPEC is tied to important local authorities; on the commission are the Chairmen of all the Executive Territorial Committees in the Soviet Far East, as well as the heads of leading industrial enterprises in the region. As a committee of high-level policy-makers and advisors from Moscow, with input from administrators from the Soviet Far East, the committee is able to shape policy for the region's economic efforts. One result of SOVNAPEC was the Vladivostok Conference in September 1988 with its symbolic, if not substantive, importance.

Gorbachev's Krasnoyarsk speech was remarkable for its delineation of policy initiatives now under consideration for the Soviet Far East. 'We are thinking', he states, in his speech (*Pravda*, 18 September 1988):

1 to establish special economic zones with all the usual incentives for direct foreign investment. Many areas have been mentioned as under consideration: Nakhodka, an area west of Vladivostok where the Chinese, North Korean and USSR borders mesh; the entire city of

Khabarovsk; an area near Yuzhno-Sakhalinsk; territory along the BAM; and the Southern Kuriles. (Interview with P.Ia. Baklanov, Director of the Institute of Economic Research, Far Eastern Branch, Academy of Sciences, Khabarovsk, 21 November 1988, Honolulu, Hawaii).

2 to create a 'Far Eastern Coefficient', to provide special investment incentives applicable to the *whole* region (in distinction to special economic zones), to make the Far East a privileged investment zone for foreign investment with tax rates more favourable than for the rest of the country, with special rules for establishing wages of personnel (agreement between foreign partner and local union), special rules for reinvestment of earnings (relaxing rule for reinvestment), and special rules for the allocation of investment shares, etc.

3 to allow Soviet Far East enterprises to keep a percentage of their hard currency earnings for social needs and imports of consumer goods (current legislation indicates that hard-currency earnings of state enterprises are to be directed for imports of machinery and equipment and technological items);

4 to give rights to direct access to the foreign market to *all* organizations, enterprises and producing co-operatives located in the territory;

5 to increase incentives of co-operatives, state enterprises and local authorities who participate in border and coastal trade by allowing saved *inputs* as well as surplus outputs to be traded in this barter trade;

6 to give to local administrative authorities the right to a percentage of the hard currency retained by enterprises located within their territory for use in developing international trade.

In addition, Gorbachev's statement in Krasnoyarsk gave credence to initiatives already *de facto* in operation: 1) economic ties with South Korea; 2) economic co-operation in Chinese–Soviet Far East border areas – establishment of Harbin/Khabarovsk airlink, (already in existence), planned ferry links, Chinese labour in construction and in agricultural activities in the region, a joint-venture pulp mill on Chinese territory (Chen Long Zhu 1988); and 3) people-to-people exchanges in the Far North where in the summer of 1988, visits between officials from Alaska and officials from Chukotsk Autonomous Region and Magadan Territory took place, each side travelling to each other's territory across the Bering Straits (*National Geographic*, October 1988).

Economic reform; decentralization of administrative decision-making; foreign trade restructuring

Coastal and border trade

Very important to the future of the Soviet Far Eastern economy and especially its role in the Asian–Pacific community are changes in foreign-trade decision-making and new rules for foreign investment. Aside from the 'Far East Coefficient' which has been described above and which gives special treatment to the region, decentralization in economic decision-making (including foreign business and trade) is important. Also significant is the policy emphasis on border and coastal trade.

A close reading of the reform legislation shows much emphasis on border and coastal trade (*On Additional Measures to Improve the Management of Foreign trade; On Measures to Improve the Management of Foreign Activity with Socialist Countries; On Improving the Management of Republic Organs of Administration; The Law on the Co-operative System*). This trade is essentially barter trade with neighbours (defined as border and coastal trade countries) which are located throughout the Asia–Pacific region and which include, for instance: China, Japan, New Zealand, Singapore and the USA (Miller 1981). Border and coastal trade allows enterprises to keep almost all of the earnings from export while normal trade gives enterprises the right to keep only a percentage (Miller 1981). Co-operatives have the right to engage in border and coastal trade. All state enterprises can engage in border and coastal trade with socialist countries, and local authorities have also been encouraged to develop this trade. For the Far Eastern enterprise, because of the special incentives in this trade and real economic possibilities for trade and co-operation with neighbours, this is not an insignificant form of business co-operation and is in evidence in USSR trade relations with practically all the Asia–Pacific countries.

Other areas of foreign trade reform

The number of state enterprises located in the Soviet Far East with the right to establish a foreign-trade firm with direct ties to foreign markets can still, at this writing, be counted only on one hand. So, too, can the number of joint-venture agreements signed and in operation. Nevertheless, the right of the local enterprise to *develop* proposals with a foreign partner (and then send to Moscow for approval), the right of the local enterprise to engage with a foreign partner in technical co-operation agreements, product development, product testing, relations during a warranty period means substantial manoeuvrability for the enterprise when compared to the previous constraints on direct contact with foreign firms (Miller 1988b).

Also significant in this category of initiatives is the fact that all territorial administrative authorities will have a Department of Foreign Economic Ties (and, if Gorbachev's Krasnoyask speech is predictive, will have hard-currency funds from local enterprises to carry out foreign economic activities). Also important is the right of the territorial executive council (on approval of the local Party Secretary) to invite two to three guests for up to one week to *any* region of the Soviet Far East (interview with Magadan officials, 7 September 1988, Anchorage, Alaska). Indeed, in the autumn of 1988, two groups of Alaskans travelled to Magadan, and Vladivostok has been receiving business visitors on an *ad hoc* basis for several years now. A State of Alaska Delegation to the Soviet Far East held meetings with officials of the Khabarovsk, Maritime *and* Magadan territories (State of Alaska, Office of International Trade, *State of Alaska Trade and Friendship Mission to the Soviet Far East, Trip Report*, November, 1988).

The new form of trade association approved in January 1988, an Association for Business Co-operation with Foreign Countries, with substantial rights to deal with the foreign market, is another example of innovation in Gorbachev's foreign-trade reform (*On the Activities of Associations for Business Co-operation with Foreign Countries*, USSR Council of Ministers Decree no. 109, 27 January 1988). In Vladivostok, the first such association in the Soviet Far East was formed: the Association for Business Co-operation with Countries in the Asia Pacific. It has a membership reaching throughout the entire region and ranging in all sectors of industry and is engaged in trade promotion for its clients (*Krasnaya znamya*, 15 July 1988).

How does it all add up? Can flexibility, enterprise and pragmatism replace rigidity, inertia, and stagnation? Is openness able to replace closedness? In our opinion, especially as the establishment of economic relations with South Korea becomes official, a watershed will have been reached. Business deals are no longer limited to fishery, forest products and coal, but can now include mining non-ferrous metals, and joint-venture possibilities in the Far North as well as in the South of the Far East. Chinese–Soviet trade and economic activity is expanding rapidly; in addition to the four border points now open, six more are being considered (Chen Long Zhu, 1988). The Amur River Steamship Company is carrying Chinese freight cargo; the Chinese have allowed a corridor for Soviet air flights to Vietnam (Shlyk 1988). The Japanese are going in small groups on winter hunting and other recreational tours into new areas of the Khabarovsk territory. Magadan is also soliciting hunting tour groups. Several crossings between Nome and Provideniya have occurred this year as if to say the border is (unofficially) open. A US airline company is proposing service across that border. A Japanese airline company is proposing an Osaka–Vladivostok service. South Korean firms are buying and selling undisguisedly and no longer only through third parties. Local officials are openly soliciting proposals for joint economic co-operation and trade.

The period of activity reminds one of the period in the 1920s where the Soviet Far East 'shifted to satisfying its own needs and to using all the available possibilities to strengthen economic relations with neighbouring countries' (quoted in Minakir 1983). If the planner (who keeps long-range goals in sight), the enterprise manager (who develops local initiative and identifies local resources) and the foreign investor (who brings technological expertise, markets and capital) can come together, and work together, the future need not be at all dim.

Yet[3] in terms of the original Gorbachev speech in 1986 and the Central Committee decision on the 'Long-term programme for the development of the Far East', these developments, though healthy and meritorious, do not truly contradict Dienes' rather pessimistic prognosis for future economic development of the Soviet Far East.

Notes

1. Research for these materials was supported in part by a grant to Elisa Miller from the International Research & Exchanges Board (IREX), with funds provided by the National Endowment for the Humanities and the United States Information Agency. None of these organizations is responsible for the views expressed.
2. These materials complement the Dienes contribution.
3. A. Rodgers.

Bibliography

Chen Long Zhu (1988) Ministry of Foreign Economic Relations and Trade, Beijing, 'Soviet-Chinese Economic Relations', paper presented at American Association for the Advancement of Slavic Studies (AAASS) Conference on *The USSR and the Pacific*, 21 November, Honolulu, Hawaii.

Koltsov, V. (1979) 'Ispolsovanie inostrannogo kapitala dlia razvitiia proizvoditelnykh sil Dalnego Vostoka v perkhodnyi period', in A. Krushanov, *Istoriaa promyshlennogo razvitiia Sovetskogo Dalnego Vostoka*, Vladivostok.

Krasnaya znamya, 15 July 1988.

Miller, Elisa (1981) 'Soviet Participation in the Emerging Pacific Basin Economy, the Role of Border Trade', *Asian Survey*, May.

—— (1988a) 'New Efforts in the Soviet Far East', *Pacific Northwest Executive*, Summer.

—— (1988b) 'The Soviet Far East and Foreign Trade Reform, Implications for Economic Integration in Asia–Pacific', unpublished manuscript, Dept. of Marketing, School of Business, University of Washington, September 1988.

Minakir, P.A. (1983) *Ekonomicheskaya razvitiya regiona: programmnyi podkhod*.

Pravda, various issues cited in the text.

Shlyk, Nadezda (1988) 'Prospects of Economic Cooperation Between the Soviet Far East and Countries in the Asian–Pacific Region', paper delivered at the AAASS Conference on *The USSR and the Pacific*, Honolulu, Hawaii, 21 November.

Index

Agafonov, S. 255, 267
Aganbegyan, A. 70, 280, 298
agricultural resettlement programme 66, 73–4
agriculture 10, 18, 63
air pollution 39, 52
Akao, N. 250, 268
Alaska 8, 17, 45
Aldan 102–3, 106, 175–7, 178
Aleksandrovsk-Sakhalinskiy 155–6
Alexander II (1855–81) 8, 12
Alexander III (1881–94) 12–13, 14
Alexandrov, V. 252, 267
Al'ter, I. 283, 298
aluminum 103
Amur Oblast 69, 75, 189, 266; forestry 121, 123, 126, 132, 134–5, 145–7, 150, 252; resources 89, 95, 102; settlement 16, 20, 22, 74; transport 73, 211
Amursk 17, 68, 145
aquaculture 156
Argentina 245
asbestos 164, 250
Australia 226, 248, 256
AYAM (Amur–Yakutsk Mainline) railway 1, 134, 146, 166–71, 177–8, 180, 204–5, 281–2, 303

Backman, C. 158, 160
Bakhrakh, M. 277, 298
Baklanov, P. Ya. 110–11, 307
Bakunin 16
Baldwin, G. S. 62, 75, 80
BAM (Baykal–Amur Mainline) railway 1, 3, 53; construction 213–14, 280–3; construction workers 68, 70; and foreign trade 272; impact of world economy on 163–4, 213–14; Little BAM 165, 166, 205, 303; role 84, 213, 234, in forestry industry 117, 119, 134, 146, 153
Bamovskaya 166
Barr, B. M. 84, 114–60, 303
Belgium 262
Belov, A. V. 53
Berkakit 166, 196
Bestyakh 93
bismuth 104
Blagoveshschensk 22, 28–9, 31
Bond, A. R. 52, 56, 159–60, 288
border trade 250, 263–6, 288, 308
Borisov, A. A. 36, 56
Borodin, A. M. 46, 49, 56
boron 108
Borts, Yu. M. 99, 113
Braden, K. 118, 139–40, 144, 151, 160
Bradshaw, M. J. 110, 114, 120, 151, 159–60, 234, 239–66, 273, 299, 303
Brezhnev, L. I. 69, 214, 243
Bunich, P. G. 102, 111
Bureya 17
Buritko, A. L. 202–3, 223

Canada 207, 226, 245, 256, 276
Chara-Tokko 178
Chelyshev, V. A. 118, 160
chemicals 176–7, 230–1, 236
Chen Long Zhu 307, 309, 310
Chernobyl 85
Chernov, V. I. 45, 57
Chernyshevskiy 95
Chichkanov, V. p. 88, 98, 112, 156, 160, 235–6, 238, 249, 277, 280
China, border trade 264–5, 288; economic co-operation 75, 78, 111,

307; relations with USSR 3, 13, 119–20, 269; trade 11, 189, 207, 243, 248, 308–9, in forestry products 115–17, 146–7, 150–1, 158

Chukchi Peninsula 104, 106

Chukotka 204–5, 208–9

Chukotsk Autonomous Okrug 105

Chul'man 90, 103, 166, 170, 172–3, 176–9, 234

Chumin, V. 153, 160

Churakov, V. Va, 73, 82

cities *see also* urbanization; company towns 62, 71; founded since the Revolution 28; growth of 28–32

Civil War (**1918–22**) 23

climate 37–40

coal 85–90, 171–5, 283; compensation agreements 254–6; costs 88, 175; imports and exports 151, 234, 245; interregional flows 229–30

coal washery 89

COCOM (Co-ordinating Committee on Multilateral Export Control) 91, 260, 273

COMECOM (Council for Mutual Economic Aid) 157

commodity movements 2–3, *see also* transport; 1950s and 1960s 225–31; 1970s and 1980s 231–7; interregional 303

communist ideology 24

compensation agreements 241–2, 249–61, 265, *see also* joint ventures

construction 40–1, 63, 71, 177, 237

consumer goods 72–3, 164–5

copper 104, 164, 250, 283

Cossak settlements 6, 20

Council for Mutual Economic Assistance (CMEA) 243, 245

Crimean War 12

Cuba 146

Dal'negorsk 104, 107–8

Danilov, S. K. 292, 299

Danilov, V. A. 91, 112

Danilova, A. 232, 238

De Kastri 259

Denisova, L. 226, 238

Desovskoye 90, 102

development, environmental protection and 45–54; natural constraints to 36–45

development strategies 271–2; export-oriented 272–4

diamonds 107–8, 237, 245, 261–2, 265

Dibb, P. 230, 238

Dienes, L. 63, 71, 83, 88, 94, 110, 120, 164, 177, 185, 190, 225, 231, 233, 236, 237, 241, 269–98, 303–4, 310

Druzenko, A. 52, 53, 56

Dzhebariki-Khaya 89

economic reform 218–20, 287–8, 303, 305, 308–10

Egyed, P. 257, 267

Elizabeth I (**1741–62**) 8

endangered species 46

energy resources 4, 45, 84–97

Enterprise Law 287–8

environment, constraints 36–45, 134, 302; degradation of 119, 155 laws on 52–3; protection 45–54

environmental impact analysis 52–4

Eronen, J. 150–1, 161

European Community 245

export trade 84, 121, 273

famine 13

Far Eastern Coefficient 307

Far Eastern Territory (**1926–38**) 23

fertilizers 108–9, 176, 230–1, 236

Finland 150

Fischer, D. 49, 56

fishing industry 66; fish breeding 156, 158; foreign trade 157, 249, 263; freshwater 155, 158; ocean 51, 154–6; off-shore 155–7; output 155–7; processed fish 118, 156, 231, 275, 280; spatial and economic impact 114–20

Five-Year Plans; 10th (**1976–80**) 63, 211; 11th (**1981–85**) 66, 68, 71, 175, 178, 211, 235, 285; 12th (**1986–90**) 72, 144, 152, 175, 178, 211, 281–3; **1990** plan 88

food products 4, 236–7, 245

foreign trade *see also* border trade; compensation agreements; commodity flows; joint ventures; natural resources, exports; Asian-Pacific 248–9; and development 239–41, 265–66; disaggregated 246–8; dynamics and structure of USSR's 243–8; effect of world economy on 239; fish 157, 249, 263;

forestry products 150–1; reforms in 240, 308–10; role of Soviet Far East 265; transport for 206–7; under central planning 239–43
Foreign Trade Organizations (FTOs) 239–40
forestry industry 121–54, 159–60, 302–3, *see also* timber; compensation agreements 250–4; national perspective 136–46; output 134–45; spatial and economic impact 114–20
forestry products 4, 51–2, 134–54, 177, 229–30, 236, 245, 249, 275
forestry project 274
forests 45, 51–2; allocation of land 130–1; allowable cut 117, 122–3, 126, 152–3; coniferous 126–7, 132; fire-control 153–4; hardwoods 126–7, 132; industrial 125; mountain 127–31; over-cutting 51, 153; quality variation 132; regional base 116, 121–2, 131–6; species composition 128, 130
France 157
Franklin, D. 249, 267
French, T. 171
friction of distance 3, 115, 117–18, *see also* transport costs
fuelwood 97
fur trade 5, 10, 16–17

Gardner, H. S. 239, 267
gas 93–4, 240–1, 244–5, 265; joint ventures 256–9
Gelman, H. 289, 294, 299
geothermal energy 96–7
Gladyshev, A. 231, 238
gold 16, 106–7, 172, 237, 245, 262–3, 265
Gorbachev, M. S. 109–10, 119–20, 122–3, 145, 152, 163, 170, 180, 239, 280, 303–4; Krasnoyarsk speech 306–7; speech in August **1986** 3–4; Valdivostok speech (July **1986**) 69–70, 83–5, 214, 270, 283, 287, 294, 306
Gorst, I. 245, 267
Goskomtrud 74
Gosplan 4, 143, 239, 284, 304
Granberg, A. G. 241, 267, 287–8
graphite 109

Great Britain 262

Ha, J. M. 293, 299
Hanson, P. 240, 267
Harrison 11, 33
Hasegawa, T. 295
Hauner, M. 270, 299
Hausladen, G. 5–33
Helgeson, A. C. 58–79, 302
Hewett, E. A. 262, 267
Hong Kong 120
housing 66, 71, 286
hydro-power sources 95–6

icebreakers, nuclear powered 199, 203–4
industrialization 12, 18, 27
infrastructure 122–3, 126, 285; social 4, 70–4
interregional linkages 225–38
Irkutsk Oblast 92, 145, 177
iron and steel 175, 229–30, 234–5
iron-ore 98–103, 175–6
Isayev, A. S. 52
Italy 243
Iultin 105
Ivanov, I. D. 240, 267
Ivanov, V. I. 78, 290, 299

Japan, economic co-operation 78, 90, 111, 120, 173–5, 242–3, 249–61, 273–4; hunting tour groups 309; relations with USSR 3, 23, 119, 189, 269, 294–6; trade 11, 84, 110, 206, 243, 245, 248, 295, 303, border 263–4, 308, in coal 89, 234, in fish 155, 157, in forestry products 115–17, 145–7, 150–1, 159, 208, 236, in gold 262, in oil and gas 90–1, 94, 230; war with Russia (**1904–5**) 13
Jensen, R. G. 98, 112
joint ventures 84, 110–11, 120–1, 195, 309, *see also* compensation agreements; with China 307; with Japan 90–1, 145, 242–3, 257–9, 273–4

Kamchatka 73–4, 156, 278; defence 289–96; employment structures 276, 290–1; environment 43–5; forestry 123, 134; resources 89, 92, 96–7; settlement 19–20; transport 73, 192, 208

Kampuchea 248
Kanevskiy, M. V. 150, 161
Kangalassy 89
Kankunskiy 177
Kaser, M. 106, 112, 262
Kavalerovo 105
Kawai 250
Khabarovsk 71, 78, 95, 111, 115, 189, 190; oil refinery 92, 234, 236; settlement 22, 28–9
Khabarovsk Kray 71, 189, 273, 309; border trade 263, 266; forestry 75, 121, 123, 126, 132, 134–5, 146–7, 150, 152; immigrants 64–5, 69; resources 89, 103–5, 109; transport 211–12
Khabarovsk Novomikhailovka 145
Khani 53
Kholmsk 155, 210–11
Kholodnaya 104, 283
Khorev, Boris 62, 80
Khotsialov, I. 306
Khrushchev, N. 69
Kibal'chich, O. A. 102, 112
Kimura, H. 294, 296, 299
Klin, P. 233, 235, 237, 238, 276–7
Knystautas, A. 50, 56
Kolbasov, O. S. 53, 56
Kolchak, Admiral 23
Kolesov, L. 231, 233–4, 238, 276
Koltsov, V. 306, 310
Kol'tsov, V. V. 120, 161
Kolyago, V. A. 37, 56
Kolyma 106–7
Komarov, 8, 36, 51, 56
Komsomol 66–8
Komsomol'sk-na-Amure 29, 31, 53, 64, 67–8, 91, 94–5, 103–4, 105, 115, 176, 212, 235, 290
Korea 75, 78, 111
Korea, North 248, 264–5
Korea, South 120, 307, 309–10
Korsakov 155, 210–11
Kostyrina, T. B. 51, 56
Kosygin 243
Kovrigin, E. B. 104, 112
Krasnoyarsk Kray 93, 170, 203
Krotova, V. M. 53, 56
Kulakov 65, 74, 80
Kuril Islands 43–5, 111, 132, 155–6, 158, 205, 257, 295
Kuznetsov, V. A. 102–3, 105, 109, 112

labour force, Chinese 181; deficits 63; foreign (guest) workers 75, 78, 158, 302; growth and migration 75; participation rates 62–3; sectoral structure 63; size of 62
labour productivity 63
Lake Baykal 43, 51, 155
Laos 248
Lawarne, S. 91, 112
lead 104, 283
Lena 85, 90
Lenin, V. I. 119
Levenson, M. 292, 300
Li Peng 84
limestone 177
lumber, 134–7, 145
Lutsenko 78
Lydolph, P. E. 36, 56

machinery 4, 231, 235, 245, 248, 260–1
McIntyre, J. 240, 268
Magadan 39, 71, 155, 209–10, 309
Magadan Oblast 71, 278; forestry 126, 132; population 31, 64; resources 89, 95, 105, 262; transport 192, 196
Mago 150, 155
Malakhovskiy, K. V. 290, 299
manganese 103
Maritime Kray 68–9, 73, 78, 89–90, 273, see also Primorskiy Kray; forestry 123, 126, 132, 134–5, 146–7, 150; resources 17, 83, 95, 104–7, 111; settlement 20, 22
Mathieson, R. S. 105, 112, 173, 183
mercury 109
Meyerhoff, A. A. 90, 93–4, 112
mica 177
migration, (1860–1910) 14–16; group 68, 74; independent 69–70; involuntary 26, 58, 64; inward 4, 63–70; and labour force growth 75; outward 65; peasant 11, 15–16, 27; rates of 64–5; resettlement cost 65; rural 73–4; to Siberia 12
migration channels, agricultural resettlement programme 66, 73–4; Komsomol 66–8; Orgnabor 66–8; university graduates 66–7
Mikheyeva, N. N. 278, 300
military, capabilities 289–90, 292; changes in command 293–4; presence 63, 270–1

Miller, E. B. 264, 268, 302–10, 305, 308, 310
Mil'ner, G., 277, 298
Milovanov, Ye. 72, 80
Minakir, P. A. 98, 112, 277–80, 310
Minakir, R. 233, 235, 238
mineral resources 5, 10, 16, 52, 83, 97–109, 231, 245, 275, 279, 282, 302
mining 63, 67, 114, *see also* coal
Minlesprom (Minlesbumprom) 119
Mirnyy 93, 266
Misko T. 102–3, 112
mix-and-share analysis (m&s) 137, 139–42
Mokhsogollokh 93
Molodenkov, L. V. 120
Molodezhnoye 164, 283
molybdenum 106
Mongolia 157, 248, 292
Mote, V. L. 36–55, 84, 102, 104, 120, 163–81, 234, 255, 282, 302–3
Mozhayev, B. 119, 161
Mozhin, V. P. 114, 161, 274–5

Nagornyy 166
Nakhodka 78, 115, 150, 155, 208–9, 211–12, 250, 260, 276
Naprasnikova, Ye, V. 53, 56
natural resources 4, 45, 83–111, *see also* mineral resources; demand for 216; exports 249, 261–4, 265–6
nature reserves 46–51
Nefedova, V. B. 42, 57
Nerchinsk 176
Neryungri 68, 88–9, 102, 166, 172–5, 178, 234, 255, 260
Nevel'sk 155
Never 165, 196
New Zealand 248, 308
Nicholas (**1894–1917**) 12–13
Nikolayevsk-na-Amure 22
Nikol'sk-Ussuriysk 22
Nikol'skiy, I. V. 292
Nirsha, T. 242, 268
Nixon, R. 243
Nizhnetambovskoye 94
Nobuhara, N. 250, 268
Noril'sk 93
North, R. N. 165, 185–221, 225, 303
Northern Sea Route (NSR) 196, 198, 202–3
Norway 91

nuclear power 85

Ogawa, K. 110, 113, 250, 268
Ognev, A. p. 242, 268
oil industry 84, 229–30, 234; development of 90–2; joint venture 257–9; Siberian 163, 241, 265; trade 240, 244–5, 248
oil shale deposits 97
Okhotsk 155
Olekminsk 8, 17
Olenicheva, M. R. 156, 162
Olyutorskoye 109
Orgnabor 66–8
Orlov, V. 156, 160
orthodox Church 11
Osetrovo 165, 196, 199–203, 205
Osleeb, J. P. 50, 103, 113
Ota Seizo 78
Ozernyy 104

paperboard 134–7, 145
peat 97
people-to-people exchanges 307, 309
permafrost 40–3
Peter the Great (**1682–1725**) 8, 11
petroleum *see* oil industry
Petropavlovsk 43, 45
Petropavlovsk-Kamchatskiy 22, 31, 155, 209–10, 290
Pevek 105
Pevzner, Ya. A. 249, 264, 268
Pionerskoye 90, 102
Pistun, N. D. 156, 162
Pokrovsk 93
Ponkratova, L. A. 286, 300
Ponomarev, A. 242, 268
population 1–2, *see also* migration; age structure 72; density 58–9, 270; distribution 275; growth 19, 25–6, 59–61; projected increase 286; regional distribution 116; rural 73–4; of working age by economic regions 76–7; Yakutian 164, 166, 180
Poronaysk 145
ports 201, 208–10
Postnikov, General 294
Pos'yet 150
preserved areas 46–51
price system 241
Primakov, Ye. 110, 306
Primorskiy Kray 189, *see also* Maritime Kray; environment 37, 43, 45–6, 50;

foreign trade 263, 266; forestry 250, 252; transport 211–12
Privalovskaya, G. A. 282, 300
Proskuryakova, A. G. 275, 300
Provideniya 210
Pryde, P. R. 36–55, 302
pulp 134–7, 145

Rasputin, V. 36, 57
Raychikhinsk 89
regional development 216–17; and foreign trade 239–41, 265–6
regional integration 189–90
reindeer 17–18, 131
resource frontiers 238, 274–6
Reut, A. 143–5
Revaykin, A. S. 156, 162
River Lena 196–202
river vessels 199–200
rivers, floods 40, 214; shallowness 197–8, 214; summer drought 199; transport 165, 193–4, 196–204, 210, 214–15; winter ice 198–9, 214
Robinson, B. V. 92, 282, 301
Rodgers, A. 1–4, 90, 225–38, 302–10
Rogachev, I. 295–6
Rudenko, P. 73, 81
Rumer, B. Z. 286, 301
Russo-Japanese War (1904–5) 13

Sagers, M. J. 88–90, 92–95, 113
Sakhalin Island 8, 43, 155, 278, 289; employment structures 289–1; forestry 121, 123, 132, 134–5, 137, 145, 150, 152; gas and oil 90–1, 94, 195, 234, 236, 257–9; settlement and development 20, 31–2, 65, 71, 74; transport 192, 195, 205, 210–12, 276
Sangar 89
Schroeder, G. E. 289, 301
sea vessels, 199, 203–4, 207–8
seismicity 43, 96–7
Seligdar 176–8, 180, 282
Semichastov 250
Semyonov, A. 85
Serebyannyy Bor 166
serfs, emancipation of 12
Shabad, T. 43, 53, 89–90, 92–4, 98, 102–3, 105–6, 108–9
Shalybkov, A. M. 50, 57
Shcheka, S. A. 98, 113
Shevernadze, Foreign Minister 296
Sheyngauz, A. 153

Shilo, N. 88, 113
Shiryayevo, P. A. 99, 113
Shlyk, N. I. 249, 268, 309, 310
Shniper, R. 226, 230, 238
Siberia 36; coal 88, 254; colonization and settlement 6–12, 14–17; forestry 117, 127–31, 131, 140, 147–50; minerals 103, 106; need for improved project management 119–21; oil and gas 90–3, 163, 241, 265, 274; prospects 287; trade with Far East 227–8, 230–4
Sidorova, V. S. 102, 113
Sinegor'ye 95
Singapore 120, 157, 207, 308
Singur, N. 72, 114, 121, 143–4, 155–6, 234–6, 284, 286
Sinitsyn, S. G. 121–2, 126–7, 131, 162
Sivagli 90, 102
Slavin, S. V. 269, 301
Smith, A. H. 245, 268
snowfall 39–40
Sobolev, Yu. A. 144, 156
Sochava, V. B. 53, 57
Sogo 89
soils 41–2
Sokolov, V. E. 45, 57
Solecki, J. J. 155, 157, 162
South Africa 261
South Yakutian Territorial-Production Complex (TPC) 166, 171–80, 266, 282, 303
Sovetskaya Gavan' 53
Soviet Far East, administration 10, 23, 119–21; area described 1; blueprint for 284–5; compared to Canada 276; conquest and settlement 5–33, 58–9; demographic description 58–62; dependence on USSR heartland 276–7; economic importance 10–11, 24; economic output 274–5; economic structure 274–80; economy (1860–1910) 16–19; impact of *perestroyka* 218–20, 288, 303, 305; imports 277; inputs under-priced 277; indicators of economic development 278–9; interregional links 277–8; investment 3–4, 119, 284–6, 305; isolation of 269–70; labour force 62–3, 75, 78, 158, 181, 302; political and psychological importance 13, 24, 120; power

consumption 280; prospects for the future 237–8, 283–89, 304–6; a resource frontier 276; spatial characteristics of the economy 189–90; strategic military importance 18–19, 24, 115, 120, 270–1, 289–96, 304
Soviet Harbour 115, 145, 155, 209
Sretensk 14
Stalin, J. 5, 24, 26, 31, 261
steamship companies 206–7
Stern, J. P. 94, 113, 257, 268
Stolyarov, Yu, S. 249, 264, 268
Storchevoy, K. 50, 57
storms and high winds 39
submarines 293
Svetlyy 95
Svobodnyy 89, 95, 103, 176
Syroyechkovskiy 49
Sysoyev, N. 154–5, 156, 162
Syuznoye 109

Taiwan 111, 120
Talakan 95
Tamginskoye 109
Tan Eng Bok, 271, 289–93, 301
tar sands 97
Tayezhnoye 90, 102, 178
Tayshet 103, 176
thorium 97
tidal power 97
Tiksi 200, 202
timber 18, 51–2, 146–54, 177
tin 104–6
Toffler, A. 163
Toffler, H. 163
Tommot 166, 178, 204
topography 37
Trainor, B. 292, 301
Trans-Siberian Container Landbridge 212, 259
Trans-Siberian Railway 10, 12, 14, 89, 185, 211–12
Transbaykalia 4, 269, 274, 285
transport see also commodity movements; air 194–6, 203; compensation agreements 259–61; economic importance 63, 73, 188–9; and foreign policy 189, 217; freight traffic patterns 192–3; functions of 190–1; future developments 215–20; impact of perestroyka 218–20;

irrational cross hauls 230, 237; in the Lena basin and the north east 196–205; maritime 194, 197–9, 202, 206–10; network density and form 191–2; new technology 217; on the Pacific coast and islands 205–11; pipelines 194–5; rail 193–4, see also AYAM; BAM; Trans-Siberian; river 37, 193–4, 196–204, 210, 214–15; road 194, 196–7, 204; sector 275–6; southern railway belt 211–15; the system 185–221
transport costs 2–3, 159, 165, 229, 231, 270, see also friction of distance
Treadgold, D. W. 15
Treml, V. 231
Treyvish, A. I. 275, 301
Tsar Mikhail 6, 11
tsunami (tidal wave) 43
tungsten 106
Tunguska 85, 90
Tynda 286

Udokan 283
Udsko-Selemdzhinsk 103
Ugol'naya 166
Ukraine 75
unemployment 286
United Nations, 'Man and the Biosphere Programme' 50
United States 11, 13, 17, 23, 115, 254, 273; economic sanctions 258; Stevenson Amendment 256; Trade Act (1974) 94; trade with USSR 157, 226, 243, 245, 308; tripartite scheme 256–7
university graduates 66
uranium 97
urbanization 20, 26–8, 59
USSR, foreign trade, with Asian-Pacific 248–9, with China 11, 16, 116–17, 150–1, disaggregated 246–8, dynamics and structure 243–8, with Finland 150, with Japan 11, 16, 94, 116–17, 150–1, 295, with US 11, 94; joint ventures with Japan, N. Korea and China 78; relations with, China 3, 115, 119, 269, Japan 3, 115, 119, 269, 294–6, USA 115
Ussuri 16
Ussuriysk 17, 29, 31
Ust-Srednekan 96

Ust'Kamchatka 150
Uzbeks 74

Vadyukhin, A. A. 102, 113
Vanino 150–1, 209–11
Vanous, J. 241, 268
Vasil'yev 50
Verkhoyansk 8
Vietnam 75, 78, 248; Cam Ranh 271, 292–3
Vilyuysk 8
Vitim 17
Vladivostok 14, 22–3, 28, 30–1, 63, 78, 85, 90, 95, 115, 150, 155, 208–9, 211, 276, 309
volcanism 43–5, 96–7
Vorob'yev, G. I. 123, 134, 162
Vorob'yev, V. V. 36, 57
Vorotnikov, V. 287
Vostochnyy 68, 151, 208–9, 211, 250, 254–5, 273, 276
Vostok 106

wage differentials 69–70
water supply 42–3
western technology 242, *see also* compensation agreements
Westwood, J. T. 292, 301
winter, cold 37–8, 255; ice 198–9, 214; lack of sunlight 38–9; roads 204

Witte, Sergei 14
Wright, A. W. 98, 112

Yakunin, A. G. 114, 119, 121–3, 126, 135, 143–5, 147, 149, 153, 162
Yakutia 278; coal 88–90, 234, 254–5, 283; forestry 123, 134, 145–6, 151; importance of railways to 163–71; oil and gas 92–4, 256–7; resources 95, 97, 102, 104–9, 261–2, 282; settlement and population 2, 8, 10, 19, 22, 29, 31, 65; South Yakutian Territorial-Production Complex (TPC) 166, 171–80; transport 190, 192, 195–6, 205
Yakutsk 8, 10, 22, 29, 31, 93, 165–7, 190, 200, 204
Yarkho, Ye. N. 99, 113
Yostochnyy 259–60
Yuzhno-Sakhalinsk 31–2, 291

Zaslavskaya, T. 70
Zeya 17
Zhebelev, G. 260, 268
Zhebrovoskiy, B. 242, 268
Zhelezko, S. N. 68, 81
zinc 104
Zolotinka 166
ZumBrunnen, C. 83–111, 234, 236, 302
Zyryanka 89